A Gentle Introduction to Optimization

Optimization is an essential technique for solving problems in areas as diverse as accounting, computer science and engineering. Assuming only basic linear algebra and with a clear focus on the fundamental concepts, this textbook is the perfect starting point for first- and second-year undergraduate students from a wide range of backgrounds and with varying levels of ability.

- Modern, real-world examples motivate the theory throughout.
- Over 140 exercises, ranging from the routine to the more advanced, give readers the opportunity to try out the skills they gain in each section.
- Solutions are available for instructors as well as algorithms for computational problems.
- Self-contained chapters allow instructors and students to tailor the material to their own needs and make the book suitable for self-study.
- Suggestions for further reading help students to take the next step to more advanced courses in optimization.
- Material has been thoroughly tried and tested by the authors, who together have 40 years of teaching experience.

B. Guenin is Professor in the Department of Combinatorics and Optimization at the University of Waterloo. He received a Fulkerson Prize awarded jointly by the Mathematical Programming Society and the American Mathematical Society in 2003. He is also the recipient of a Premier's Research Excellence Award in 2001 from the Government of Ontario, Canada. Guenin currently serves on the Editorial Board of the *SIAM Journal of Discrete Mathematics*.

J. Könemann is Professor in the Department of Combinatorics and Optimization at the University of Waterloo. He received an IBM Corporation Faculty Award in 2005, and an Early Researcher Award from the Government of Ontario, Canada, in 2007. He served on the program committees of several major conferences in Mathematical Optimization and Computer Science, and is a member of the editorial board of *Elsevier's Surveys in Operations Research and Management Science*.

L. Tunçel is Professor in the Department of Combinatorics and Optimization at the University of Waterloo. In 1999 he received a Premier's Research Excellence Award from the Government of Ontario, Canada. More recently, he received a Faculty of Mathematics Award for Distinction in Teaching from the University of Waterloo in 2012. Tunçel currently serves on the Editorial Board of the *SIAM Journal on Optimization* and as an Associate Editor of *Mathematics of Operations Research*.

A Gentle Introduction to Optimization

B. GUENIN

J. KÖNEMANN

L. TUNÇEL

University of Waterloo, Ontario

CAMBRIDGE
UNIVERSITY PRESS

CAMBRIDGE
UNIVERSITY PRESS

University Printing House, Cambridge CB2 8BS, United Kingdom

One Liberty Plaza, 20th Floor, New York, NY 10006, USA

477 Williamstown Road, Port Melbourne, VIC 3207, Australia

314-321, 3rd Floor, Plot 3, Splendor Forum, Jasola District Centre, New Delhi - 110025, India

79 Anson Road, #06-04/06, Singapore 079906

Cambridge University Press is part of the University of Cambridge.

It furthers the University's mission by disseminating knowledge in the pursuit of
education, learning and research at the highest international levels of excellence.

www.cambridge.org
Information on this title: www.cambridge.org/9781107658790

© B. Guenin, J. Könemann and L. Tunçel 2014

First published 2014
Reprinted 2015

A catalogue record for this publication is available from the British Library

Library of Congress Cataloging in Publication data
Guenin, B. (Bertrand)
A gentle introduction to optimization / B. Guenin, J. Könemann, L. Tunçel, University of Waterloo, Ontario.
    pages   cm
Includes bibliographical references.
ISBN 978-1-107-05344-1 (Hardback)
ISBN 978-1-107-65879-0 (Paperback)
1. Mathematical optimization.   I. Könemann, J. (Jochen)   II. Tuncel, Levent, 1965-   III. Title.
IV. Title: Introduction to optimization.
QA402.5.G84 2014
519.6–dc23   2014008067

ISBN 978-1-107-05344-1 Hardback
ISBN 978-1-107-65879-0 Paperback

Additional resources for this publication at www.cambridge.org/9781107658790

Contents

Preface

Desire to improve drives many human activities. Optimization can be seen as a means for identifying better solutions by utilizing a scientific and mathematical approach. In addition to its widespread applications, optimization is an amazing subject with very strong connections to many other subjects and deep interactions with many aspects of computation and theory. The main goal of this textbook is to provide an attractive, modern, and accessible route to learning the fundamental ideas in optimization for a large group of students with varying backgrounds and abilities. The only background required for the textbook is a first-year linear algebra course (some readers may even be ready immediately after finishing high school). However, a course based on this book can serve as a header course for all optimization courses. As a result, an important goal is to ensure that the students who successfully complete the course are able to proceed to more advanced optimization courses.

Another goal of ours was to create a textbook that could be used by a large group of instructors, possibly under many different circumstances. To a degree, we tested this over a four-year period. Including the three of us, 12 instructors used the drafts of the book for two different courses. Students in various programs (majors), including accounting, business, software engineering, statistics, actuarial science, operations research, applied mathematics, pure mathematics, computational mathematics, computer science, combinatorics and optimization, have taken these courses. We believe that the book will be suitable for a wide range of students (mathematics, mathematical sciences including computer science, engineering including software engineering, and economics). To accomplish our goals, we operated with the following rules:

1. Always motivate the subject/algorithm/theorem (leading by modern, relatable examples which expose important aspects of the subject/algorithm/theorem).
2. Keep the text as concise and as focused as possible (this meant, that some of the more advanced or tangential topics are either treated in advanced sections or in starred exercises).
3. Make sure that some of the pieces are *modular* so that an instructor or a reader can choose to skip certain parts of the text smoothly. (Please see the potential usages of the book below.)

In particular, for the derivation and overall presentation of the simplex method, we focused on the main ideas rather than gritty details (which in our opinion and experience, distract from the beauty and power of the method as well as the upcoming generalizations of the underlying ideas).

We emphasized the unifying notion of *relaxation* in our discussion of duality, integer programming, and combinatorial optimization as well as nonlinear optimization. We also emphasized the power and usefulness of *primal–dual approaches* as well as *convexity* in deriving algorithms, understanding the theory, and improving the usage of optimization in applications.

We strived to enhance understanding by weaving in *geometric notions, interpretations, and ideas* starting with the first chapter, Introduction, and all the way through to the last chapter (Nonlinear optimization) in a cohesive and consistent manner.

We made sure that the themes of *efficiency of algorithms* and *good certificates of correctness* as well as their relevance were present. We included a brief introduction to the relevant parts of computational complexity in the appendix.

All of these ideas come to a beautiful meeting point in the last chapter, Nonlinear optimization. First of all, we develop the ideas only based on linear algebraic and geometric notions, capitalizing on the strength built through linear programming (geometry, halfspaces, duality) and discrete optimization (relaxation). We arrive at the powerful Karush–Kuhn–Tucker Theorem without requiring more background in continuous mathematics and real analysis.

We thank Yu Hin (Gary) Au, Joseph Cheriyan, Bill Cook, Bill Cunningham, Ricardo Fukasawa, Konstantinos Georgiou, Stephen New, Clinton Reddekop, Patrick Roh, Laura Sanita and Nick Wormald for very useful suggestions, corrections and ideas for exercises. We also thank the Editor, David Tranah, for very useful suggestions, and for his support and patience.

Some alternative ways of using the book

We designed the textbook so that starred sections/chapters can be skipped without any trouble. In Chapter 3 it is sufficient to pick only one of the two motivating problems (the shortest path or the minimum cost matching problem). Moreover, there are many seamless ways of using the textbook, we outline some of them below.

- For a high-paced, academically demanding course, cover the material from beginning to end by inserting the Appendix (Computational complexity) between Chapter 2 or 3, or 4 or 5.
- Cover in order Chapters 1,2,3,4,5,6,7 (do not cover the Appendix).
- For an audience mostly interested in modeling and applications, cover Chapters 1,2,3,6,7.
- For an audience with prior knowledge of the simplex method, cover Chapters 1,3,4,5,6,7.
- For a slow-paced course based only on linear programming, cover Chapters 1,2,3,4.
- For a course based only on linear programming, cover Chapters 1,2,3,4,5 (possibly with the Appendix included).

- For a course based only on linear programming and discrete optimization, cover chapters 1,2,3,4,5,6 (possibly with the Appendix included). This version may be particularly suitable for an introductory course offered in computer science departments.
- For a course based only on linear programming and discrete optimization (but at a slower pace than the last one above), cover Chapters 1,2,3,4,6.
- For a course based only on linear programming and nonlinear optimization, cover Chapters 1,2,3,4,5,7.
- For an audience with some prior course in elementary linear programming, cover Chapters 1,5,6,7 (insert the Appendix after Chapter 5).
- Our book can also be used for independent study and by undergraduate research assistants to quickly build up the required background for research studies.

1 Introduction

Broadly speaking, optimization is the problem of minimizing or maximizing a function subject to a number of constraints. Optimization problems are ubiquitous. Every chief executive officer (CEO) is faced with the problem of maximizing profit given limited resources. In general, this is too general a problem to be solved exactly; however, many aspects of decision making can be successfully tackled using optimization techniques. This includes, for instance, production, inventory, and machine-scheduling problems. Indeed, the overwhelming majority of Fortune 500 companies make use of optimization techniques. However, optimization problems are not limited to the corporate world. Every time you use your GPS, it solves an optimization problem, namely how to minimize the travel time between two different locations. Your hometown may wish to minimize the number of trucks it requires to pick up garbage by finding the most efficient route for each truck. City planners may need to decide where to build new fire stations in order to efficiently serve their citizens. Other examples include: how to construct a portfolio that maximizes its expected return while limiting volatility; how to build a resilient tele-communication network as cheaply as possible; how to schedule flights in a cost-effective way while meeting the demand for passengers; or how to schedule final exams using as few classrooms as possible.

Suppose that you are a consultant hired by the CEO of the WaterTech company to solve an optimization problem. Say for simplicity that it is a maximization problem. You will follow a two-step process:

(1) find a *formulation* of the optimization problem,
(2) use a suitable *algorithm* to solve the formulation.

A formulation is a mathematical representation of the optimization problem. The various parameters that the WaterTech CEO wishes to determine are represented as *variables* (unknowns) in your formulations. The *objective function* will represent the quantity that needs to be maximized. Finally, every constraint to the problem is expressed as a *mathematical constraint*.

Now given a mathematical formulation of an appropriate form, you need to develop (or use an existing) algorithm to solve the formulation. By an algorithm, we mean in this

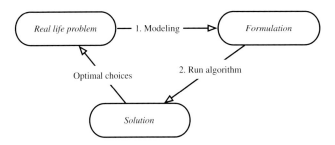

Figure 1.1 Modeling and solving optimization problems.

case a finite procedure (something that can be coded as a computer program) that will take as input the formulation, and return an assignment of values to each of the variables such that all constraints are satisfied, and, subject to these conditions, maximizes the objective function. The values assigned to the variables indicate the optimal choices for the parameters that the CEO of WaterTech wishes to determine.

This two-step process is summarized in Figure 1.1. In this chapter, we will focus our attention on the first step, namely how to formulate optimization problems.

1.1   A first example

To clarify these ideas, let us consider a simple example. Suppose WaterTech manufactures four products, requiring time on two machines and two types (skilled and unskilled) of labor. The amount of machine time and labor (in hours) needed to produce a unit of each product and the sales prices in dollars per unit of each product are given in the following table:

Product	Machine 1	Machine 2	Skilled labor	Unskilled labor	Unit sale price
1	11	4	8	7	300
2	7	6	5	8	260
3	6	5	5	7	220
4	5	4	6	4	180

Each month, 700 hours are available on machine 1 and 500 hours on machine 2. Each month, WaterTech can purchase up to 600 hours of skilled labor at $8 per hour and up to 650 hours of unskilled labor at $6 per hour. The company wants to determine how much of each product it should produce each month and how much labor to purchase in order to maximize its profit (i.e. revenue from sales minus labor costs).

1.1.1 The formulation

We wish to find a formulation for this problem, i.e. we need to determine the variables, the objective function, and the constraints.

Variables. WaterTech must decide how much of each product to manufacture; we capture this by introducing a variable x_i for each $i \in \{1, 2, 3, 4\}$ for the number of units of product i to manufacture. As part of the planning process, the company must also decide on the number of hours of skilled and unskilled labor that it wants to purchase. We therefore introduce variables y_s and y_u for the number of purchased hours of skilled and unskilled labor, respectively.

Objective function. Deciding on a production plan now amounts to finding values for variables $x_1, \ldots, x_4$, y_s and y_u. Once the values for these variables have been found, WaterTech's profit is easily expressed by the following function:

$$\underbrace{300x_1 + 260x_2 + 220x_3 + 180x_4}_{\text{Profit from sales}} - \underbrace{(8y_s + 6y_u)}_{\text{Labor costs}},$$

and the company wants to maximize this quantity.

Constraints. We manufacture x_1 units of product 1 and each unit of product 1 requires 11 hours on machine 1. Hence, product 1 will use $11x_1$ hours on machine 1. Similarly, for machine 1, product 2 will use $7x_2$ hours, product 3 will use $6x_3$ hours, and product 4 will use $5x_4$ hours. Hence, the total amount of time needed on machine 1 is given by

$$11x_1 + 7x_2 + 6x_3 + 5x_4,$$

and this must not exceed the available 700 hours of time on that machine. Thus

$$11x_1 + 7x_2 + 6x_3 + 5x_4 \leq 700. \tag{1.1}$$

In a similar way, we derive a constraint for machine 2

$$4x_1 + 6x_2 + 5x_3 + 4x_4 \leq 500. \tag{1.2}$$

Analogously, once we decide how much of each product should be produced, we know how much skilled and unskilled labor is needed. Naturally, we need to make sure that enough hours of each type of labor are purchased. The following two constraints enforce this:

$$8x_1 + 5x_2 + 5x_3 + 6x_4 \leq y_s, \tag{1.3}$$
$$7x_1 + 8x_2 + 7x_3 + 4x_4 \leq y_u. \tag{1.4}$$

Finally, we need to add constraints that force each of the variables to take on only nonnegative values as well as constraints that limit the number of hours of skilled and unskilled labor purchased. Combining the objective function with (1.1)–(1.4) gives the following formulation:

$$\max \quad 300x_1 + 260x_2 + 220x_3 + 180x_4 - 8y_s - 6y_u$$

subject to

$$
\begin{aligned}
11x_1 + 7x_2 + 6x_3 + 5x_4 &\leq 700 \\
4x_1 + 6x_2 + 5x_3 + 4x_4 &\leq 500 \\
8x_1 + 5x_2 + 5x_3 + 6x_4 &\leq \ y_s \\
7x_1 + 8x_2 + 7x_3 + 4x_4 &\leq \ y_u \\
y_s &\leq 600 \\
y_u &\leq 650 \\
x_1, x_2, x_3, x_4, y_u, y_s &\geq \ 0.
\end{aligned}
\tag{1.5}
$$

1.1.2   Correctness

Is the formulation given by (1.5) correct, i.e. does this formulation capture exactly the WaterTech problem? We will argue that it does and outline a procedure to verify whether a given formulation is correct. Let us introduce a bit of terminology. By the *word description* of the optimization problem, we mean a description of the optimization problem in plain English. This is the description that the CEO of WaterTech would give you. By a *formulation*, we mean the mathematical formulation, as in (1.5). A *solution* to the formulation is an assignment of values to each of the variables of the formulation. A solution is *feasible* if it has the property that all the constraints are satisfied. An *optimal solution* to the formulation is a feasible solution that maximizes the objective function (or minimizes it if the optimization problem is a minimization problem). Similarly, we define a solution to the word description of the optimization problem to be a choice for the unknowns, and a feasible solution to be such a choice that satisfies all the constraints.

To construct a formulation for an optimization problem, there are many approaches. Not all of them may apply to a given problem. Conceptually, an easy approach is to make sure that there is a mapping between feasible solutions of the word description and feasible solutions of the formulation and vice versa (between feasible solutions of the formulation and feasible solutions of the word description). For instance, a feasible solution for WaterTech is to produce 10 units of product 1, 50 of product 2, 0 units of product 3, and 20 of product 4, and buy 600 hours of both skilled and unskilled labor. This corresponds to the following feasible solution of the formulation:

$$x_1 = 10, \ x_2 = 50, \ x_3 = 0, \ x_4 = 20, \ y_s = 600, \ y_u = 600. \tag{1.6}$$

Conversely, given the feasible solution (1.6), we can construct a feasible solution for the word description of the WaterTech problem. Note that this works for every feasible solution. When constructing a formulation using this approach, you need to make sure that through the map that you defined:

(1) every feasible solution of the word description gives a feasible solution of the math-
    ematical formulation, and
(2) every feasible solution of the mathematical formulation gives a feasible solution of
    the word description.

If (2) does not hold, feasible solutions to the formulation may violate constraints of the
word description, and if (1) does not hold, then the formulation is more restrictive than
the word description. A common mistake is to violate (2) by forgetting some constraint
when writing down the formulation. For instance, we may forget to write down the condition for WaterTech that y_s is nonnegative. In this case, when solving the formulation
using an algorithm, we may end up with a negative value for y_s, i.e. we buy a negative
amount of skilled labor or equivalently we sell skilled labor; the latter is not allowed in
our word description. Thus far, we have only discussed feasible solutions. Clearly, we
also need to verify that the objective function in the word description and the formulation are the same. This is clearly the case for the WaterTech formulation, and is usually
straightforward to verify.
    We used an algorithm to find an optimal solution to (1.5) and obtained

$$x_1 = 16 + \frac{2}{3}, \; x_2 = 50, \; x_3 = 0, \; x_4 = 33 + \frac{1}{3}, \; y_s = 583 + \frac{1}{3}, \; y_u = 650, \quad (1.7)$$

achieving a total profit of $\$15,433 + \frac{1}{3}$. Thus, the optimal strategy for WaterTech is to
manufacture $16 + \frac{2}{3}$ units of product 1, 50 units of product 2, 0 units of product 3, $33 + \frac{1}{3}$
units of product 4, and to buy 583 hours of skilled labor and 650 units of unskilled labor.
    Since constructing the formulation is only our first step (see Figure 1.1) and we need
to use an algorithm to find an optimal solution to the formulation, we will strive to get,
among all possible formulations, one that is as simple as possible. In the remainder of
the chapter, we will introduce three types of formulation:

- linear programs (Section 1.2),
- integer programs (Section 1.3), and
- nonlinear programs (Section 1.6).

There are efficient techniques to solve linear programs and we will see some of these
in Chapters 2 and 7. Integer programs and nonlinear programs can be hard to solve
however. Thus, we will always attempt to formulate our problem as a linear program.
Unfortunately, this may not always be possible, and sometimes the only valid formulation is an integer program or a nonlinear program.

1.2   Linear programs

A function $f: \mathbb{R}^n \to \mathbb{R}$ is an *affine* function if $f(x) = a^\top x + \beta$, where x and a are vectors
with n entries and β is a real number. If $\beta = 0$, then f is a *linear function*. Thus, every
linear function is affine, but the converse is not true.

Example 1 Suppose $x = (x_1, x_2, x_3, x_4)^\top$. Then:

1. $f(x) = x_1 - x_3 + x_4$ is a linear function,
2. $f(x) = 2x_1 - x_3 + x_4 - 6$ is an affine function, but not a linear function, and
3. $f(x) = 3x_1 + x_2 - 6x_3x_4$ is not an affine function (because of the product x_3x_4).

A *linear constraint* is any constraint that is of one of the following forms (after moving all variables to the left-hand side and all constants to the right-hand side):

$$f(x) \le \beta \quad \text{or} \quad f(x) \ge \beta \quad \text{or} \quad f(x) = \beta,$$

where $f(x)$ is a linear function, and β is a real number. A *linear program* (LP) is the problem of maximizing or minimizing an affine function subject to a finite number of linear constraints. We will abbreviate the term linear program by LP, throughout this book.

Example 2

(a) The following is an LP:

$$\min \quad x_1 - 2x_2 + x_4$$
$$\text{subject to}$$
$$\begin{aligned} x_1 - x_3 &\le 3 \\ x_2 + x_4 &\ge 2 \\ x_1 + x_2 &= 4 \\ x_1, x_2, x_3, x_4 &\ge 0. \end{aligned}$$

(b) The following is not an LP, as $x_2 + x_3 < 3$ is not a linear constraint:[1]

$$\max \quad x_1$$
$$\text{subject to}$$
$$\begin{aligned} x_2 + x_3 &< 3 \\ x_1 - x_4 &\ge 1. \end{aligned}$$

(c) The following is not an LP, as the objective function is not affine:

$$\min \quad \frac{1}{x_1}$$
$$\text{subject to}$$
$$x_1 \ge 3.$$

---

[1] Strict inequalities are not allowed.

(d) The following is not an LP, as there are an infinite number of constraints:

$$\max \qquad x_1$$

subject to

$$x_1 + \alpha x_2 \leq 5 \qquad \text{(for all real numbers } \alpha \in [2, 3]).$$

Observe also that (1.5) is an example of an LP. In that example, the constraint $8x_1 + 5x_2 + 5x_3 + 6x_4 \leq y_s$ can be rewritten as $8x_1 + 5x_2 + 5x_3 + 6x_4 - y_s \leq 0$.

1.2.1   Multiperiod models

In this section, we present another example of a type of optimization problem that can be formulated as an LP. In the problem discussed in Section 1.1, we were asked to make a one-time decision on a production plan. Often times, the decision-making process has a temporal component; time is split into so-called *periods* and we have to make certain decisions at the beginning or end of each of them. Each of these decisions will naturally determine the final outcome at the end of all periods. We introduce this area with an example. KWOil is a local supplier of heating oil. The company has been around for many years, and knows its home turf. In particular, KWOil has developed a dependable model to forecast future demand for oil. For each of the following four months, the company expects the following amounts of demand for heating oil.

Month	1	2	3	4
Demand (litres)	5000	8000	9000	6000

At the beginning of each of the four months, KWOil may purchase heating oil from a regional supplier at the current market rate. The following table shows the projected price per litre at the beginning of each of these months:

Month	1	2	3	4
Price ($/litres)	0.75	0.72	0.92	0.90

KWOil has a small storage tank on its facility. The tank can hold up to 4000 litres of oil, and currently (at the beginning of month 1) contains 2000 litres. The company wants to know how much oil it should purchase at the beginning of each of the four months such that it satisfies the projected customer demand at the minimum possible total cost. Note, oil that is delivered at the beginning of each month can be delivered directly to the customer, it does not need to be first put into storage; only oil that is left over at the end of the month goes into storage. We wish to find an LP formulation for this problem. Thus, we need to determine the variables, the objective function, and the constraints.

Variables. KWOil needs to decide how much oil to purchase at the beginning of each of the four months. We therefore introduce variables p_i for $i \in \{1, 2, 3, 4\}$ denoting the number of litres of oil purchased at the beginning of month i for $i \in \{1, 2, 3, 4\}$. We also introduce variables t_i for each $i \in \{1, 2, 3, 4\}$ to denote the number of litres of heating oil in the company's tank at the beginning of month i (we already know that $t_1 = 2000$ – we can substitute this value later as we finish constructing our mathematical formulation). Thus, while every unknown of the word description always needs to be represented as a variable in the formulation, it is sometimes useful, or necessary, to introduce additional variables to keep track of various parameters.

Objective function. Given the variables defined above, it is straightforward to write down the cost incurred by KWOil. The objective function of KWOil's problem is

$$\min \quad 0.75p_1 + 0.72p_2 + 0.92p_3 + 0.90p_4. \tag{1.8}$$

Constraints. In each month i, the company needs to have enough heating oil available to satisfy customer demand. The amount of available oil at the beginning of month 1, for example, is comprised of two parts: the p_1 litres of oil purchased in month 1, and the t_1 litres contained in the tank. The sum of these two quantities needs to cover the demand in month 1, and the excess is stored in the local tank. Hence, we obtain the following constraint:

$$p_1 + t_1 = 5000 + t_2. \tag{1.9}$$

We obtain similar constraints for months 2 and 3

$$p_2 + t_2 = 8000 + t_3, \tag{1.10}$$

$$p_3 + t_3 = 9000 + t_4. \tag{1.11}$$

Finally, in order to satisfy the demand in month 4, we need to satisfy the following constraint:

$$p_4 + t_4 \geq 6000. \tag{1.12}$$

Notice that each of the variables t_i for $i \in \{2, 3, 4\}$ appears in two of the constraints (1.9) and (1.12). The constraints are therefore linked by the variables t_i. Such linkage is a typical feature in multiperiod models. Constraints (1.9) and (1.12) are sometimes called *balance constraints* as they *balance* demand and inventory between periods.

We now obtain the entire LP for the KWOil problem by combining (1.8)–(1.12), and by adding upper bounds and initialization constraints for the tank contents, as well as non-negativity constraints

$$\text{min} \quad 0.75p_1 + 0.72p_2 + 0.92p_3 + 0.90p_4$$

subject to

$$
\begin{aligned}
p_1 + t_1 &= 5000 + t_2 \\
p_2 + t_2 &= 8000 + t_3 \\
p_3 + t_3 &= 9000 + t_4 \\
p_4 + t_4 &\geq 6000 \\
t_1 &= 2000 \\
t_i &\leq 4000 \quad (i = 2, 3, 4) \\
t_i, p_i &\geq 0 \quad (i = 1, 2, 3, 4).
\end{aligned}
$$

Solving this LP yields

$$p_1 = 3000, \ p_2 = 12000, \ p_3 = 5000, \ p_4 = 6000, \ t_1 = 2000, \ t_2 = 0, \ t_3 = 4000, \ t_4 = 0,$$

corresponding to a total purchasing cost of $20\,890. Not surprisingly, this solution suggests to take advantage of the low oil prices in month 2, while no oil should be stored in month 3 when prices are higher.

Exercises

1 Consider the following table indicating the nutritional value of different food types:

Foods	Price ($) per serving	Calories per serving	Fat (g) per serving	Protein (g) per serving	Carbohydrate (g) per serving
Raw carrots	0.14	23	0.1	0.6	6
Baked potatoes	0.12	171	0.2	3.7	30
Wheat bread	0.2	65	0	2.2	13
Cheddar cheese	0.75	112	9.3	7	0
Peanut butter	0.15	188	16	7.7	2

You need to decide how many servings of each food to buy each day so that you minimize the total cost of buying your food while satisfying the following daily nutritional requirements:

- calories must be at least 2000,
- fat must be at least 50g,
- protein must be at least 100g,
- carbohydrates must be at least 250g.

Write an LP that will help you decide how many servings of each of the aforementioned foods are needed to meet all the nutritional requirement, while minimizing the total cost of the food (you may buy fractional numbers of servings).

2   MUCOW (Milk Undertakings, C and O, Waterloo) owns a herd of Holstein cows and a herd of Jersey cows. For each herd, the total number of litres produced each day, and milk-fat content (as a *percentage*) are as follows:

	Litres produced	Percent milk-fat
Holstein	500	3.7
Jersey	250	4.9

The fat is split off and blended again to create various products. For each product, the volume, required milk-fat percentage, and profit are as follows. In particular, the milk-fat percentage must exactly equal what is specified.

	Skimmed milk	2%	Whole milk	Table cream	Whipping cream
Volume (litres)	2	2	2	0.6	0.3
Milk-fat requirement	0%	2%	4%	15%	45%
Profit ($)	0.1	0.15	0.2	0.5	1.2

(a) Formulate as an LP the problem of deciding how many items of each type to produce, so that the total profit is maximized. Don't worry about fractional numbers of items. Write your formulation in matrix notation.

(b) MUCOW is told of a regulation change: 'skimmed milk' can now contain anything up to 0.1% milk-fat, but no more. How does the formulation of the problem change? Note the resulting formulation should also be an LP.

3   The director of the CO-Tech startup needs to decide what salaries to offer its employees for the coming year. In order to keep the employees satisfied, she needs to satisfy the following constraints:

- Tom wants at least $20 000 or he will quit;
- Peter, Nina, and Samir each want to be paid at least $5000 more than Tom;
- Gary wants his salary to be at least as high as the combined salary of Tom and Peter;
- Linda wants her salary to be $200 more than Gary;
- the combined salary of Nina and Samir should be at least twice the combined salary of Tom and Peter;
- Bob's salary is at least as high as that of Peter and at least as high as that of Samir;
- the combined salary of Bob and Peter should be at least $60 000;
- Linda should not make more money than the combined salary of Bob and Tom.

(a) Write an LP that will determine salaries for the employees of CO-tech that satisfy each of these constraints while minimizing the total salary expenses.
(b) Write an LP that will determine salaries for the employees of CO-tech that satisfy each of these constraints while minimizing the salary of the highest paid employee.
(c) What is the relation between the solutions for (a) and (b)?

4   You wish to build a house and you have divided the process into a number of tasks, namely:
B. excavation and building the foundation,
F. raising the wooden frame,
E. electrical wiring,
P. indoor plumbing,
D. dry walls and flooring,
L. landscaping.
You estimate the following duration for each of the tasks (in weeks):

Task	B	F	E	P	D	L
Duration	3	2	3	4	1	2

Some of the tasks can only be started when some other tasks are completed. For instance, you can only build the frame once the foundation has been completed, i.e. F can start only after B is completed. All the precedence constraints are summarized as follows:

- F can start only after B is completed,
- L can start only after B is completed,
- E can start only after F is completed,
- P can start only after F is completed,
- D can start only after E is completed,
- D can start only after P is completed.

The goal is to schedule the starting time of each task such that the entire project is completed as soon as possible.
   As an example, here is a feasible schedule with a completion time of ten weeks.

Tasks	B	F	E	P	D	L
Starting time	0	3	6	5	9	6
End time	3	5	9	9	10	8

Formulate this problem as an LP. Explain your formulation. Note, that there is no limit on the number of tasks that can be done in parallel.
   HINT: Introduce variables to indicate the times that the tasks start.

5 The CRUD chemical plant produces as part of its production process a noxious compound called chemical X. Chemical X is highly toxic and needs to be disposed of properly. Fortunately, CRUD is linked by a pipe system to the FRESHAIR recycling plant that can safely reprocess chemical X. On any give day, the CRUD plant will produce the following amount of Chemical X (the plant operates between 9am and 3pm only):

Time	9–10 am	10–11 am	11am–12pm	12–1 pm	1–2 pm	2–3 pm
Chemical X (in litres)	300	240	600	200	300	900

Because of environmental regulation, at no point in time is the CRUD plant allowed to keep more than 1000 litres on site and no chemical X is allowed to be kept overnight. At the top of every hour, an arbitrary amount of chemical X can be sent to the FRESHAIR recycling plant. The cost of recycling chemical X is different for every hour:

Time	10am	11am	12pm	1pm	2pm	3pm
Price ($ per litre)	30	40	35	45	38	50

You need to decide how much chemical to send from the CRUD plant to the FRESHAIR recycling plant at the top of each hour, so that you can minimize the total recycling cost but also meet all the environmental constraints. Formulate this problem as an LP.

6 We are given an m by n matrix A and a vector b in $\mathbb{R}^m$, for which the system $Ax = b$ has no solution. Here is an example:

$$2x_1 - x_2 = -1$$
$$x_1 + x_2 = 1$$
$$x_1 + 3x_2 = 4$$
$$-2x_1 + 4x_2 = 3.$$

We are interested in finding a vector $x \in \mathbb{R}^n$ that "comes close" to solving the system. Namely, we want to find an $x \in \mathbb{R}^n$ whose *deviation* is minimum, and where the deviation of x is defined to be

$$\sum_{i=1}^{n} |b_i - \sum_{j=1}^{n} a_{ij}x_j|.$$

(For the example system above, the vector $x = (1, 1)^\top$ has deviation $2 + 1 + 0 + 1 = 4$.)

(a) Show that a solution to the optimization problem

$$\text{minimize} \sum_{i=1}^{m} y_i$$

subject to

$$|\sum_{j=1}^{n} a_{ij}x_j - b_i| \le y_i \qquad\qquad (i = 1, \ldots, m)$$

will give a vector x of minimum deviation.

(b) The problem of part (a) is not an LP. (Why?) Show that it can be formulated as an LP.

(c) Suppose that we had instead defined the deviation of x as

$$\max_{1 \le i \le m} |b_i - \sum_{j=1}^{n} a_{ij}x_j|.$$

(According to this definition, in the example above $x = (1, 1)^\top$ would have deviation $\max(2, 1, 0, 1) = 2$.) With this new definition, write the problem of finding a vector of minimum deviation as an optimization problem, and show that this problem can also be formulated as an LP.

7   Consider the following set up: we have factories 1 through m and stores 1 through n. Each factory i produces u_i units of a commodity and each store j requires ℓ_j units of that commodity. Note, each factory produces the same commodity, and each store requires the same commodity. The goal is to transfer the commodity from the factories to the stores. All the commodities going from the factories to the stores are first sent to one of two central storage hubs A and B. The cost of transporting one unit of commodity from factory i to hub A (resp. B) is given by a_i (resp. b_i). The cost of transporting one unit of commodity from hub A (resp. B) to store j is given by a'_j (resp. b'_j).[2] In the figure on top of the next page, we illustrate the case of three factories and four stores. The problem is to decide how much to send from each factory to each hub and how much to send from each hub to each store so that each store receives the amount of commodity it requires, no factory sends out more commodity than it produces, and such that the total transportation cost is minimized. Formulate this problem as an LP (we may assume that the number of units of commodity sent may be fractional).

8   We are given a matrix $A \in \mathbb{R}^{m \times n}$ and a matrix $B \in \mathbb{R}^{p \times n}$ so that the rows of A denote observations for healthy human tissue and the rows of B denote observations for unhealthy human tissue. We would like to find $a \in \mathbb{R}^n$ and $\alpha, \beta \in \mathbb{R}$ such that all rows of A are in the set $\{x \in \mathbb{R}^n : a^\top x \le \alpha\}$, all rows of B are in the set $\{x \in \mathbb{R}^n : a^\top x \ge \beta\}$,

---

[2] Quantities $u_i, \ell_j, a_i, b_i, a'_j, b'_j$ are constants for all i and j.

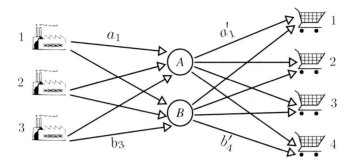

and such that the distance between the sets $\{x \in \mathbb{R}^n : a^\top x = \alpha\}$ and $\{x \in \mathbb{R}^n : a^\top x = \beta\}$ is maximized. The following figure illustrates the situation for $n = 2$; circles correspond to rows in A, and squares to rows in B.

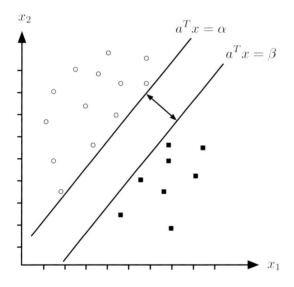

Formulate the problem of computing a, α, and β achieving the above-mentioned goals as an LP.

1.3   Integer programs

An *integer program* is obtained by taking a linear program and adding the condition that a nonempty subset of the variables be required to take integer values. When all variables are required to take integer values, the integer program is called a *pure integer program* otherwise it is called a *mixed integer program*. We will abbreviate the term integer program by IP, throughout this book.

Example 3 The following is a mixed IP, where variables x_1 and x_3 are required to take integer values:

$$\max \quad x_1 + x_2 + 2x_4$$

subject to

$$
\begin{aligned}
x_1 + x_2 &\leq 1 \\
-x_2 - x_3 &\geq -1 \\
x_1 + x_3 &= 1
\end{aligned}
$$

$$x_1, x_2, x_3 \geq 0 \text{ and } x_1, x_3 \text{ integer.}$$

In Section 1.1, we introduced the WaterTech production problem. We gave an LP formulation (1.5) and a solution to that formulation in (1.7). This solution told us to manufacture, $16 + \frac{2}{3}$ units of product 1. Depending on the nature of product 1, it may not make sense to produce a fractional number of units of this product. Thus, we may want to add the condition that each of x_1, x_2, x_3, x_4 is an integer. The resulting program would be an IP. In this example, we could try to ignore the integer condition, and round down the solution, hoping to get a reasonably good approximation to the optimal solution.

1.3.1   Assignment problem

Our friends at WaterTech are once again looking to us for help. The company faces the following problem: there is a set of J jobs that need to be handled urgently. The company has selected I of its most trusted employees to handle these jobs. Naturally, the skill sets of these employees differ, and not all of the jobs are equally well handled by all of the employees. From past experience, management knows the number of hours c_{ij} each worker $i \in I$ is expected to take in order to complete any of the jobs $j \in J$. The following table gives an example for a case with $|J| = 4$ jobs and $|I| = 4$ employees:

Employees	Jobs			
	1	2	3	4
1	3	5	1	7
2	8	2	2	4
3	2	1	6	8
4	8	3	3	2

For instance, the table says that $c_{3,4} = 8$, i.e. employee 3 would take eight hours to finish job 4. WaterTech wants to assign jobs to employees with the conditions that:

(1) each employee $i \in I$ is assigned exactly one job $j \in J$,
(2) each job $j \in J$ is assigned to exactly one employee $i \in I$.

Both of these conditions can only be satisfied when $|I| = |J|$. Naturally, we want to find such an assignment that minimizes the total expected amount of time needed to process all jobs J. A feasible solution would be to assign job k to employee k, for $k = 1, 2, 3, 4$. The total amount of time required for this assignment is $3 + 2 + 6 + 2 = 13$ hours. This not an optimal solution, however. We wish to find an IP formulation for this problem. Thus, we need to determine the variables, the constraints, and the objective function.

Variables. In this case, we need to decide for each employee $i \in I$ and each job $j \in J$ whether employee i is assigned job j. We will introduce for every such pair i, j a variable x_{ij} that we restrict to take values 0 or 1, where $x_{ij} = 1$ represents the case where employee i is assigned job j, and $x_{ij} = 0$ means that employee i is not assigned job j. Thus, we have $|I||J|$ variables. In the example given by the above table we end up with 16 variables.

Constraints. We need to encode condition (1) as a mathematical constraint. Let $i \in I$ be an employee, then $\sum_{j \in J} x_{ij}$ is the number of jobs employee i is assigned to (we do the sum over all jobs). We want this quantity to be one, thus the following should hold:

$$\sum_{j \in J} x_{ij} = 1. \tag{1.13}$$

In the example given by the table, this says that $x_{i1} + x_{i2} + x_{i3} + x_{i4} = 1$ for all jobs $i \in \{1, 2, 3, 4\}$. We need to encode condition (2) as a mathematical constraint. Let $j \in J$ be a job then $\sum_{i \in I} x_{ij}$ is the number of employees job j is assigned to (we do the sum over all employees). We want this quantity to be one, thus the following should hold:

$$\sum_{i \in I} x_{ij} = 1. \tag{1.14}$$

Objective function. The objective function should calculate the total amount of time spent to complete the jobs. For every employee $i \in I$ and job $j \in J$, if employee i is assigned job j, then we should contribute c_{ij} to the objective function, otherwise we should contribute 0. Thus, we should contribute $c_{ij} x_{ij}$. Therefore, the objective function is given by

$$\sum_{i \in I} \sum_{j \in J} c_{ij} x_{ij}. \tag{1.15}$$

For instance, in our specific example, the objective function is

$$3x_{11} + 5x_{12} + 1x_{13} + 7x_{14} + 8x_{21} + 2x_{22} + 2x_{23}$$
$$+ 4x_{24} + 2x_{31} + 1x_{32} + 6x_{33} + 8x_{34} + 8x_{41} + 3x_{42} + 3x_{43} + 2x_{44}.$$

Thus, the IP formulation is given by objective function (1.15), and constraints (1.13) and (1.14), as well as the condition that each variable x_{ij} can only take values $0, 1$

$$\min \qquad \sum_{i \in I} \sum_{j \in J} c_{ij} x_{ij}$$

subject to

$$\begin{aligned}
&\textstyle\sum_{j \in J} x_{ij} = 1 && (i \in I) \\
&\textstyle\sum_{i \in I} x_{ij} = 1 && (j \in J) \\
&x_{ij} \in \{0, 1\} && (i \in I, j \in J).
\end{aligned} \qquad (1.16)$$

Note, formally speaking, that (1.16) is not an IP, because of the constraints $x_{ij} \in \{0, 1\}$. However, we can clearly replace these constraints by the constraints $x_{ij} \geq 0$, $x_{ij} \leq 1$ and x_{ij} integer. If we do this for all $i \in I$ and $j \in J$, then the resulting optimization formulation is an IP. Hence, we will abuse notation slightly and call the formulation (1.16) an IP.

Solving this IP for the special case given in the table, yields

$$x_{11} = 1, \ x_{23} = 1, \ x_{32} = 1, \ x_{44} = 1$$

and all other values $x_{ij} = 0$. Thus, an optimal solution is to assign job 1 to employee 1, job 3 to employee 2, job 2 to employee 3, and job 4 to employee 4.

Note, in this example we represent a binary choice (whether to assign job j to employee i) by a variable taking values 0 or 1. We call such a variable a *binary variable*. When using a binary variable y, we can express $y \in \{0, 1\}$ by $0 \leq y \leq 1$ and y integer, but the condition that y is integer cannot easily be omitted, as we may otherwise get any value between 0 and 1 (say $\frac{1}{2}$ for instance) and this value gives no information as to what our binary choice should be.

1.3.2 Knapsack problem

The company KitchTech wishes to ship a number of crates from Toronto to Kitchener in a freight container. The crates are of six possible types, say type 1 through type 6. Each type of crate has a given weight in kilograms and has a particular retail value in $, as indicated in the following table:

Type	1	2	3	4	5	6
Weight (kg)	30	20	30	90	30	70
Value ($)	60	70	40	70	20	90

In addition you have the following constraints:

(1) you cannot send more than ten crates of the same type in the container;
(2) you can only send crates of type 3, if you send at least one crate of type 4;

(3) at least one of the following two conditions has to be satisfied:
    i. a total of at least four crates of type 1 or type 2 is selected or
    ii. a total of at least four crates of type 5 or type 6 is selected.

Finally, the total weight allowed on the freight container is 1000 kilograms. Your goal is to decide how many crates of each type to place in the freight container so that the value of the crates in the container is maximized. We wish to find an IP formulation for this problem. Thus, we need to determine the variables, the constraints, and the objective function.

Variables. We will have variables x_i for $i = 1, \ldots, 6$ to indicate how many crates of type i we place in the container. We will also require an additional binary variable y to handle condition (3).

Constraints. The total weight of the crates selected in kilograms is given by

$$30x_1 + 20x_2 + 30x_3 + 90x_4 + 30x_5 + 70x_6.$$

This weight should not exceed 1000 kilograms. Thus

$$30x_1 + 20x_2 + 30x_3 + 90x_4 + 30x_5 + 70x_6 \leq 1000. \qquad (1.17)$$

Condition (1) simply says that for all $i = 1, \ldots, 6$

$$x_i \leq 10. \qquad (1.18)$$

We claim that condition (2) can be stated as

$$x_3 \leq 10x_4. \qquad (1.19)$$

If no crates of type 4 is sent, then $x_4 = 0$, which implies by (1.19) that $x_3 = 0$ as well, i.e. no crates of type 3 are sent. On the other hand, if at least one crate of type 4 is sent, then $x_4 \geq 1$ and (1.19) says that $x_3 \leq 10$, which is the maximum number of crates of type 3 we can send anyway.

It remains to express condition (3). The binary variable y will play the following role. If $y = 1$, then we want (i) to be true, and if $y = 0$, then we want (ii) to be true. This can be achieved by adding the following two constraints:

$$x_1 + x_2 \geq 4y$$
$$x_5 + x_6 \geq 4(1 - y). \qquad (1.20)$$

Let us verify that (1.20) behaves as claimed. If $y = 1$, then the conditions become $x_1 + x_2 \geq 4$ and $x_5 + x_6 \geq 0$, which implies that (i) holds. If $y = 0$, then the conditions become $x_1 + x_2 \geq 0$ and $x_5 + x_6 \geq 1$, which implies that (ii) holds.

Objective function. The total value of the crates selected for the container is given by

$$60x_1 + 70x_2 + 40x_3 + 70x_4 + 20x_5 + 90x_6. \qquad (1.21)$$

Thus, the IP formulation is given by objective function (1.21), and constraints (1.17), (1.18), (1.19), (1.20) as well as the condition that each variable x_i is integer, and $y \in \{0, 1\}$. We obtain

$$\text{max} \quad 60x_1 + 70x_2 + 40x_3 + 70x_4 + 20x_5 + 90x_6$$

subject to

$$30x_1 + 20x_2 + 30x_3 + 90x_4 + 30x_5 + 70x_6 \leq 10000$$

$$
\begin{aligned}
x_i &\leq 10 & (i = 1, \ldots, 6) \\
x_3 &\leq 10x_4 \\
x_1 + x_2 &\geq 4y \\
x_5 + x_6 &\geq 4(1 - y) \\
x_i &\geq 0 & (i = 1, \ldots, 6) \\
x_i &\quad \text{integer} & (i = 1, \ldots, 6) \\
y &\in \{0, 1\}.
\end{aligned}
$$

Exercises

1 You are about to trek across the desert with a vehicle having 3.6 cubic metres (3.6m³) of cargo space for goods. There are various types of items available for putting in this space, each with a different volume and a different net value for your trip, shown as follows:

Item type i	1	2	3	4	5	6	7
Volume v_i (m³)	0.55	0.6	0.7	0.75	0.85	0.9	0.95
Net value n_i	250	300	500	700	750	900	950

(a) You need to decide which items to take, not exceeding the volume constraint. You can take at most one of any item. No item can be split into fractions. The total net value must be maximized. Formulate this problem as an LP or IP. (You may use the notation v_i and n_i for volume and net value of item i.)

(b) Your two friends have decided to come as well, each with an identical vehicle. There are exactly two items of each type. The question is, can you fit all 14 items in the vehicles without exceeding the volume constraints? No cutting items into pieces is permitted! Each item taken must be placed entirely in one of the vehicles. Formulate an LP or IP that has a feasible solution if and only if the items can be packed as desired. Describe how to determine from a feasible solution how to pack the items into the vehicles. Note that net value is ignored for part (b).

2 Consider a public swimming pool. In the following table, we give a list of seven potential lifeguards. For each lifeguard, we have the time he/she wants to start and finish work and how much he/she wishes to be paid for the work. The problem is to find a selection of lifeguards so that there is at least one (but possibly more than one) lifeguard present at each time between 1pm. and 9pm. An example of a possible selection would be Joy, Tim, and Beth. This selection has a total cost of $30 + 21 + 20$.

Lifeguards	Joy	Dean	Tim	Celicia	Beth	Ivar	Eilene
Hours	1–5	1–3	4–7	4–9	6–9	5–8	8–9
Amount required	30	18	21	38	20	22	9

Formulate this problem as an IP.

3   You have gone to an exotic destination during the summer vacation and decided to do your part to stimulate the economy by going on a shopping spree. Unfortunately, the day before your return you realize that you can only take 20 kilograms of luggage on your flight which is less than the total weight of the items that you purchased. The next table gives the value and the weight in kilograms of each item:

	A	B	C	D	E	F
Weight (kg)	6	7	4	9	3	8
Value ($)	60	70	40	70	16	100

The problem is to decide which subset of the items to put in your luggage so that you maximize the total value of the items selected without exceeding the weight requirement (i.e. that the total weight of the items selected is no more than 20 kilograms). For instance, you could select items A, C, and D for a total value of $170 and a total weight of 19 kilograms.

(a) Formulate this problem as an IP.

(b) Suppose that you only want to pack item D when item A is selected, but it is ok to pack item A without item D. Add constraints to your formulation that impose this additional condition.

(c) Suppose that the airline allows you exceed the 20 kilogram weight limit at a cost of $15 per additional kilogram. For instance, you could select items A, B and D for a total value of $200 and a total weight of 22 kilogram and pay $2 \times \$15$ to the airline for exceeding the maximum capacity by 2 kilograms. Modify your formulation so that you are allowed to go over the 20 kilogram capacity and such that you maximize the total value of the items packed minus the cost paid to the airline.

Note, for (b) and (c) the resulting formulation should remain an IP.

4   The Waterloo hotel wants to rent rooms 1, 2, and 3 for New Year's night. Abby is willing to pay $60 for room 1, $50 for room 2, but is not interested in room 3. Bob is willing to pay $40 for room 1, $70 for room 2, and $80 for room 3. Clarence is not interested in room 1, but is willing to pay $55 for room 2 and $75 for room 3. Donald is willing to pay $65 for room 1, $90 for room 2, but is not interested in room 3. The information is summarized in the following table:

Room number	Abby's offer	Bob's offer	Clarence's offer	Donald's offer
1	$60	$40	not interested	$65
2	$50	$70	$55	$90
3	not interested	$80	$75	not interested

The hotel wants to fill up rooms 1,2,3 with some of the potential clients (Abby, Bob, Clarence, and Donald) in a way that maximizes the total revenue. Each room is to be assigned to *exactly one* potential client, and each potential client is to be assigned *at most one* room. As an example, Room 1 could be assigned to Bob, room 2 to Abby, and room 3 to Clarence (while Donald would not get to stay in the hotel). This would yield a revenue of $40+$50+$75=$165.

(a) Formulate this problem as an IP. Your solution should be easy to modify if we change the values in the table.

(b) Abby and Bob have a history of loud and rude behavior when celebrating together. In the interest of keeping the New Year's eve party orderly, the hotel management decides that it does not wish to rent rooms to *both* Abby and Bob. Add a constraint to the IP in (a) that will enforce this condition (the resulting formulation should still be an IP).

5 You wish to find out how to pack crates on a transport plane in an optimal way. The crates are of five possible types, namely A, B, C, D, E. For each crate type, the next table gives its weight (in kg), its volume (in cubic meters), and its value (in dollars):

Type	A	B	C	D	E
Weight	500	1500	2100	600	400
Volume	25	15	13	20	16
Value	50000	60000	90000	40000	30000

The transport plane is divided into three segments: Front, Middle, and Back. Each segment has a limited volume (in cubic meters), and a limit on the weight of the cargo in that segment (in kg):

Segment	Front	Middle	Back
Available volume	200	500	300
Weight capacity	8000	20000	6000

Finally, to keep the plane balanced we need to satisfy the following constraints:

weight of Middle cargo $\geq$ weight of Front cargo $+$ weight Back cargo,

weight of Middle cargo $\leq 2 \times$ (weight of Front cargo $+$ weight Back cargo).

Suppose that there are 12 crates of type A, eight crates of type B, 22 crates of type C, 15 crates of type D, and 11 crates of type E that are waiting to be transported. Your goal is to maximize the total value of the crates on the plane. You need to decide how many crates of each type are going in what segment of the plane. Formulate your problem as an IP.

6 Consider an LP with variables x_1, x_2, x_3, x_4. Suppose that the LP includes the constraints $x_1, x_2, x_3, x_4 \geq 0$.

(a) Consider the constraint:

$$x_4 \geq |x_3 - 2x_1|. \tag{1.22}$$

Suppose that we want to add to the LP the condition that (1.22) is satisfied. Show how to satisfy this requirement so that the resulting formulation is an LP.

HINT: rewrite (1.22) as a pair of linear inequalities.

(b) Consider the following inequalities:

$$6x_1 + 2x_2 + 3x_3 + 3x_4 \geq 3, \tag{1.23}$$

$$2x_1 + 4x_2 + 2x_3 + 7x_4 \geq 9. \tag{1.24}$$

Suppose that we want to add to an IP the condition that at *least one* of constraints (1.23) or (1.24) is satisfied. Show how to satisfy this requirement so that the resulting formulation is an IP.

HINT: add a binary variable indicating whether (1.23) or (1.24) must be satisfied. Note that the left-hand side of either (1.23) or (1.24) is always nonnegative.

(c) Suppose that for $i = 1, \ldots, k$ we have a non-negative vector a^i with four entries and a number β_i (both a^i and β_i are constants). Let r be any number between 1 and k. Consider the following set of inequalities:

$$(a^i)^\top x \geq \beta_i \qquad (i = 1, \ldots, k). \tag{1.25}$$

We want to add to an IP the condition that at *least r* of the constraints are satisfied. Show how to satisfy this requirement so that the resulting formulation is an IP.

HINT: add a binary variable for each constraint in (1.25).

(d) Consider the following set of values:

$$\mathscr{S} := \{3, 9, 17, 19, 36, 67, 1893\}.$$

Suppose that we want to add to an IP the condition that the variable x takes only one of the values in $\mathscr{S}$. Show how to satisfy this requirement so that the resulting formulation is an IP.

HINT: add a binary variables for each number in the set $\mathscr{S}$.

7 The company C & O operates an oil pipeline pumping oil from Alberta to various states in the Northwestern USA. Figure 1.2 shows the direction of flow, four input lines, and the three output lines. Note for instance that State A can only get its oil from either Input 1 or from the Yukon input line.

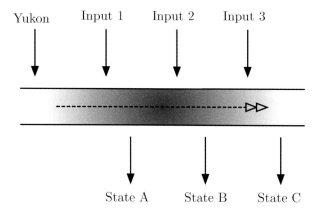

Figure 1.2 The structure of the oil pipeline, inputs and outputs.

Each input line has a capacity (barrels/day) and a cost per barrel:

Input line	1	2	3	Yukon
Capacity	4000	2000	3000	10000
Cost per barrel ($)	70	50	30	60

Each state has a daily demand (barrels/day) that must be met exactly:

State	A	B	C
Demand	3500	3000	4000

The input from the Yukon is not owned by the company and activating that line has a fixed cost of $11 000 per day.

Write an IP that plans the activities of the company C & O for a day (how many barrels of oil to pump from each input line) by minimizing the total daily cost of the company while meeting all the demand.

8 A company won a government bid to meet the yearly demands $d_1, d_2, \ldots, d_n$ in the areas $j \in \{1, 2, \ldots, n\}$. Now the company has to decide where to build its factories and how much of each factory's output will be shipped to which of these n areas.

There are m potential locations for building the factories. If the company decides to build at location $i \in \{1, 2, \ldots, m\}$, then the fixed cost of building the factory (yearly amortized version) is f_i and the yearly capacity of the factory will be s_i. The cost of transporting one unit of the product from location i to area j is given as c_{ij}.

Construct an IP whose solution indicates where to build the factories, how many units of product to ship from each factory to each demand area so that the demand is met and the total yearly cost of the company is minimized.

9 Your boss asks you to purchase b_i units of product i for each i in a set P of products. (These products are all divisible, i.e. they can be obtained in fractional amounts.) Of course, your boss wants you to spend as little money as possible. You call up all the stores in a set S of stores, and store j gives you a per-unit price c_{ij} for product i for all i, j.

(a) You decide to just order all b_i units of product i from the store that gives the cheapest per-unit price for each i. Show that this is optimal.

(b) Actually, there is another constraint. Your boss forgot to tell you that he does not want you to buy from too many different stores – he wants you to keep the number of stores from which you buy to at most (integer) k. Modify your formulation in (a), the resulting formulation should be an IP.

(c) It turns out that the stores have special deals. If the total value of your order from store j is at least t_j dollars, it will give you d_j cash back. (All stores j offer such a deal, with perhaps different values of t_j and d_j.) Modify your formulation in (b), the resulting formulation should be an IP.

10 KW mining has an open-pit mine with 12 blocks of ore as shown in the figure below. The *mineability condition* says that no block can be mined without mining all blocks which lie at least partially above it. For example, block 7 cannot be mined unless blocks 2 and 3 are mined, and block 12 requires blocks 8 and 9, and hence 3, 4, and 5 too.

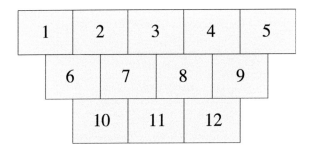

(a) The net value of mining block i is given by m_i (in \$) for $i = 1, \dots, 12$. Formulate as an IP the problem of deciding which blocks should be mined, in order to maximize the total value of blocks mined, and satisfy the mineability condition if at most seven blocks can be mined.

(b) The volume of mining block i is given by v_i (in m^3) for $i = 1, \dots, 12$. What extra constraint would you add if, in addition to all constraints needed in (a), it is required that the total volume mined must not exceed $10\,000\ m^3$?

(c) Mines often have a minimum short-term return requirement. For this mine, the board of the company requires that the total value of blocks mined in the first two years must total at least \$1\,000\,000. Each block takes one year to mine, at most two blocks can be mined at once, and a block cannot be mined in a given year unless all blocks lying at least partially above it were mined by the year before. Besides this new condition, the mineability condition still applies, and at most seven blocks can be mined. Formulate as an IP the problem of deciding which blocks should be mined, subject to these constraints, in order to maximize total value of blocks mined.

11 Suppose that you are given an $N \times N$ chessboard. We wish to place N queens on the board such that no two queens share any row, column, or diagonal. Figure 1.3 shows a solution for $N = 8$.

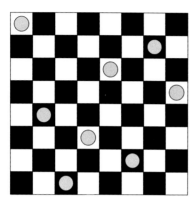

Figure 1.3

Formulate this problem as an integer feasibility problem (i.e. an IP without an objective function).

12 A 9×9 matrix A is partitioned into nine 3×3 submatrices $A_1, \ldots, A_9$ (of consecutive elements). Certain entries of A contain numbers from the set $\{1, \ldots, 9\}$. An example of such a pre-assignment is shown in Figure 1.4. A solution to the Sudoku game is an assignment of integers from 1 to 9 to each (unassigned) entry of the matrix such that:

- each row of A,
- each column of A,
- each 3 by 3 submatrix $A_1, \ldots, A_9$

contains every number from $\{1, \ldots, 9\}$ exactly once.

Formulate the problem of checking whether there is a solution to the Sudoku game as an integer feasibility problem (i.e. an IP without an objective function).

HINT: define a binary variable x_{ijk} that takes value 1 when entry i, j is assigned value k.

13 (Advanced) In *Conway's Game of Life*, we are given a chess board of size $n \times n$ for some positive integer n. Each cell (i, j) of this board has up to eight neighboring cells $N_{i,j}$. A cell (i, j) can be either *alive* or *dead* and a configuration of the game consists of a set of cells that are alive: $\mathcal{L} = \{(i_1, j_1), \ldots, (i_k, j_k)\}$. We use a set of simple rules to compute the successor configuration $\text{succ}(\mathcal{L})$ to $\mathcal{L}$:

(a) if there is at most one living cell in the neighborhood of (i, j), then (i, j) will be dead in the next iteration;

(b) if there are exactly two living cells in $N_{i,j}$, then we do not change the status of (i, j);

(c) if (i,j) has exactly three living neighbors in $N_{i,j}$, then its status in the next configuration will be alive;

(d) if there are at least four living cells in the neighborhood of (i,j), then (i,j) will be dead in the next iteration.[3]

A *still life* is a configuration $\mathcal{L}$ such that $\text{succ}(\mathcal{L}) = \mathcal{L}$ and the density of $\mathcal{L}$ is defined as $|\mathcal{L}|/n^2$. Given an $n \times n$ board, we are interested in finding a still life of maximum density.

Formulate the problem of finding a maximum density still life as an IP.

		5						3
				4	6			
		7						2
	1				3		6	9
	4		6		9		5	
9	8		2				7	
2						9		
			8	1				
6						4		

Figure 1.4

1.4   Optimization problems on graphs

A *graph* G is a pair (V, E), where V is a finite set and E is a set of *unordered* pairs of elements of V (i.e. the pair uv and vu is the same). Elements of V are called *vertices* and elements of E *edges*. We can represent a graph as a drawing where vertices correspond to points and a pair u, v of vertices is joined by a line when $uv \in E$. Observe that different drawings may correspond to the same graph.

Example 4 Two drawings of $G = (V, E)$ with

$$V = \{1, 2, 3, 4, 5\} \qquad E = \{12, 23, 34, 14, 15, 25, 35, 45\}.$$

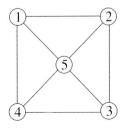   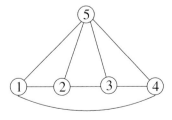

---

[3] Check out http://www.bitstorm.org/gameoflife/ for a Java applet for LIFE.

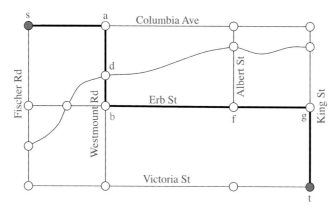

Figure 1.5 Map of Watertown.

Let $G = (V, E)$ be a graph. Suppose $uv \in E$. Then u and v are *adjacent* vertices, u, v are the *endpoints* of edge uv, and edge uv is *incident* to vertices u and v. In this section, we present several optimization problems that can be expressed as optimization problems on graphs.

1.4.1   Shortest path problem

The shortest path problem is a fundamental optimization problem that most of us encounter (and solve) frequently: starting in geographic location s, we wish to travel to location t. Since we are frugal travellers, we wish to choose a route of minimum total length. Let us illustrate this problem using a concrete example.

Suppose you are visiting the city of Watertown (see Figure 1.5). You are staying with a friend who lives at the intersection of Fischer Road and Columbia Avenue (location s), and you wish to visit the Tannery District which is located at the intersection of King Street and Victoria Street (location t). A shortest route from s to t is indicated by the thick lines in Figure 1.5. The problem of finding such a shortest route is known as the *shortest path problem*.

Let us rephrase the problem in the language of graph theory. We represent in Figure 1.6 the street network by a graph, where vertices correspond to intersections and edges to roads. For instance, the graph $G = (V, E)$ corresponding to the WaterTown map is given in Figure 1.6, where the number $c_e \geq 0$ next to each edge e corresponds to the length (in meters) of the corresponding road on the map.

An *st-path* P is a sequence of edges

$$v_1 v_2, v_2 v_3, \ldots, v_{k-2} v_{k-1}, v_{k-1} v_k$$

such that $v_1 = s$, $v_k = t$, and $v_i \neq v_j$ for all $i \neq j$.[4] The length $c(P)$ of an st-path is defined as the sum of the lengths of the edges of P, i.e. as $\sum (c_e : e \in P)$.

---

[4] It is often useful to think of an st-path as a *set* of edges that can be ordered to form an st-path. We will alternate between the representation of an st-path as a set of edges and a sequence of edges.

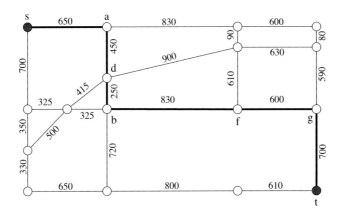

Figure 1.6 Graph representing map of Watertown.

Example 5 In Figure 1.6, the thick black edges in the graph form an *st*-path

$$P = sa, ad, db, bf, fg, gt$$

of total length

$$c(P) = c_{sa} + c_{ad} + c_{db} + c_{bf} + c_{fg} + c_{gt}$$
$$= 650 + 450 + 250 + 830 + 600 + 700 = 3480.$$

We can now define formally the shortest path problem. We are given a graph $G = (V, E)$ with nonnegative weights c_e for every edge $e \in E$, and distinct vertices s, t. We wish to find among all possible *st*-paths, one that is of minimum length. Thus, the optimization problem we wish to solve is

$$\min \quad c(P)$$
$$\text{subject to} \qquad\qquad (1.26)$$
$$P \text{ is an } st\text{-path.}$$

We return to this problem in Section 1.5.2, where we will see how it can be formulated as an IP.

1.4.2   Minimum cost perfect matching

Recall our assignment problem from Section 1.3.1. WaterTech wishes to assign a set I of employees to do a set J of jobs with the condition that every employee is assigned exactly one job and every job is assigned to exactly one employee. For every $i \in I$ and $j \in J$, c_{ij} is the number of hours it takes employee i to do job j. The goal is to find among all such assignments one that minimizes the total completion time of the jobs. To make the example more realistic, we have to consider the fact that some employees are not qualified to do certain jobs, and obviously we do not wish to assign these jobs to those employees. The following table gives an example for a case with $|J| = 4$ jobs and $|I| = 4$ employees. Entries – indicate that the employee is not qualified for the corresponding job.

Employees	Jobs			
	$1'$	$2'$	$3'$	$4'$
1	–	5	–	7
2	8	–	2	4
3	–	1	–	8
4	8	3	3	–

For instance, employee 3 is not qualified to do job 1. We can represent these data by a graph $G = (V, E)$ where $V = I \cup J$, and $ij \in E$, where $i \in I, j \in J$ if employee i is qualified to do job j. Moreover, edge ij is assigned weight c_{ij}. We represent the weighted graph corresponding to the table in Figure 1.7.

A graph $G = (V, E)$ is *bipartite* if we can partition the vertices into two sets, say I and J, such that every edge has one endpoint in I and one endpoint in J. Given a graph, a subset of edges M is called a *matching* if no vertex is incident to more than one edge of M. A matching is called *perfect* if every vertex is incident to exactly one edge in the matching. The weight $c(M)$ of a matching M is defined as the sum of the weights of the edges of M, i.e. as $\sum(c_e : e \in M)$.

Example 6 The graph in Figure 1.7 is bipartite as every edge has an endpoint in $\{1, 2, 3, 4\}$ and one endpoint in $\{1', 2', 3', 4'\}$. The set of edges $\{12', 24', 41'\}$ is a matching. However, it is not a perfect matching as no edge of M is incident to $3'$ for instance. The set of edges $M = \{14', 21', 32', 43'\}$ is a perfect matching. The matching M has weight

$$c(M) = c_{14'} + c_{21'} + c_{32'} + c_{43'} = 7 + 8 + 1 + 3 = 19.$$

Going back to our optimization problem, since we wish to assign every employee to exactly one job, and every job to exactly one employee, we are looking for a perfect

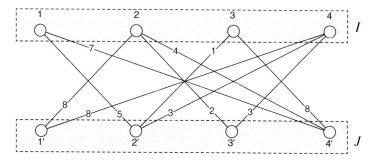

Figure 1.7 Graph representing assignment problem data.

matching in the graph given in Figure 1.7. As we are trying to minimize the total completion time of all the jobs, we wish to find among all possible perfect matchings one that is of minimum weight. Thus, given a bipartite graph $G = (V, E)$ with nonnegative edge weights c_e for all $e \in E$, the optimization problem we wish to solve is

$$\min \quad c(M)$$
$$\text{subject to} \qquad\qquad\qquad\qquad\qquad (1.27)$$
$$M \text{ is a perfect matching.}$$

In the following section, we will see how to formulate this optimization problem as an IP.

Exercises

1 (Advanced) Let $G = (V, E)$ be a graph with distinct vertices s, t. An even st-path is an st-path that has an even number of edges. Show that we can formulate the problem of finding an even st-path with as few edges as possible, as a minimum cost perfect matching problem.

HINT: Make a copy of the graph G and remove vertices s and t. Call the resulting graph G'. Construct a new graph H starting with the union of graph G and G' and by joining every vertex $v \notin \{s, t\}$ in G with its copy v' in G'.

1.5   Integer programs continued

In the previous section, we presented two optimization problems on graphs, namely the shortest path and minimum weight perfect matching problems. We will see in this section that both of these problems can be formulated as IPs. Furthermore, in Chapter 3 we will use these formulations to guide us in devising efficient algorithms to solve these problems. We consider the minimum cost perfect matching first as the formulation is simpler.

1.5.1   Minimum cost perfect matching

Let $G = (V, E)$ be a graph and let $v \in V$. We denote by $\delta(v)$ the set of edges that have v as one of the endpoints, i.e. $\delta(v)$ is the set of edges incident to v. For instance, for the graph in Figure 1.8, we have $\delta(a) = \{ad, ag, ab\}$ and $\delta(b) = \{ab, bd, bg\}$.

We are now ready to find an IP formulation for the minimum cost perfect matching problem. We are given a graph $G = (V, E)$ and edge weight c_e for every edge e. We need to determine the variables, the objective function, and the constraints.

Variables. We introduce a binary variable x_e for every edge e, where $x_e = 1$ indicates that edge e is selected to be part of our perfect matching and $x_e = 0$ indicates that edge e is not selected. Note, here we indicate how to interpret our variables, we of course need to define the constraints so that the edges in the matching are given by $\{e \in E : x_e = 1\}$.

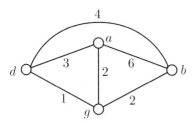

Figure 1.8

Constraints. Let v be a vertex. The number of edges incident to v that are selected is

$$\sum \left(x_e \; : \; e \in \delta(v) \right).$$

Since we want a perfect matching, we need that number to be equal to 1, i.e.

$$\sum \left(x_e \; : \; e \in \delta(v) \right) = 1. \tag{1.28}$$

Objective function. We wish to minimize the total weight of the edges in the matching M we selected. Thus, the objective function should return the weight of M. If e is an edge of M, then we will have $x_e = 1$ and we should contribute c_e to the objective function, otherwise $x_e = 0$ and we should contribute 0 to the objective function. This can be modeled by the term $c_e x_e$. Thus, the objective function should be

$$\sum \left(c_e x_e \; : \; e \in E \right). \tag{1.29}$$

The formulation for the minimum weight matching problem is obtained by combining (1.28) and (1.29) as well as the condition that x_e be a binary variables. Thus, we obtain

$$\begin{aligned}
\min \quad & \sum \left(c_e x_e \; : \; e \in E \right) \\
\text{subject to} \quad & \\
& \sum \left(x_e \; : \; e \in \delta(v) \right) = 1 \qquad (v \in V) \\
& x_e \geq 0 \qquad (e \in E) \\
& x_e \text{ integer} \qquad (e \in E).
\end{aligned} \tag{1.30}$$

It is not necessary to include the constraints $x_e \leq 1$ for $e \in E$ since this condition will be satisfied by every feasible solution of (1.30). Indeed, consider an arbitrary edge f and let v be one of its endpoints. Then $f \in \delta(v)$, and thus $x_f \leq \sum (x_e \; : \; e \in \delta(v)) = 1$. Thus, $0 \leq x_f \leq 1$. As x_f is integer, $x_f \in \{0, 1\}$, as required.

As an example let us write the formulation (1.30) for the graph in Figure 1.8. Let x denote the vector $(x_{ab}, x_{bg}, x_{dg}, x_{ad}, x_{ag}, x_{bd})^\top$ and let $\mathbb{0}$ and $\mathbb{1}$ denote the vector of all zeros and ones (of appropriate dimension) respectively. The IP in that case is,

$$\min \qquad (6, 2, 1, 3, 2, 4)x$$

$$\text{subject to}$$

$$\begin{array}{c} & ab \quad bg \quad dg \quad ad \quad ag \quad bd \\ \begin{array}{c} a \\ b \\ d \\ g \end{array} & \left(\begin{array}{cccccc} 1 & 0 & 0 & 1 & 1 & 0 \\ 1 & 1 & 0 & 0 & 0 & 1 \\ 0 & 0 & 1 & 1 & 0 & 1 \\ 0 & 1 & 1 & 0 & 1 & 0 \end{array} \right) x = \mathbb{1} \end{array}$$

$$x \geq \mathbb{0} \quad \text{integer.}$$

Each of the first four constraints corresponds to one of the four vertices of G. For instance, the second constraint, corresponding to vertex b, states that $x_{ab} + x_{bg} + x_{bd} = 1$. It says that we should select exactly one of the edges incident to b.

Let us verify the correctness of the formulation (1.30). We have a graph $G = (V, E)$ and edge weights c_e for every edge e. Let M be a perfect matching. We need to verify (see Section 1.1.2) that M corresponds to a feasible solution $\bar{x}$ of (1.30). To do this, we set $\bar{x}_e = 1$ if e is an edge of M, and $\bar{x}_e = 0$ otherwise. Conversely, suppose that $\bar{x}$ is a feasible solution of (1.30), then $M := \{e : \bar{x}_e = 1\}$ is a perfect matching. Finally, observe that the objective function of the formulation computes the weight of the corresponding perfect matching.

1.5.2   Shortest path problem

The IP formulation will rely on a characterization of st-paths. We first require a few definitions. Let $G = (V, E)$ be a graph and let $U \subseteq V$ be a subset of the vertices. Generalizing from the definition for singletons from the previous section, we let $\delta(U)$ denote the set of edges that have exactly one endpoint in U, i.e.

$$\delta(U) = \{uv \in E : u \in U, v \notin U\}.$$

Consider the following graph given in Figure 1.9. Then $\delta(\{s, a\})$ is the set of all edges with exactly one end equal to either a or s, i.e. it is the set $\{sb, ab, at\}$. Suppose a graph G has a distinct pair of vertices s and t. Then an st-cut is a set of edges of the form $\delta(U)$, where $s \in U$ and $t \notin U$. The set of all st-cuts in the graph in Figure 1.9 is given by

$$\delta(\{s\}) = \{sa, sb\}$$
$$\delta(\{s, a\}) = \{at, ab, sb\}$$
$$\delta(\{s, b\}) = \{sa, ab, bt\}$$
$$\delta(\{s, a, b\}) = \{at, bt\}. \tag{1.31}$$

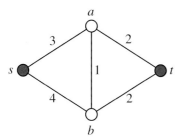

Figure 1.9

These st-cuts are obtained by considering all possible sets of vertices U, where $s \in U$ and $t \notin U$. Note, the term st-cut arises from the fact that if we remove all the edges in an st-cut, the resulting graph breaks into (at least) two parts that are not joined by any edge and where one part contains s and another contains t.

Consider a graph $G = (V, E)$ with distinct vertices s and t and let P be an st-path of G. Let $\delta(U)$ be an arbitrary st-cut of $G = (V, E)$. Follow the path P, starting from s and denote by u the last vertex of P in U, and denote by u' the vertex that follows u in P. Note that u exists since $s \in U$ and $t \notin U$. Then by definition uu' is an edge that is in $\delta(U)$. Since $\delta(U)$ was an arbitrary st-cut, we have shown the following remark:

Remark 1.1 *An st-path P intersects every st-cut.*

In fact the converse, also holds, namely:

Remark 1.2 *Let S be a set of edges that contains at least one edge from every st-cut. Then there exists an st-path P that is contained in the edges of S.*

Proof Let U denote the set of vertices that contain s as well as all vertices u for which there exists a path from s to u that only uses edges in S. We need to show that there exists an st-path that is contained in S, i.e. that $t \in U$. Suppose for a contradiction that $t \notin U$. 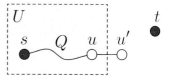 Then by definition $\delta(U)$ is an st-cut. By hypothesis, there exists an edge of S in $\delta(U)$, say $uu' \in S$, where $u \in U$ and $u' \notin U$ (see figure on the right). By construction, there exists an su-path Q that is contained in S. Then the path obtained from Q by adding the edge uu' is an su'-path contained in S. It follows by definition of U that $u' \in U$, a contradiction. $\qquad\square$

We are now ready to find an IP formulation for the shortest path problem. We are given a graph $G = (V, E)$, distinct vertices s and t and edge lengths $c_e \geq 0$ for every edge e. We need to determine the variables, the objective function, and the constraints.

Variables. We will introduce a binary variable x_e for every edge e, where $x_e = 1$ represents the case where edge e is selected to be part of our st-path, and $x_e = 0$, means that the edge e is not selected.

Constraints. Remark 1.2 says that in order to construct an st-path, it suffices to select one edge from every st-cut. We will use this idea for our formulation. Let $\delta(U)$

be an arbitrary st-cut of G. The number of edges we selected from the st-cut $\delta(U)$ is given by

$$\sum (x_e : e \in \delta(U))$$

and since we wish to select at least one edge from $\delta(U)$, the constraint should be

$$\sum (x_e : e \in \delta(U)) \geq 1. \tag{1.32}$$

Objective function. We wish to minimize the total length of the edges in the st-path P we selected. Thus, the objective function should return the length of P. If e is an edge of P then we will have $x_e = 1$ and we should contribute c_e to the objective function, otherwise $x_e = 0$ and we should contribute 0 to the objective function. This can be modeled by the term $c_e x_e$. Thus, the objective function should be

$$\sum (c_e x_e : e \in E). \tag{1.33}$$

The formulation for the shortest path problem is obtained by combining (1.32) and (1.33) as well as the condition that x_e be a binary variable for all $e \in E$. Thus, we obtain

$$\min \quad \sum (c_e x_e : e \in E)$$

subject to

$$
\begin{aligned}
\sum (x_e : e \in \delta(U)) \geq 1 \quad & (U \subseteq V, s \in U, t \notin U) \\
x_e \geq 0 \quad & (e \in E) \\
x_e \text{ integer} \quad & (e \in E).
\end{aligned}
\tag{1.34}
$$

Observe in this formulation that the condition that $x_e \in \{0, 1\}$ has been replaced by the condition that $x_e \geq 0$ and x_e is integer, i.e. we removed the condition that $x_e \leq 1$. This is correct, for if $\bar{x}$ is a feasible solution to (1.34), where $\bar{x}_e \geq 2$, then x', obtained from $\bar{x}$ by setting $x'_e = 1$, is also a feasible solution to (1.34), moreover, the objective value for x' is smaller or equal to the objective value for $\bar{x}$ (as $c_e \geq 0$).

As an example, let us write the formulation (1.34) for the graph in Figure 1.9. Let x denote the vector $(x_{sa}, x_{sb}, x_{ab}, x_{at}, x_{bt})^\top$ and let $\mathbb{1}$ denote the vector of all ones (of appropriate dimension). The IP in that case is

$$\min \quad (3, 4, 1, 2, 2)x$$

subject to

	sa	sb	ab	at	bt	
$\{s\}$	1	1	0	0	0	
$\{s, a\}$	0	1	1	1	0	$x \geq \mathbb{1}$
$\{s, b\}$	1	0	1	0	1	
$\{s, a, b\}$	0	0	0	1	1	

$$x \geq 0 \qquad x \text{ integer.}$$

Each of the first four constraints correspond to one of the four distinct st-cuts $\delta(U)$, given by (1.31). For instance, the second constraint, corresponding to $\delta(\{s, a\}) = \{sb, ab, at\}$, states that $x_{sb} + x_{ab} + x_{at} \geq 1$. It says that we should select at least one edge from the st-cut $\delta(\{s, a\})$.

Let us verify the correctness of the formulation (1.34). We have a graph $G = (V, E)$ with distinct vertices s and t and will assume that $c_e > 0$ for every $e \in E$, i.e. every edge has positive length.[5] Let P be an st-path. We need to verify (see Section 1.1.2) that P corresponds to a feasible solution $\bar{x}$ of (1.34). To do this, we set $\bar{x}_e = 1$ if e is an edge of P, and $\bar{x}_e = 0$ otherwise. By Remark 1.1, every st-cut $\delta(U)$ contains an edge of P, hence, $\sum(\bar{x}_e : e \in \delta(U)) \geq 1$. It follows that $\bar{x}$ is feasible for the formulation. It is not true, however, that every feasible solution to the formulation corresponds to an st-path. This is seemingly in contradiction with the strategy outlined in Section 1.1.2. However, it suffices to show that every optimal solution $\bar{x}$ to the formulation corresponds to an st-path. Let $S = \{e \in E : \bar{x} = 1\}$. We know from Remark 1.2 that the edges of S contain an st-path P. If $S = P$, then we are done. Otherwise, define x' by setting $x'_e = 1$ if e is an edge of P, and $x'_e = 0$ otherwise. Then x' is also a feasible solution to the formulation, but it has an objective value that is strictly smaller than $\bar{x}$, a contradiction. It follows that S is an st-path, as required.

Let us re-evaluate our simple strategy for constructing mathematical formulations from word descriptions as given in Section 1.1.2. When dealing with the shortest path problem, instead of forcing all solutions to the mathematical formulation to correspond to solutions to the word formulation, we found it advantageous to utilize the objective function and argue that optimal solutions of the word problem are correctly formulated by our mathematical formulation. This approach will be very handy when we deal with many other problems, including those where we try to minimize the maximum of finitely many affine functions (or maximize the minimum of finitely many affine functions).

Exercises

1 Let $G = (V, E)$ be a graph with nonnegative edge weights c_e for every edge e. The *maximum weight matching* problem is the problem of finding a matching of maximum weight. An *edge-cover* is a set of edges S that has the property that every vertex is the endpoint of some edge of S. The *minimum weight edge-cover* problem is the problem of finding an edge-cover of minimum weight.
(a) Formulate as an IP the problem of finding a maximum weight matching.
(b) Formulate as an IP the problem of finding a minimum edge-cover.

2 A *vertex-cover* of a graph G is a set $\mathscr{S}$ of vertices of G such that each edge of G is incident with at least one vertex of $\mathscr{S}$. For instance, in Figure 1.10 the set $\mathscr{S} = \{2, 3, 4, 6, 7\}$ is a vertex-cover. The size of a vertex-cover is the number of vertices it contains. So the above set is a vertex-cover of size 5.
(a) Formulate the following problem as an IP: find a vertex-cover in the graph G given in Figure 1.10 of smallest possible size.

---

[5] The formulation does not work if $c_e < 0$ for some edge e and there are technical difficulties when $c_e = 0$.

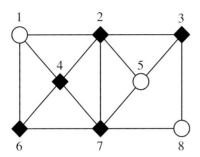

Figure 1.10

(b) Now do the same for an arbitrary graph G. That is, you are given a graph $G = (V, E)$. Formulate the problem of finding a minimum size vertex-cover as an IP.

3 Consider a graph $G = (V, E)$. A *triangle T* is a set of three distinct vertices i, j, k, where ij, jk and ik are all edges. A *packing of triangles* is a set $\mathscr{S}$ of triangles with the property that no two triangles in $\mathscr{S}$ have a common vertex. The maximum triangle packing problem (MTP) is the problem of finding a packing of triangles $\mathscr{S}$ with as many triangles as possible. As an example consider the graph in Figure 1.11. Then triangles

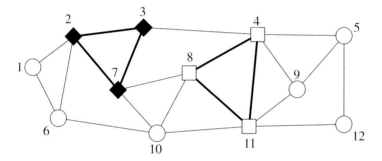

Figure 1.11

$T_1 = \{2, 3, 7\}$, $T_2 = \{4, 8, 11\}$, form a packing of triangles. Formulate the (MTP) problem for this graph G as an IP.

HINT: Define one variable for every triangle and write one constraint for every vertex.

4 Let $G = (V, E)$ be a graph, let s, t be vertices of G, and suppose each edge e has a nonnegative weight c_e. Remark 1.1 states that an *st*-cut $\delta(U)$ intersects every *st*-path.

(a) Suppose S is a set of edges that contains at least one edge from every *st*-path. Show that there exists an *st*-cut $\delta(U)$ that is contained in the edges of S.

(b) The weight of an *st*-cut $\delta(U)$ is defined as $c(\delta(U)) := \sum(c_e : e \in \delta(U))$. Find an IP formulation for the problem of finding an *st*-cut of minimum weight, where we have a binary variable for every edge and a constraint for every *st*-path.

HINT: This is analogous to the formulation given in (1.34).

(c) Show that the formulation given in (b) is correct.

1.6   Nonlinear programs

Consider functions $f : \mathbb{R}^n \to \mathbb{R}$, and $g_i : \mathbb{R}^n \to \mathbb{R}$, for every $i \in \{1, 2, \ldots, m\}$. A *nonlinear program* is an optimization problem of the form

$$
\begin{aligned}
\min \quad & z = f(x) \\
\text{s.t.} \quad & \\
& g_1(x) \leq 0, \\
& g_2(x) \leq 0, \\
& \quad \vdots \quad \vdots \quad \vdots \\
& g_m(x) \leq 0.
\end{aligned}
\tag{1.35}
$$

We will abbreviate the term nonlinear program by NLP throughout this book. The reader might be troubled by the fact that NLP are always minimization problems. What if we wish to maximize the objective function $f(x)$? Then we could replace in (1.35) $z = f(x)$ by $z = -f(x)$, as making the function $-f(x)$ as small as possible is equivalent to making $f(x)$ as large as possible. Note that when each of the functions $f(x)$ and $g_1(x), \ldots, g_m(x)$ are affine functions in (1.35), then we obtain an LP.

1.6.1   Pricing a tech gadget

Company Dibson is a local producer of tech gadgets. Its newest model, the *Dibson BR-1* will be sold in three regions: 1, 2, and 3. Dibson is considering a price between $50 and $70 for the device, but is uncertain as to which exact price it should choose. Naturally, the company wants to maximize its revenue, and that depends on the demand. The demand on the other hand is a function of the price. Dibson has done a considerable amount of market research, and has reliable estimates for the demand (in thousands) for the new product for three different prices in each of the three regions. The following table summarizes this:

Price ($)	Demand		
	Region 1	Region 2	Region 3
50	130	90	210
60	125	80	190
70	80	20	140

The company wants to model the demand in region i as a function $d_i(p)$ of the unit's price. Dibson believes that the demand within a region can be modeled by a quadratic function, and uses the data from the above table to obtain the following quadratic demand functions:

$$d_1(p) = -0.2p^2 + 21.5p - 445$$
$$d_2(p) = -0.25p^2 + 26.5p - 610 \tag{1.36}$$
$$d_3(p) = -0.15p^2 + 14.5p - 140.$$

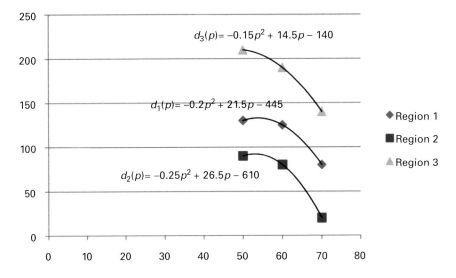

The previous figure shows the demand functions for the three regions. Dibson wants to restrict the price to be between \$50 and \$70. What price should Dibson choose in order to maximize revenue? We will formulate this problem as an NLP. In this case, there is a single variable, namely the price p of the unit and the constraints state that $50 \le p \le 70$.

Objective function. For each region i, the demand for region i is given by $d_i(p)$. Since the price is p, the revenue from region i is given by $pd_i(p)$. Thus, the total revenue from each of the three regions is given by

$$p\sum_{i=1}^{3} d_i(p) = p\left(-0.2p^2 + 21.5p - 445 - 0.25p^2 + 26.5p - 610 - 0.15p^2 + 14.5p - 140\right)$$
$$= -0.6p^3 + 62.5p^2 - 1195p,$$

where the first equality follows by (1.36). Thus, we obtain the following NLP:

$$\min \quad -0.6p^3 + 62.5p^2 - 1195p$$

subject to

$$p \ge 50$$
$$p \le 70.$$

Using an algorithm to solve NLPs, we find that a price of $p = 57.9976$ maximizes Dibson's revenue. The revenue at this price is approximately \$23 873.

We should comment that using a single quadratic function is not a good way to model demand. We used this overly simplified model for the sake of a clear presentation. It would be better to model the demand in region i, either by a piecewise quadratic function or by exponential functions $\beta_i \exp(-\gamma_i p)$, for suitable choices of positive β_i, γ_i.

1.6.2   Finding a closest point feasible in an LP

Suppose we formulated the set of feasible production plans for a company as the feasible solution set of an LP. We then have an m-by-n matrix A and $b \in \mathbb{R}^m$ and the set of feasible production plans is represented by

$$\left\{ x \in \mathbb{R}^n : Ax \leq b \right\}.$$

Moreover, a decision-making body gave us some goal values for the variables. Let $\bar{x} \in \mathbb{R}^n$ represent these goals.

Our job is to find a production plan that is feasible and gets as close to the stated goals as possible. Therefore, we are asked to find a point in the set

$$\left\{ x \in \mathbb{R}^n : Ax \leq b \right\}$$

that is closest to $\bar{x}$. This problem can be formulated as

$$\begin{aligned} \min \quad & \|x - \bar{x}\| \\ \text{subject to} \quad & Ax \leq b, \end{aligned}$$

where $\|x\|$ denotes the L_2-norm $\sqrt{x_1^2 + x_2^2 + \cdots + x_n^2}$ of vector x. Even though the above problem is equivalent to

$$\begin{aligned} \min \quad & \|x - \bar{x}\|^2 \\ \text{subject to} \quad & Ax \leq b \end{aligned}$$

the objective function is still nonlinear.

1.6.3   Finding a "central" feasible solution of an LP

Suppose we are operating a nuclear reactor. We have a mathematical model, based on a system of linear inequalities, for safe operation of the reactor, i.e. we have an m-by-n matrix A and $b \in \mathbb{R}^m$. We are interested in the solutions x satisfying

$$Ax \leq b.$$

However, since these constraints model safe operation modes, we are worried about not only violating the constraints but also about getting close to violating the constraints. As a result, we want a solution that is as far away as possible from getting close to violating any of the constraints. Then one option is to solve the optimization problem

$$\max \quad \sum_{i=1}^{m} \ln \left(b_i - \sum_{j=1}^{n} A_{ij}x_j \right)$$

subject to
$$Ax \le b.$$

Another option is to solve the optimization problem

$$\min \quad \sum_{i=1}^{m} \frac{1}{b_i - \sum_{j=1}^{n} A_{ij}x_j}$$

subject to
$$Ax \le b.$$

Indeed, we can pick any, suitable function which is finite valued and monotone decreasing for positive arguments, and tends to infinity as its argument goes to zero with positive values. Note that the functions used above as building blocks $f_1 : \mathbb{R} \to \mathbb{R}$, $f_1(x) := \frac{1}{x}$ and $f_2 : \mathbb{R} \to \mathbb{R}, f_2(x) := -\ln(x)$ both have these desired properties.

Further note that the objective functions of both of these problems are nonlinear. Therefore, they are both NLP problems but not LP problems.

Exercises

1 UW Smokestacks has a chimney which needs cleaning regularly. Soot builds up on the inside of the chimney at a rate of 1cm per month. The cost of cleaning when the soot has thickness x cm is $13 + x^2$ hundred dollars. Cleaning can occur at the end of any month.

The chimney is clean at present (i.e. at the start of month 1). According to the contract with the cleaners, it must be cleaned precisely nine times in the next 48 months, one of which times must be at the end of month 48. Also, at least three cleanings must occur in the first 15 months.

It is needed to determine which months the cleaning should occur, to minimize total cost. Formulate a nonlinear mathematical program to solve this problem, such that *all constraints are either linear or integrality constraints*. (Hint: Consider a variable giving the time of the *i*th cleaning.)

2 A number of remote communities want to build a central warehouse. The location of each of the communities is described by an (x, y)-coordinate as given in the next table:

	x	y
Dog River	3	9
Springfield	10	6
Empire Falls	0	12
Worthing	4	9

If the warehouse is located at the coordinate $(7, 6)$, then the distance from the warehouse to Dog River will be

$$\sqrt{(7 - 3)^2 + (6 - 9)^2} = 5.$$

Goods from the warehouse will be delivered to the communities by plane. The price of a single delivery to a community is proportional to the distance between the warehouse and the community. Each community will need a number of deliveries from the warehouse which depends on the size of the community. Dog River will require 20 deliveries, Springfield 14, Empire Falls 8, and Worthing 24. The goal is to locate the warehouse, i.e. find (x, y)-coordinates, so that the total cost of all deliveries is minimized. Formulate this problem as an NLP.

3 A $0, 1$ *program* is an integer program where every variable is a binary variable, i.e. can only take values 0 or 1. Show that every $0, 1$ program can be expressed as an NLP.

HINT: For every variable x_j, consider including a nonlinear equation with left-hand side $x_j(x_j - 1)$.

4 (Note, this exercise requires basic concepts in probability.) Another important application of nonlinear optimization is *portfolio optimization*. In a fundamental version of such a problem, we usually have a fixed amount of capital that we wish to invest into several investment options. The usual goal is to maximize the return from this investment while ensuring that the risk of the constructed portfolio is small. Usually, stocks that have the highest potential returns are the most volatile, and hence the most risky stocks to invest in at the same time. The risk-averse investor will therefore have to strike a balance between return and risk. One way to achieve this balance is to attempt to minimize the overall volatility while guaranteeing a minimum expected return.

More concretely, imagine you had $500 available for investment into three different stocks $1, 2$, and 3. In the following, we let S_i be the random variable for the annual return on $1 invested into stock i. Assume that we know the expected annual return and its variance for each of the given stocks. The following table shows expected values and variances for the given variables:

i	$E[S_i]$	$\text{var}[S_i]$
1	0.1	0.2
2	0.13	0.1
3	0.08	0.15

In addition to the information above, we are also given the following covariances:

$$\mathrm{cov}(S_1, S_2) = 0.03, \quad \mathrm{cov}(S_1, S_3) = 0.04, \text{ and } \mathrm{cov}(S_2, S_3) = 0.01.$$

Recall, for a random variable ξ

$$\mathrm{var}(\xi) := E\left([\xi - E(\xi)]^2\right),$$

and for a pair of random variables ξ_1, ξ_2

$$\mathrm{cov}(\xi_1, \xi_2) := E\left([\xi_1 - E(\xi_1)][\xi_2 - E(\xi_2)]\right).$$

The goal is now to invest the given \$500 into the three stocks such that the expected return is at least \$50, and such that the total variance of our investment is minimized.

Introduce a variable x_i denoting the amount of money invested into stock i for all $i \in \{1, 2, 3\}$. Using these variables, formulate the problem as a nonlinear optimization problem such that the objective function is a quadratic function of x_i and all the constraints are linear. (Such problems are called *quadratic optimization problems.*)

1.7   Overview of the book

In this chapter, we have seen a number of examples of optimization problems all put in a practical context. In each of these cases, we have developed a concise mathematical model that captures the respective problem. Clearly, modeling a problem in mathematical terms is nice and perhaps theoretically satisfying to a degree; however, in most cases, a mathematical model by itself is not enough. After all, in most cases, most of us are interested in solutions! For many of the given examples in this chapter, we also provided a solution. How do we find these solutions? The techniques to solve these problems depend on the type of mathematical model considered. The simplest and most widespread of these models is linear programming and an algorithm to solve linear programs is the simplex algorithm that we describe in Chapter 2.

In Chapter 3, we will develop efficient algorithms to solve two classical optimization problems, namely the problem of finding a shortest path between a fixed pair of vertices in an undirected graph and the problem of finding a minimum cost perfect matching in a bipartite graph. We show in the process that the natural framework for understanding these problems is the theory of linear programming duality. This theory will be further developed in Chapter 4. Further applications to this theory are given in Chapter 5. General techniques to solve integer programs are studied in Chapter 6. Finally, in Chapter 7 we explain how some of the concepts introduced for linear programs extend to nonlinear programs. One of the guiding principles in designing algorithms is the consideration of efficiency. The appendix provides an introduction to a foundational treatment of this issue, under the title of Computational Complexity.

1.8   Further reading and notes

A recent survey (see also Winston [70]) of Fortune 500 firms shows that 85% of all respondents use linear programming in their operations. The roots of linear programming can be traced back at least a couple of hundred years to the work of Fourier on solutions of systems of linear inequalities. The name *linear programming*, however, was coined only in the late 1930s and early 1940s when the Russian mathematician Leonid Kantorovich and the American mathematician George B. Dantzig formally defined the underlying techniques. George Dantzig also developed the *simplex algorithm* which to this date remains one of the most popular methods to solve linear programs. Dantzig who worked as a mathematical advisor for the US Air Force initially applied linear programming to solve logistical problems arising in the military. It did, however, not take long for industry to realize the technique's potential, and its use is widespread today.

We just saw some examples of modeling a given problem as an optimization problem. An important part of the mathematical modeling process is to prove the correctness of the mathematical model constructed. In this book, typically, our starting point will be a well-described statement of the problem with clearly stated data. Then we take this description of the problem, define variables, the objective function, and the constraints. Once the mathematical model is constructed, we prove that the mathematical model we constructed is correct, i.e. it exactly represents the given problem statement. For many examples of such mathematical models, see the book [70].

In real applications, we have to go through a lot of preparation to arrive at the clear description of the problem and the data. In some applications (actually in almost all applications), some of the data will be uncertain. There are more advanced tools to deal with such situations (see, for instance, the literature on robust optimization starting with Ben-Tal, Ghaoui and Nemirovski [4]).

Many of the subjects that we introduced in this chapter have a whole course dedicated to them. The portfolio optimization example (see Exercise 4, Section 1.6.1) is an instance of Markowitz model. It was proposed by Harry Markowitz in the 1950s. For his work in the area, Markowitz received the Nobel Prize in Economics in 1990. For applications of optimization in financial mathematics, see Best [6] and Cornuéjols and Tütüncü [17]. For further information on scheduling, see Pinedo [55].

2 Solving linear programs

2.1 Possible outcomes

Consider an LP (P) with variables $x_1, \ldots, x_n$. Recall that an assignment of values to each of $x_1, \ldots, x_n$ is a feasible solution if the constraints of (P) are satisfied. We can view a feasible solution to (P) as a vector $x = (x_1, \ldots, x_n)^\top$. Given a vector x, by the *value of x* we mean the value of the objective function of (P) for x. Suppose (P) is a maximization problem. Then recall that we call a vector x an *optimal solution* if it is a feasible solution and no feasible solution has larger value. The value of the optimal solution is the *optimal value*. By definition, an LP has only one optimal value; however, it may have many optimal solutions. When solving an LP, we will be satisfied with finding any optimal solution. Suppose (P) is a minimization problem. Then a vector x is an optimal solution if it is a feasible solution and no feasible solution has *smaller* value.

If an LP (P) has a feasible solution, then it is said to be *feasible*, otherwise it is *infeasible*. Suppose (P) is a maximization problem and for every real number α there is a feasible solution to (P) which has value greater than α, then we say that (P) is *unbounded*. In other words, (P) is unbounded if we can find feasible solutions of arbitrarily high value. Suppose (P) is a minimization problem and for every real number α there is a feasible solution to (P) which has value *smaller* than α, then we say that (P) is unbounded. Unbounded LPs are easy to construct. Try to construct one yourself. (HINT: You can do this for an LP with a single variable.)

We have identified three possible outcomes for an LP (P), namely:

(1) it is infeasible,
(2) it has an optimal solution,
(3) it is unbounded.

Clearly, each of these outcomes are mutually exclusive (i.e. no two can occur at the same time). We will show that in fact exactly one of these outcomes must occur (this is a form of the *fundamental theorem of linear programming* which will be established at the end of this chapter). We will now illustrate each of these outcomes with a different example.

2.1.1 Infeasible linear programs

If we are interested in knowing whether an LP (P) is infeasible, it suffices to consider the constraints, as this does not depend on the objective function. Suppose that the constraints of (P) are as follows:

$$4x_1 + 10x_2 - 6x_3 - 2x_4 = 6, \tag{2.1}$$
$$-2x_1 + 2x_2 - 4x_3 + x_4 = 5, \tag{2.2}$$
$$-7x_1 - 2x_2 \qquad + 4x_4 = 3, \tag{2.3}$$
$$x_1, \quad x_2, \quad x_3, \quad x_4 \geq 0. \tag{2.4}$$

Suppose further that you are asked to solve (P). After spending a substantial amount of time on the problem, you are starting to become more and more convinced that (P) is in fact infeasible. But how can you be certain? How will you convince a third party that (P) does indeed have no solution? You certainly cannot claim to have tried all possible sets of values for x_1, x_2, x_3, x_4 as there are an infinite number of possibilities.

We now present a concise way of proving that the system (2.1)–(2.4) has no solution. The first step is to create a new equation by combining equations (2.1), (2.2), and (2.3). We pick some values y_1, y_2, y_3. Then we multiply equation (2.1) by y_1, equation (2.2) by y_2, and equation (2.3) by y_3, and add each of the resulting equations together. If we choose the values $y_1 = 1, y_2 = -2$ and $y_3 = 1$, we obtain the equation

$$x_1 + 4x_2 + 2x_3 = -1. \tag{2.5}$$

Let us proceed by contradiction and suppose that there is in fact a solution $\bar{x}_1, \bar{x}_2, \bar{x}_3, \bar{x}_4$ to (2.1)–(2.4). Then clearly $\bar{x}_1, \bar{x}_2, \bar{x}_3, \bar{x}_4$ must satisfy (2.5) as it satisfies each of (2.1), (2.2), and (2.3). As $\bar{x}_1, \bar{x}_2, \bar{x}_3, \bar{x}_4$ are all nonnegative, it follows that $\bar{x}_1 + 4\bar{x}_2 + 2\bar{x}_3 \geq 0$. But this contradicts constraint (2.5). Hence, our hypothesis that there was in fact a solution to (2.1)–(2.4) must be false.

The vector $y = (1, -2, 1)^\top$ is the kind of proof that should satisfy a third party; given y, anyone can now easily and quickly check that there is no solution to (P). Of course, this proof will only work for an appropriate choice of y and we have not told you how to find such a vector at this juncture. We will derive an algorithm that either finds a feasible solution, or finds a vector y that proves that no feasible solution exists.

We wish to generalize this argument, but before this can be achieved, we need to become comfortable with matrix notation. Equations (2.1)–(2.3) can be expressed as,

$$Ax = b,$$

where

$$A = \begin{pmatrix} 4 & 10 & -6 & -2 \\ -2 & 2 & -4 & 1 \\ -7 & -2 & 0 & 4 \end{pmatrix} \quad b = \begin{pmatrix} 6 \\ 5 \\ 3 \end{pmatrix} \quad x = \begin{pmatrix} x_1 \\ x_2 \\ x_3 \\ x_4 \end{pmatrix}. \tag{2.6}$$

Then Ax is a vector with three components. Components 1, 2, and 3 of the vector Ax correspond to respectively the left-hand side of equations (2.1), (2.2), and (2.3). For a vector $y = (y_1, y_2, y_3)^\top$, $y^\top (Ax)$ is the scalar product of y and Ax and it consists of multiplying the left-hand side of equation (2.1) by y_1, the left-hand side of equation (2.2) by y_2, the left-hand side of equation (2.3) by y_3, and adding each of resulting expressions. Also, $y^\top b$ is the scalar product of y and b and it consists of multiplying the right-hand side of equation (2.1) by y_1, the right-hand side of equation (2.2) by y_2, and the right-hand side of equation (2.3) by y_3, and adding each of the resulting values. Thus

$$y^\top (Ax) = y^\top b$$

is the equation obtained by multiplying equation (2.1) by y_1, (2.2) by y_2, (2.3) by y_3, and by adding each of the resulting equations together. Note that in the previous relation, we may omit the parentheses because of associativity, i.e. $y^\top (Ax) = (y^\top A)x$. For instance, if we choose $y_1 = 1, y_2 = -2$ and $y_3 = 1$, then we obtain

$$(1, -2, 1) \begin{pmatrix} 4 & 10 & -6 & -2 \\ -2 & 2 & -4 & 1 \\ -7 & -2 & 0 & 4 \end{pmatrix} x = (1, -2, 1) \begin{pmatrix} 6 \\ 5 \\ 3 \end{pmatrix},$$

and after simplifying

$$(1, 4, 2, 0)(x_1, x_2, x_3, x_4)^\top = -1,$$

which is equation (2.5). We then observed that all the coefficients in the left-hand side of (2.5) are nonnegative, i.e. that $(1, 4, 2, 0) \geq 0^\top$ or equivalently that $y^\top A \geq 0^\top$. We also observed that the right-hand side of (2.5) is negative or equivalently that $y^\top b < 0$. (Note, 0 denotes the number zero and 0 the column vector whose entries are all zero.) These two facts implied that there is no solution to (2.1)–(2.4), i.e. that $Ax = b, x \geq 0$ has no solution.

Let us generalize this argument to an arbitrary matrix A and vector b. We will assume that the matrices and vectors have appropriate dimensions so that the matrix relations make sense. This remark holds for all subsequent statements.

PROPOSITION 2.1   *Let A be a matrix and b be a vector. Then the system*

$$Ax = b \quad x \geq 0$$

has no solution if there exists a vector y such that:

(1) $y^\top A \geq 0^\top$, *and*
(2) $y^\top b < 0$.

Note, if A has m rows, then the vector y must have m components. Then the equation $y^\top Ax = y^\top b$ is obtained by multiplying for every $i \in \{1, \ldots, m\}$ row i of A by y_i and adding all the resulting equations together.

Proof of Proposition 2.1   Let us proceed by contradiction and suppose that there exists a solution $\bar{x}$ to $Ax = b, x \geq 0$ and that we can find y such that $y^\top A \geq 0^\top$ and $y^\top b < 0$. Since $A\bar{x} = b$ is satisfied, we must also satisfy $y^\top A\bar{x} = y^\top b$. Since $y^\top A \geq 0^\top$ and $\bar{x} \geq 0$, it follows that $y^\top A\bar{x} \geq 0$. Then $0 \leq y^\top A\bar{x} = y^\top b < 0$, a contradiction.   □

We call a vector y which satisfies conditions (1) and (2) of Proposition 2.1 a *certificate of infeasibility*. To convince a third party that a particular system $Ax = b, x \geq 0$ has no solution, it suffices to exhibit a certificate of infeasibility. Note, while we have argued that such a certificate will be sufficient to prove infeasibility, it is not clear at all that for *every* infeasible system there exists a certificate of infeasibility. The fact that this is indeed so is a deep result which is known as *Farkas' lemma*; (see Theorem 4.8).

2.1.2 Unbounded linear programs

Consider the LP

$$\max\{z(x) = c^{\top}x : Ax = b, x \geq 0\},$$

where

$$A = \begin{pmatrix} 1 & 1 & -3 & 1 & 2 \\ 0 & 1 & -2 & 2 & -2 \\ -2 & -1 & 4 & 1 & 0 \end{pmatrix} \quad b = \begin{pmatrix} 7 \\ -2 \\ -3 \end{pmatrix} \quad c = \begin{pmatrix} -1 \\ 0 \\ 3 \\ 7 \\ -1 \end{pmatrix} \quad x = \begin{pmatrix} x_1 \\ x_2 \\ x_3 \\ x_4 \\ x_5 \end{pmatrix}. \quad (2.7)$$

This LP is unbounded. Just like in the previous case of infeasibility, we are looking for a concise proof that establishes this fact. Specifically, we will define a family of feasible solutions $x(t)$ for all real numbers $t \geq 0$ and show that as t tends to infinity, so does the value of $x(t)$. This will show that (2.7) is indeed unbounded.

We define for every $t \geq 0$

$$x(t) = \bar{x} + td,$$

where

$$\bar{x} = (2, 0, 0, 1, 2)^{\top} \quad \text{and} \quad d = (1, 2, 1, 0, 0)^{\top}.$$

For instance, when $t = 0$, then $x(t) = (2, 0, 0, 1, 2)^{\top}$, and when $t = 2$, then $x(t) = (4, 4, 2, 1, 2)^{\top}$. We claim that for every $t \geq 0$, $x(t)$ is feasible for (2.7). Let us first check that the equation $Ax = b$ holds for every $x(t)$. You can verify that $A\bar{x} = b$ and that $Ad = 0$. Then we have

$$Ax(t) = A(\bar{x} + td) = A\bar{x} + A(td) = \underbrace{A\bar{x}}_{=b} + t\underbrace{Ad}_{=0} = b$$

as required. We also need to verify that $x(t) \geq 0$ for every $t \geq 0$. Note that $\bar{x}, d \geq 0$ hence, $x(t) = \bar{x} + td \geq td \geq 0$, as required. Let us investigate what happens to the objective function for $x(t)$ as t increases. Observe that $c^{\top}d = 2 > 0$, then

$$c^{\top}x(t) = c^{\top}(\bar{x} + td) = c^{\top}\bar{x} + c^{\top}(td) = c^{\top}\bar{x} + tc^{\top}d = c^{\top}\bar{x} + 2t,$$

and as t tends to infinity so does $c^{\top}x(t)$. Hence, we have proved that the LP is in fact unbounded.

Given $\bar{x}$ and d, anyone can now easily verify that the linear program is unbounded. We have not told you how to find such a pair of vectors $\bar{x}$ and d. We will want an algorithm that detects if an LP is unbounded and when it is, provides us with the vectors $\bar{x}$ and d.

Let us generalize the previous argument. Let A be an $m \times n$ matrix, b a vector with m components, and c a vector with n components. The vector of variables x has n components. Consider the LP

$$\max\{c^\top x : Ax = b, x \geq 0\}. \tag{P}$$

We leave the proof of the following proposition to the reader, as the argument is essentially the one which we outlined in the above example.

PROPOSITION 2.2   *Suppose there exists a feasible solution $\bar{x}$ and a vector d such that:*

(1) *$Ad = 0$,*
(2) *$d \geq 0$,*
(3) *$c^\top d > 0$.*

Then (P) is unbounded.

We call a pair of vectors $\bar{x}, d$ as in the previous proposition a *certificate of unboundedness*. We will show that there exists a certificate of unboundedness for every unbounded LP.

2.1.3   Linear programs with optimal solutions

Consider the LP

$$\max \qquad z(x) = (0, -1, -2, 0, -3)x + 3$$

$$\text{subject to}$$

$$\begin{pmatrix} 1 & -2 & 1 & 0 & 2 \\ 0 & 1 & -1 & 1 & 3 \end{pmatrix} x = \begin{pmatrix} 2 \\ 4 \end{pmatrix} \tag{2.8}$$

$$x \geq 0.$$

Suppose you are being told that $(2, 0, 0, 4, 0)^\top$ is an optimal solution to (2.8). Clearly, you will not simply believe such a claim but rightfully request a proof. It is easy enough to verify that $(2, 0, 0, 4, 0)^\top$ is feasible and that the corresponding value is 3. However, once again, trying all feasible solutions, and comparing their values to that of $(2, 0, 0, 4, 0)^\top$ is impossible.

In order to construct a concise proof of optimality, we will prove that $z(\bar{x}) \leq 3$ for every feasible solution $\bar{x}$, i.e. no feasible solution has a value that exceeds 3. Since $(2, 0, 0, 4, 0)^\top$ has value equal to 3, it will imply that it is optimal. We seemingly traded one problem for another however: how to show that $z(\bar{x}) \leq 3$ for every feasible solution $\bar{x}$. It suffices in this particular case to analyze the objective function as

$$z(\bar{x}) = \underbrace{(0, -1, -2, 0, -3)}_{\leq 0^\top} \underbrace{\bar{x}}_{\geq 0} + 3 \leq 3,$$

where the inequality follows from the fact that the scalar product of two vectors, one where all entries are nonpositive and one where all entries are nonnegative, is always nonpositive. Hence, $(2, 0, 0, 4, 0)^\top$ is indeed optimal.

Consider now the following LP:

$$\max \quad z(x) = (-1, 3, -5, 2, 1)x - 3$$

subject to

$$\begin{pmatrix} 1 & -2 & 1 & 0 & 2 \\ 0 & 1 & -1 & 1 & 3 \end{pmatrix} x = \begin{pmatrix} 2 \\ 4 \end{pmatrix} \tag{2.9}$$

$$x \geq 0.$$

Observe that the only difference between (2.8) and (2.9) is the objective function. In particular, $(2, 0, 0, 4, 0)^\top$ is a feasible solution to (2.9). We claim that it is in fact an optimal solution. To do this, we construct a new constraint by multiplying the first constraint of the LP by -1, the second by 2, and by adding the two constraints together. Then we obtain

$$(-1, 2) \begin{pmatrix} 1 & -2 & 1 & 0 & 2 \\ 0 & 1 & -1 & 1 & 3 \end{pmatrix} x = (-1, 2) \begin{pmatrix} 2 \\ 4 \end{pmatrix}$$

and after simplifying

$$(-1, 4, -3, 2, 4)x = 6 \qquad \text{or equivalently} \qquad 0 = -(-1, 4, -3, 2, 4)x + 6.$$

This equation holds for every feasible solution of (2.8). Thus, adding this equation to the objective function $z(x) = (-1, 3, -5, 2, 1)x - 3$ will not change the value of the objective function for any of the feasible solutions. The resulting objective function is

$$z(x) = (-1, 3, -5, 2, 1)x - 3 - (-1, 4, -3, 2, 4)x + 6 = (0, -1, -2, 0, -3)x + 3.$$

However, this is the same objective function as in the LP (2.8). It means that the LP (2.9) is essentially the same LP written differently. In particular, this proves that $(2, 0, 0, 4, 0)^\top$ is an optimal solution to (2.9).

More generally for LPs of the form

$$\max\{z(x) = c^\top x + \bar{z} : Ax = b, x \geq 0\}, \tag{P}$$

we will always be able to show that an optimal solution $\bar{x}$ of value $z(\bar{x})$ is indeed optimal by proving that $z(\bar{x})$ is an upper bound for (P). We will see later in this chapter that it is always possible to prove such an upper bound by combining the objective function with a linear combination of the constraints. The coefficients used in constructing such a proof is called a *certificate of optimality*.

Exercises

1 (a) Prove that the following LP problem is infeasible:

$$
\begin{array}{rrrrrrl}
\max & 3x_1 & + & 4x_2 & + & 6x_3 & \\
\text{s.t.} & 3x_1 & + & 5x_2 & - & 6x_3 & = & 4 \\
& x_1 & + & 3x_2 & - & 4x_3 & = & 2 \\
& -x_1 & + & x_2 & - & x_3 & = & -1 \\
& x_1 & , & x_2 & , & x_3 & \geq & 0.
\end{array}
$$

(b) Prove that the following LP problem is unbounded:

$$
\begin{array}{rrrrrrl}
\max & & & & -x_3 & + & x_4 & \\
\text{s.t.} & x_1 & & & + & x_3 & - & x_4 & = & 1 \\
& & x_2 & + & 2x_3 & - & x_4 & = & 2 \\
& x_1 & , & x_2 & , & x_3 & , & x_4 & \geq & 0.
\end{array}
$$

(c) Prove that the LP problem

$$\max\{c^\top x : Ax = b, \ x \geq 0\}$$

is unbounded, where

$$
A = \begin{pmatrix} 4 & 2 & 1 & -6 & -1 \\ -1 & 1 & -4 & 1 & 3 \\ 3 & -6 & 5 & 3 & -5 \end{pmatrix}, \ b = \begin{pmatrix} 11 \\ -2 \\ -8 \end{pmatrix}, \ c = \begin{pmatrix} 1 \\ -2 \\ 1 \\ 1 \\ 1 \end{pmatrix}.
$$

HINT: Consider the vectors $\hat{x} = (1, 3, 1, 0, 0)^\top$ and $d = (1, 1, 1, 1, 1)^\top$.

(d) For each of the problems in parts (b) and (c), give a feasible solution having objective value exactly 5000.

2 Let A be an $m \times n$ matrix and let b be a vector with m entries. Prove or disprove each of the following statements (in both cases y is a vector with m entries):

(i) If there exists y such that $y^\top A \geq 0^\top$ and $b^\top y < 0$, then $Ax \leq b, x \geq 0$ has no solution.

(ii) If there exists $y \geq 0$ such that $y^\top A \geq 0^\top$ and $b^\top y < 0$, then $Ax \leq b, x \geq 0$ has no solution.

2.2   Standard equality form

An LP is said to be in *standard equality form* (SEF) if it is of the form

$$\max\{c^\top x + \bar{z} : Ax = b, x \geq 0\},$$

where $\bar{z}$ denotes some constant. In other words, an LP is in SEF if it satisfies the following conditions:

(1) it is a maximization problem,
(2) other than the nonnegativity constraints, all constraints are equations,
(3) every variable has a nonnegativity constraint.

Here is an example

$$\max \qquad (1, -2, 4, -4, 0, 0)x + 3$$

subject to

$$\begin{pmatrix} 1 & 5 & 3 & -3 & 0 & -1 \\ 2 & -1 & 2 & -2 & 1 & 0 \\ 1 & 2 & -1 & 1 & 0 & 0 \end{pmatrix} x = \begin{pmatrix} 5 \\ 4 \\ 2 \end{pmatrix} \qquad (2.10)$$

$$x \geq 0.$$

We will develop an algorithm that given an LP (P) in SEF will either prove that (P) is infeasible by exhibiting a certificate of infeasibility, or prove that (P) is unbounded by exhibiting a certificate of unboundedness, or find an optimal solution and show that it is indeed optimal by exhibiting a certificate of optimality.

However, not every LP is in SEF. Given an LP (P) which is not in SEF, we wish to "convert" (P) into an LP (P′) in SEF and apply the algorithm to (P′) instead. Of course, we want the answer for (P′) to give us some meaningful answer for (P). More precisely what we wish is for (P) and (P′) to satisfy the following relationships:

(1) (P) is infeasible if and only if (P′) is infeasible;
(2) (P) is unbounded if and only if (P′) is unbounded;
(3) given any optimal solution of (P′), we can construct an optimal solution of (P), and given any optimal solution of (P), we can construct an optimal solution of (P′).

Linear programs that satisfy the relationships (1), (2), and (3) are said to be *equivalent*.

We now illustrate on an example how to convert an arbitrary LP into an equivalent LP in SEF. Note, we will leave it to the reader to verify that at each step we do indeed get an equivalent LP

$$\min \qquad (-1, 2, -4)(x_1, x_2, x_3)^{\top}$$

subject to

$$\begin{pmatrix} 1 & 5 & 3 \\ 2 & -1 & 2 \\ 1 & 2 & -1 \end{pmatrix} \begin{pmatrix} x_1 \\ x_2 \\ x_3 \end{pmatrix} \begin{matrix} \geq \\ \leq \\ = \end{matrix} \begin{pmatrix} 5 \\ 4 \\ 2 \end{pmatrix} \qquad (2.11)$$

$$x_1, x_2 \geq 0.$$

The notation used here means that $\geq$, $\leq$ and $=$ refer to the first, second, and third constraints respectively. We want to convert (2.11) into an LP in SEF. We will proceed step by step. Note that (2.11) is a minimization problem. We can replace $\min(-1, 2, -4)(x_1, x_2, x_3)^\top$ by $\max(1, -2, 4)(x_1, x_2, x_3)^\top$, or more generally $\min c^\top x$ by $\max -c^\top x$. The resulting LP is as follows:

$$\max \quad (1, -2, 4)(x_1, x_2, x_3)^\top$$

$$\text{subject to}$$

$$\begin{pmatrix} 1 & 5 & 3 \\ 2 & -1 & 2 \\ 1 & 2 & -1 \end{pmatrix} \begin{pmatrix} x_1 \\ x_2 \\ x_3 \end{pmatrix} \begin{matrix} \geq \\ \leq \\ = \end{matrix} \begin{pmatrix} 5 \\ 4 \\ 2 \end{pmatrix} \tag{2.12}$$

$$x_1, x_2 \geq 0.$$

We do not have the condition that $x_3 \geq 0$ as part of the formulation in (2.12). We call such a variable *free*. We might be tempted to simply add the constraint $x_3 \geq 0$ to the formulation. However, by doing so we may change the optimal solution as it is possible for instance that all optimal solutions to (2.12) satisfy $x_3 < 0$. The idea here is to express x_3 as the difference of two nonnegative variables, say $x_3 = x_3^+ - x_3^-$ where $x_3^+, x_3^- \geq 0$. Let us rewrite the objective function with these new variables

$$\begin{aligned} (1, -2, 4)(x_1, x_2, x_3)^\top &= x_1 - 2x_2 + 4x_3 \\ &= x_1 - 2x_2 + 4(x_3^+ - x_3^-) \\ &= x_1 - 2x_2 + 4x_3^+ - 4x_3^- \\ &= (1, -2, 4, -4)(x_1, x_2, x_3^+, x_3^-)^\top. \end{aligned}$$

Let us rewrite the left-hand side of the constraints with these new variables

$$\begin{pmatrix} 1 & 5 & 3 \\ 2 & -1 & 2 \\ 1 & 2 & -1 \end{pmatrix} \begin{pmatrix} x_1 \\ x_2 \\ x_3 \end{pmatrix} = x_1 \begin{pmatrix} 1 \\ 2 \\ 1 \end{pmatrix} + x_2 \begin{pmatrix} 5 \\ -1 \\ 2 \end{pmatrix} + x_3 \begin{pmatrix} 3 \\ 2 \\ -1 \end{pmatrix}$$

$$= x_1 \begin{pmatrix} 1 \\ 2 \\ 1 \end{pmatrix} + x_2 \begin{pmatrix} 5 \\ -1 \\ 2 \end{pmatrix} + (x_3^+ - x_3^-) \begin{pmatrix} 3 \\ 2 \\ -1 \end{pmatrix}.$$

$$= x_1 \begin{pmatrix} 1 \\ 2 \\ 1 \end{pmatrix} + x_2 \begin{pmatrix} 5 \\ -1 \\ 2 \end{pmatrix} + x_3^+ \begin{pmatrix} 3 \\ 2 \\ -1 \end{pmatrix} + x_3^- \begin{pmatrix} -3 \\ -2 \\ 1 \end{pmatrix}$$

$$= \begin{pmatrix} 1 & 5 & 3 & -3 \\ 2 & -1 & 2 & -2 \\ 1 & 2 & -1 & 1 \end{pmatrix} \begin{pmatrix} x_1 \\ x_2 \\ x_3^+ \\ x_3^- \end{pmatrix}.$$

In general, for an LP with variables $x = (x_1, \ldots, x_n)^\top$, if we have a variable x_i where $x_i \geq 0$ is not part of the formulation (i.e. a free variable), we introduce variables $x_i^+, x_i^- \geq 0$ and define $x' = (x_1, \ldots, x_{i-1}, x_i^+, x_i^-, x_{i+1}, \ldots, x_n)^\top$. Replace the objective function $c^\top x$ by $c'^\top x'$ where $c' = (c_1, \ldots, c_{i-1}, c_i, -c_i, c_{i+1}, \ldots, c_n)^\top$. If the left-hand side of the constraints is of the form Ax where A is a matrix with columns $A_1, \ldots, A_n$, replace Ax by $A'x'$ where A' is the matrix which consists of columns $A_1, \ldots, A_{i-1}, A_i, -A_i, A_{i+1}, \ldots, A_n$.

The new LP is as follows:

$$\max \quad (1, -2, 4, -4)(x_1, x_2, x_3^+, x_3^-)^\top$$

subject to

$$\begin{pmatrix} 1 & 5 & 3 & -3 \\ 2 & -1 & 2 & -2 \\ 1 & 2 & -1 & 1 \end{pmatrix} \begin{pmatrix} x_1 \\ x_2 \\ x_3^+ \\ x_3^- \end{pmatrix} \begin{matrix} \geq \\ \leq \\ = \end{matrix} \begin{pmatrix} 5 \\ 4 \\ 2 \end{pmatrix} \qquad (2.13)$$

$$x_1, x_2, x_3^+, x_3^- \geq 0.$$

Let us replace the constraint $2x_1 - x_2 + 2x_3^+ - 2x_3^- \leq 4$ in (2.13) by an equality constraint. We introduce a new variable x_4 where $x_4 \geq 0$ and we rewrite the constraint as $2x_1 - x_2 + 2x_3^+ - 2x_3^- + x_4 = 4$. The variable x_4 is called a *slack variable*. More generally, given a constraint of the form $\sum_{i=1}^{n} a_i x_i \leq \beta$, we can replace it by $\sum_{i=1}^{n} a_i x_i + x_{n+1} = \beta$ where $x_{n+1} \geq 0$.

The resulting LP is as follows:

$$\max \quad (1, -2, 4, -4, 0)(x_1, x_2, x_3^+, x_3^-, x_4)^\top$$

subject to

$$\begin{pmatrix} 1 & 5 & 3 & -3 & 0 \\ 2 & -1 & 2 & -2 & 1 \\ 1 & 2 & -1 & 1 & 0 \end{pmatrix} \begin{pmatrix} x_1 \\ x_2 \\ x_3^+ \\ x_3^- \\ x_4 \end{pmatrix} \begin{matrix} \geq \\ = \\ = \end{matrix} \begin{pmatrix} 5 \\ 4 \\ 2 \end{pmatrix} \qquad (2.14)$$

$$x_1, x_2, x_3^+, x_3^-, x_4 \geq 0.$$

Let us replace the constraint $x_1 + 5x_2 + 3x_3^+ - 3x_3^- \geq 5$ in (2.14) by an equality constraint. We introduce a new variable x_5, where $x_5 \geq 0$ and we rewrite the constraint as $x_1 + 5x_2 + 3x_3^+ - 3x_3^- - x_5 = 5$. The variable x_5 is also called a *slack variable*. More generally, given a constraint of the form $\sum_{i=1}^{n} a_i x_i \geq \beta$, we can replace it by $\sum_{i=1}^{n} a_i x_i - x_{n+1} = \beta$, where $x_{n+1} \geq 0$.

The resulting LP is as follows:

$$\max \qquad (1, -2, 4, -4, 0, 0)(x_1, x_2, x_3^+, x_3^-, x_4, x_5)^\top$$

subject to

$$\begin{pmatrix} 1 & 5 & 3 & -3 & 0 & -1 \\ 2 & -1 & 2 & -2 & 1 & 0 \\ 1 & 2 & -1 & 1 & 0 & 0 \end{pmatrix} \begin{pmatrix} x_1 \\ x_2 \\ x_3^+ \\ x_3^- \\ x_4 \\ x_5 \end{pmatrix} = \begin{pmatrix} 5 \\ 4 \\ 2 \end{pmatrix}$$

$$x_1, x_2, x_3^+, x_3^- x_4, x_5 \geq 0.$$

Note that after relabeling the variables $x_1, x_2, x_3^+, x_3^-, x_4, x_5$ by $x_1, x_2, x_3, x_4, x_5, x_6$, we obtain the LP (2.10) modulo, a constant in the objective function. We leave to the reader to verify that the aforementioned transformations are sufficient to convert any LP into an LP in SEF.

Exercises

1 Convert the following LPs into SEF:

(a)

$$\min \qquad (2, -1, 4, 2, 4)(x_1, x_2, x_3, x_4, x_5)^\top$$

subject to

$$\begin{pmatrix} 1 & 2 & 4 & 7 & 3 \\ 2 & 8 & 9 & 0 & 0 \\ 1 & 1 & 0 & 2 & 6 \\ -3 & 4 & 3 & 1 & -1 \end{pmatrix} \begin{pmatrix} x_1 \\ x_2 \\ x_3 \\ x_4 \\ x_5 \end{pmatrix} \begin{matrix} \leq \\ = \\ \geq \\ \geq \end{matrix} \begin{pmatrix} 1 \\ 2 \\ 3 \\ 4 \end{pmatrix}$$

$$x_1 \geq 0, x_2 \geq 0, x_4 \geq 0.$$

(b) Let A, B, D be matrices and b, c, d, f be vectors (all of suitable dimensions). Convert the following LP with variables x and y (where x, y are vectors) into SEF:

$$\min \qquad c^\top x + d^\top y$$

subject to

$$\begin{aligned} Ax &\geq b \\ Bx + Dy &= f \\ x &\geq 0. \end{aligned}$$

Note, the variables y are free.

2.3   A simplex iteration

Consider the following LP in SEF:

$$\text{max} \qquad z(x) = (2, 3, 0, 0, 0)x$$

subject to

$$\begin{pmatrix} 1 & 1 & 1 & 0 & 0 \\ 2 & 1 & 0 & 1 & 0 \\ -1 & 1 & 0 & 0 & 1 \end{pmatrix} x = \begin{pmatrix} 6 \\ 10 \\ 4 \end{pmatrix} \tag{2.15}$$

$$x_1, x_2, x_3, x_4, x_5 \geq 0,$$

where $x = (x_1, x_2, x_3, x_4, x_5)^\top$. Because (2.15) has a special form, it is easy to verify that $\bar{x} = (0, 0, 6, 10, 4)^\top$ is a feasible solution with value $z(\bar{x}) = 0$. Let us try to find a feasible solution x with larger value. Since the objective function is $z(x) = 2x_1 + 3x_2$, by increasing the value of $\bar{x}_1$ or $\bar{x}_2$ we will increase the value of the objective function. Let us try to increase the value of $\bar{x}_1$ while keeping $\bar{x}_2$ equal to zero. In other words, we look for a new feasible solution x, where $x_1 = t$ for some $t \geq 0$ and $x_2 = 0$. The matrix equation will tell us which values we need for x_3, x_4, x_5. We have

$$\begin{pmatrix} 6 \\ 10 \\ 4 \end{pmatrix} = \begin{pmatrix} 1 & 1 & 1 & 0 & 0 \\ 2 & 1 & 0 & 1 & 0 \\ -1 & 1 & 0 & 0 & 1 \end{pmatrix} x$$

$$= x_1 \begin{pmatrix} 1 \\ 2 \\ -1 \end{pmatrix} + x_2 \begin{pmatrix} 1 \\ 1 \\ 1 \end{pmatrix} + x_3 \begin{pmatrix} 1 \\ 0 \\ 0 \end{pmatrix} + x_4 \begin{pmatrix} 0 \\ 1 \\ 0 \end{pmatrix} + x_5 \begin{pmatrix} 0 \\ 0 \\ 1 \end{pmatrix}$$

$$= t \begin{pmatrix} 1 \\ 2 \\ -1 \end{pmatrix} + 0 \begin{pmatrix} 1 \\ 1 \\ 1 \end{pmatrix} + \begin{pmatrix} x_3 \\ x_4 \\ x_5 \end{pmatrix}.$$

Thus

$$\begin{pmatrix} x_3 \\ x_4 \\ x_5 \end{pmatrix} = \begin{pmatrix} 6 \\ 10 \\ 4 \end{pmatrix} - t \begin{pmatrix} 1 \\ 2 \\ -1 \end{pmatrix}. \tag{2.16}$$

The larger we pick $t \geq 0$, the more we will increase the objective function. How large can we choose t to be? We simply need to make sure that $x_3, x_4, x_5 \geq 0$, i.e. that

$$t \begin{pmatrix} 1 \\ 2 \\ -1 \end{pmatrix} \leq \begin{pmatrix} 6 \\ 10 \\ 4 \end{pmatrix}.$$

Thus, $t \leq 6$ and $2t \leq 10$, i.e. $t \leq \frac{10}{2}$. Note that $-t \leq 4$ does not impose any bound on how large t can be. Picking the largest possible t yields 5. We can summarize this computation as

$$t = \min\left\{\frac{6}{1}, \frac{10}{2}, -\right\}.$$

Replacing $t = 5$ in (2.16) yields $x' := (5, 0, 1, 0, 9)^\top$ with $z(x') = 10$.

What happens if we try to apply the same approach again? We could try to increase the value of x_2', but in order to do this we would have to decrease the value of x_1', which might *decrease* the objective function. The same strategy no longer seems to work! We were able to carry out the computations the first time around because the LP (2.15) was in a suitable form for the original solution $\bar{x}$. However, (2.15) is not in a suitable form for the new solution x'. In the next section, we will show that we can rewrite (2.15) so that the resulting LP is in a form that allows us to carry out the kind of computations outlined for $\bar{x}$. By repeating this type of computation and rewriting the LP at every step, this will lead to an algorithm for solving LPs which is know as the *simplex algorithm*.

Exercises

1   In this exercise, you are asked to repeat the argument in Section 2.3 with different examples.

(a) Consider the following LP:

$$\begin{array}{ll} \max & (-1, 0, 0, 2)x \\ \text{subject to} & \end{array}$$

$$\begin{pmatrix} -1 & 1 & 0 & 2 \\ 1 & 0 & 1 & -3 \end{pmatrix} x = \begin{pmatrix} 2 \\ 3 \end{pmatrix}$$

$$x \geq 0.$$

Observe that $\bar{x} = (0, 2, 3, 0)^\top$ is a feasible solution. Starting from $\bar{x}$, construct a feasible solution x' with value larger than that of $\bar{x}$ by increasing as much as possible the value of exactly one of $\bar{x}_1$ or $\bar{x}_4$ (keeping the other variable unchanged).

(b) Consider the following LP:

$$\begin{array}{ll} \max & (0, 0, 4, -6)x \\ \text{subject to} & \end{array}$$

$$\begin{pmatrix} 1 & 0 & -1 & 1 \\ 0 & 1 & -3 & 2 \end{pmatrix} x = \begin{pmatrix} 2 \\ 1 \end{pmatrix}$$

$$x \geq 0.$$

Observe that $\bar{x} = (2, 1, 0, 0)^\top$ is a feasible solution. Starting from $\bar{x}$, construct a feasible solution x' with value larger than that of $\bar{x}$ by increasing as much as possible the value of exactly one of $\bar{x}_3$ or $\bar{x}_4$ (keeping the other variable unchanged). What can you deduce in this case?

2.4   Bases and canonical forms

2.4.1   Bases

Consider an $m \times n$ matrix A, where the rows of A are linearly independent. We will denote column j of A by A_j. Let J be a subset of the column indices (i.e. $J \subseteq \{1, \ldots, n\}$), we define A_J to be the matrix formed by columns A_j for all $j \in J$ (where the columns appear in the order given by their corresponding indices). We say that a set of column indices B forms a *basis* if the matrix A_B is a square nonsingular matrix. Equivalently, a basis corresponds to a maximal subset of linearly independent columns. Consider for instance

$$A = \begin{pmatrix} 2 & 1 & 2 & -1 & 0 & 0 \\ 1 & 0 & -1 & 2 & 1 & 0 \\ 3 & 0 & 3 & 1 & 0 & 1 \end{pmatrix}.$$

Then $B = \{2, 5, 6\}$ is a basis as the matrix A_B is the identity matrix. Note that $B = \{1, 2, 3\}$ and $\{1, 5, 6\}$ are also bases, while $B = \{1, 3\}$ is not a basis (as A_B is not square in this case) and neither is $B = \{1, 3, 5\}$ (as A_B is singular in this case). We will denote by N the set of column indices not in B. Thus, B and N will always denote a partition of the column indices of A.

Suppose that in addition to the matrix A we have a vector b with m components, and consider the system of equations $Ax = b$. For instance

$$\begin{pmatrix} 2 & 1 & 2 & -1 & 0 & 0 \\ 1 & 0 & -1 & 2 & 1 & 0 \\ 3 & 0 & 3 & 1 & 0 & 1 \end{pmatrix} x = \begin{pmatrix} 2 \\ 1 \\ 1 \end{pmatrix}. \tag{2.17}$$

Variables x_j are said to be *basic* when $j \in B$ and *nonbasic* otherwise. The vector which is formed by the basic variables is denoted by x_B and the vector which is formed by the nonbasic variables is x_N. We assume that the components in x_B appear in the same order as A_B and that the components in x_N appear in the same order as A_N. For instance, for (2.17) $B = \{1, 5, 6\}$ is a basis, then $N = \{2, 3, 4\}$ and $x_B = (x_1, x_5, x_6)^\top$, $x_N = (x_2, x_3, x_4)^\top$.

The following easy observation will be used repeatedly:

$$Ax = \sum_{j=1}^{n} x_j A_j = \sum_{j \in B} x_j A_j + \sum_{j \in N} x_j A_j = A_B x_B + A_N x_N.$$

A vector $\bar{x}$ is a *basic solution* of $Ax = b$ for a basis B if the following conditions hold:

(1) $A\bar{x} = b$, and
(2) $\bar{x}_N = \mathbb{0}$.

Suppose $\bar{x}$ is such a basic solution, then

$$b = A\bar{x} = A_B \bar{x}_B + A_N \underbrace{\bar{x}_N}_{= \mathbb{0}} = A_B \bar{x}_B.$$

Since A_B is nonsingular, it has an inverse and we have $\bar{x}_B = A_B^{-1} b$. In particular, it shows:

Remark 2.3 *Every basis is associated with a unique basic solution.*

For (2.17) and basis $B = \{1, 5, 6\}$, the unique basic solution $\bar{x}$ is $\bar{x}_2 = \bar{x}_3 = \bar{x}_4 = 0$ and

$$\begin{pmatrix} \bar{x}_1 \\ \bar{x}_5 \\ \bar{x}_6 \end{pmatrix} = \begin{pmatrix} 2 & 0 & 0 \\ 1 & 1 & 0 \\ 3 & 0 & 1 \end{pmatrix}^{-1} \begin{pmatrix} 2 \\ 1 \\ 1 \end{pmatrix} = \begin{pmatrix} 1 \\ 0 \\ -2 \end{pmatrix}.$$

Thus, $\bar{x} = (1, 0, 0, 0, 0, -2)^\top$. A basic solution $\bar{x}$ is *feasible* if $\bar{x} \geq 0$ (in which case we refer to $\bar{x}$ as a *basic feasible solution*). A basis B is feasible if the corresponding basic solution is feasible. If a basic solution (resp. a basis) is not feasible, then it is *infeasible*. For instance, the basis $B = \{1, 5, 6\}$ is infeasible as the corresponding basic solution $\bar{x}$ has negative entries such as $\bar{x}_6 = -2$. When $B = \{2, 5, 6\}$, then as A_B is the identity matrix, the corresponding basic solution is $\bar{x} = (0, 2, 0, 0, 1, 1)^\top$ which is feasible as all entries are nonnegative.

The simplex algorithm will solve LPs in SEF by considering the bases of the matrix A where $Ax = b$ are the equations that define all the equality constraints of the LP. Of course, if the rows of A are not linearly independent, then A will not have any basis. We claim however that we may assume without loss of generality that the rows of A are indeed linearly independent. For otherwise, you can prove using elementary linear algebra that one of the following two possibilities must occur:

1. The system $Ax = b$ has no solution,
2. The system $Ax = b$ has a redundant constraint.

If (1) occurs, then the LP is infeasible and we can stop. If (2) occurs, then we can eliminate a redundant constraint. We repeat the procedure until all rows of A are linearly independent. Hence, throughout this chapter we will always assume (without stating it explicitly) that the rows of the matrix defining the left-hand side of the equality constraints in an LP in SEF are linearly independent.

2.4.2   Canonical forms

Let us restate the LP (2.15) we were trying to solve in Section 2.3:

$$\begin{array}{ll} \max & z(x) = (2, 3, 0, 0, 0)x \\ \text{subject to} & \end{array}$$

$$\begin{pmatrix} 1 & 1 & 1 & 0 & 0 \\ 2 & 1 & 0 & 1 & 0 \\ -1 & 1 & 0 & 0 & 1 \end{pmatrix} x = \begin{pmatrix} 6 \\ 10 \\ 4 \end{pmatrix} \tag{2.15}$$

$$x_1, x_2, x_3, x_4, x_5 \geq 0.$$

Observe that $B = \{3, 4, 5\}$ is a basis. The corresponding basic solution is given by $\bar{x} = (0, 0, 6, 10, 4)^\top$. Note that $\bar{x}$ is the feasible solution with which we started the iteration. The basis B has the property that A_B is an identity matrix. In addition, the objective function has the property that $c_B = (c_3, c_4, c_5)^\top = (0, 0, 0)^\top$. These are the two properties that allowed us to find a new feasible solution $x' = (5, 0, 1, 0, 9)^\top$ with larger value. We let the readers verify that x' is a basic solution for the basis $B = \{1, 3, 5\}$. Clearly, for the new basis B, A_B is not an identity matrix, and c_B is not the $\mathbb{0}$ vector. Since these properties are no longer satisfied for x', we could not carry the computation further.

This motivates the following definition.

Consider the following LP in SEF:

$$\max\{c^\top x + \bar{z} : Ax = b, x \geq \mathbb{0}\}, \tag{P}$$

where $\bar{z}$ is a constant and let B be a basis of A. We say that (P) is in *canonical form for* B if the following conditions are satisfied:

(C1) A_B is an identity matrix,
(C2) $c_B = \mathbb{0}$.

We will show that given any basis B, we can rewrite the LP (P) so that it is in canonical form. For instance, $B = \{1, 2, 4\}$ is a basis of the LP (2.15) that can be rewritten as the following equivalent LP:

$$\max \qquad z(x) = 17 + \left(0, 0, -\frac{5}{2}, 0, -\frac{1}{2}\right) x$$

subject to

$$\begin{pmatrix} 1 & 0 & 1/2 & 0 & -1/2 \\ 0 & 1 & 1/2 & 0 & 1/2 \\ 0 & 0 & -3/2 & 1 & 1/2 \end{pmatrix} x = \begin{pmatrix} 1 \\ 5 \\ 3 \end{pmatrix} \tag{2.18}$$

$$x \geq \mathbb{0}.$$

How did we get this LP?

Let us first rewrite the equations of (2.15) so that condition (C1) is satisfied for $B = \{1, 2, 4\}$. We left multiply the equations by the inverse of A_B (where $Ax = b$ denote the equality constraints of (2.15)) and where $B = \{1, 2, 4\}$):

$$\begin{pmatrix} 1 & 1 & 0 \\ 2 & 1 & 1 \\ -1 & 1 & 0 \end{pmatrix}^{-1} \begin{pmatrix} 1 & 1 & 1 & 0 & 0 \\ 2 & 1 & 0 & 1 & 0 \\ -1 & 1 & 0 & 0 & 1 \end{pmatrix} x = \begin{pmatrix} 1 & 1 & 0 \\ 2 & 1 & 1 \\ -1 & 1 & 0 \end{pmatrix}^{-1} \begin{pmatrix} 6 \\ 10 \\ 4 \end{pmatrix}.$$

The resulting equation is exactly the set of equations in the LP (2.18).

Let us rewrite the objective function of (2.15) so that condition (C2) is satisfied (still for basis $B = \{1, 2, 4\}$). We generate an equation obtained by multiplying the first equation of (2.15) by y_1, multiplying the second equation of (2.15) by y_2, the third equation by y_3, and adding each of the corresponding constraints together (where the values of y_1, y_2, y_3 are yet to be decided). The resulting equation can be written as

$$(y_1, y_2, y_3) \begin{pmatrix} 1 & 1 & 1 & 0 & 0 \\ 2 & 1 & 0 & 1 & 0 \\ -1 & 1 & 0 & 0 & 1 \end{pmatrix} x = (y_1, y_2, y_3) \begin{pmatrix} 6 \\ 10 \\ 4 \end{pmatrix},$$

or equivalently as

$$0 = (y_1, y_2, y_3) \begin{pmatrix} 6 \\ 10 \\ 4 \end{pmatrix} - (y_1, y_2, y_3) \begin{pmatrix} 1 & 1 & 1 & 0 & 0 \\ 2 & 1 & 0 & 1 & 0 \\ -1 & 1 & 0 & 0 & 1 \end{pmatrix} x.$$

Since this equation holds for every feasible solution x, we can add this previous constraint to the objective function of (2.15), namely $z(x) = (2, 3, 0, 0, 0)x$. The resulting objective function is

$$z(x) = (y_1, y_2, y_3) \begin{pmatrix} 6 \\ 10 \\ 4 \end{pmatrix} + \left[(2, 3, 0, 0, 0) - (y_1, y_2, y_3) \begin{pmatrix} 1 & 1 & 1 & 0 & 0 \\ 2 & 1 & 0 & 1 & 0 \\ -1 & 1 & 0 & 0 & 1 \end{pmatrix} \right] x, \quad (\star)$$

which is of the form $z(x) = \bar{z} + \bar{c}^\top x$. For (C2) to be satisfied, we need $\bar{c}_1 = \bar{c}_2 = \bar{c}_4 = 0$. We need to choose y_1, y_2, y_3 accordingly. Namely, we need

$$(2, 3, 0) - (y_1, y_2, y_3) \begin{pmatrix} 1 & 1 & 0 \\ 2 & 1 & 1 \\ -1 & 1 & 0 \end{pmatrix} = \mathbb{0}^\top$$

or equivalently

$$(y_1, y_2, y_3) \begin{pmatrix} 1 & 1 & 0 \\ 2 & 1 & 1 \\ -1 & 1 & 0 \end{pmatrix} = (2, 3, 0)$$

and by taking the transpose on both sides of the equation, we get

$$\begin{pmatrix} 1 & 1 & 0 \\ 2 & 1 & 1 \\ -1 & 1 & 0 \end{pmatrix}^\top \begin{pmatrix} y_1 \\ y_2 \\ y_3 \end{pmatrix} = \begin{pmatrix} 2 \\ 3 \\ 0 \end{pmatrix}.$$

By solving the system, we get the unique solution $y = \left(\frac{5}{2}, 0, \frac{1}{2} \right)^\top$. By substituting y in $(\star)$, we obtain

$$z(x) = 17 + \left(0, 0, -\frac{5}{2}, 0, -\frac{1}{2} \right) x,$$

which is the objective function of (2.18). Let us formalize these observations. Consider an LP in SEF

$$\max\{c^\top x + \bar{z} : Ax = b, x \geq \mathbb{0}\},$$

where $\bar{z}$ is a constant. By definition, A_B is nonsingular, hence A_B^{-1} exists.

We claim that to achieve condition (C1), it suffices to replace $Ax = b$ by

$$A_B^{-1} Ax = A_B^{-1} b. \tag{2.19}$$

Observe that $Ax = A_B x_B + A_N x_N$, thus

$$A_B^{-1} Ax = A_B^{-1}(A_B x_B + A_N x_N)$$
$$= A_B^{-1} A_B x_B + A_B^{-1} A_N x_N$$
$$= x_B + A_B^{-1} A_N x_N.$$

In particular, the columns corresponding to B on the left-hand side of (2.19) form an identity matrix as required. Moreover, we claim that the set of solutions to $Ax = b$ is equal to the set of solutions to (2.19). Clearly, every solution to $Ax = b$ is also a solution to $A_B^{-1} Ax = A_B^{-1} b$ as these equations are linear combinations of the equations $Ax = b$. Moreover, every solution to $A_B^{-1} Ax = A_B^{-1} b$ is also a solution to $A_B A_B^{-1} Ax = A_B A_B^{-1} b$, but this equation is simply $Ax = b$, proving the claim.

Let us consider condition (C2). Let B be a basis of A. Suppose A has m rows, then for any vector $y = (y_1, \ldots, y_m)^\top$ the equation

$$y^\top Ax = y^\top b$$

can be rewritten as

$$0 = y^\top b - y^\top Ax.$$

Since this equation holds for every feasible solution, we can add this constraint to the objective function $z(x) = c^\top x + \bar{z}$. The resulting objective function is

$$z(x) = y^\top b + \bar{z} + (c^\top - y^\top A)x. \tag{2.20}$$

Let $\bar{c}^\top := c^\top - y^\top A$. For (C2) to be satisfied, we need $\bar{c}_B = \mathbb{0}$ and also to choose y accordingly. Namely, we want that

$$\bar{c}_B^\top = c_B^\top - y^\top A_B = \mathbb{0}^\top$$

or equivalently that

$$y^\top A_B = c_B^\top.$$

By taking the transpose on both sides, we get

$$A_B^\top y = c_B.$$

Note that B is a basis, and thus matrices A_B and its transpose are nonsingular. Furthermore, the inverse and transpose operations commute, and hence $(A_B^{-1})^\top = (A_B^\top)^{-1}$. Therefore, we will write $A_B^{-\top}$ for $(A_B^{-1})^\top$. Hence, the previous relation can be rewritten as

$$y = A_B^{-\top} c_B. \tag{2.21}$$

We have shown the following, see (2.19), (2.20), (2.21):

PROPOSITION 2.4 *Suppose an LP*

$$\max\{z(x) = c^\top x + \bar{z} : Ax = b, x \geq \mathbb{0}\}$$

and a basis B of A are given. Then the following LP is an equivalent LP in canonical form for the basis B:

$$\max \qquad z(x) = y^{\mathsf{T}} b + \bar{z} + (c^{\mathsf{T}} - y^{\mathsf{T}} A)x$$

subject to

$$A_B^{-1} Ax = A_B^{-1} b$$
$$x \geq 0,$$

where $y = A_B^{-\mathsf{T}} c_B$.

Exercises

1 Consider the system $Ax = b$ where the rows of A are linearly independent. Let $\bar{x}$ be a solution to $Ax = b$. Let J be the column indices j of A for which $\bar{x}_j \neq 0$.
(a) Show that if $\bar{x}$ is a basic solution, then the columns of A_J are linearly independent.
(b) Show that if the columns of A_J are linearly independent, then $\bar{x}$ is a basic solution for some basis $B \supseteq J$.

Note that (a) and (b) give you a way of checking whether $\bar{x}$ is a basic solution, namely you simply need to verify whether the columns of A_J are linearly independent.
(c) Consider the system of equations

$$\begin{pmatrix} 1 & 1 & 0 & 2 & 1 & 1 & 1 \\ 0 & 2 & 2 & 0 & 0 & -2 & 1 \\ 1 & 2 & 1 & 5 & 4 & 3 & 3 \end{pmatrix} x = \begin{pmatrix} 2 \\ 2 \\ 6 \end{pmatrix}$$

and the following vectors:
  (i) $(1, 1, 0, 0, 0, 0, 0)^{\mathsf{T}}$,
  (ii) $(2, -1, 2, 0, 1, 0, 0)^{\mathsf{T}}$,
  (iii) $(1, 0, 1, 0, 1, 0, 0)^{\mathsf{T}}$,
  (iv) $(0, 0, 1, 1, 0, 0, 0)^{\mathsf{T}}$,
  (v) $(0, \frac{1}{2}, 0, 0, \frac{1}{2}, 0, 1)^{\mathsf{T}}$.
For each vector in (i)–(v), indicate if it is a basic solution or not. *You need to justify your answers.*
(d) Which of the vectors in (i)–(v) are basic *feasible* solutions?

2 The following LP is in SEF:

$$\max \qquad (1, -2, 0, 1, 3)x$$

subject to

$$\begin{pmatrix} 1 & -1 & 2 & -1 & 0 \\ 2 & 0 & 1 & -1 & 1 \end{pmatrix} x = \begin{pmatrix} 1 \\ -1 \end{pmatrix}$$
$$x \geq 0.$$

Find an equivalent LP in canonical form for:
(a) the basis $\{1, 4\}$,
(b) the basis $\{3, 5\}$.

In each case, state whether the basis (i.e. the corresponding basic solution) is feasible.

3   Let A be an $m \times n$ matrix and consider the following LP (P):

$$\max \quad c^\top x$$

subject to

$$Ax = b$$

$$x_j \geq 0 \quad \text{for all } j = 1, \ldots, n - 1.$$

Convert (P) into an LP (P') in SEF by replacing the free variable x_n by two variables x_n^+ and x_n^-. Show then that no basic solution $\bar{x}$ of (P') satisfies $\bar{x}_n^+ > 0$ and $\bar{x}_n^- > 0$.

4   Consider the LP

$$\max\{c^\top x : Ax = b, x \geq \mathbb{0}\}. \tag{P}$$

Assume that the rows of A are linearly independent. Let $x^{(1)}$ and $x^{(2)}$ be two distinct feasible solutions of (P) and define

$$\bar{x} := \frac{1}{2}x^{(1)} + \frac{1}{2}x^{(2)}.$$

(a) Show that $\bar{x}$ is a feasible solution to (P).
(b) Show that if $x^{(1)}$ and $x^{(2)}$ are optimal solutions to (P), then so is $\bar{x}$.
(c) Show that $\bar{x}$ is not a *basic feasible solution*.
      HINT: Proceed by contradiction and suppose that $\bar{x}$ is a basic feasible solution for some basis B. Denote by N the indices of the columns of A not in B. Then show that $\bar{x}_N = x_N^{(1)} = x_N^{(2)} = \mathbb{0}$ and that $x_B^{(1)} = x_B^{(2)}$.
(d) Deduce from (b) and (c) that if (P) has two optimal solutions, then (P) has an optimal solution that is NOT a basic feasible solution.

5   Consider the following LP (P):

$$\max \quad (0, 0, 0, -1, -2, -6)x + 17$$

subject to

$$\begin{pmatrix} 1 & 0 & 0 & -9 & 12 & 17 \\ 0 & 1 & 0 & 39 & 0 & 14 \\ 0 & 0 & 1 & -2 & 13 & -66 \end{pmatrix} x = \begin{pmatrix} 4 \\ 8 \\ 5 \end{pmatrix}$$

$$x \geq \mathbb{0}.$$

(a) Find the basic solution $\bar{x}$ of (P) for the basis $B = \{1, 2, 3\}$.
(b) Show that $\bar{x}$ is an optimal solution of (P).
(c) Show that $\bar{x}$ is the *unique* optimal solution of (P).
      HINT: Show that for every feasible solution x' of value 17 $\{j : x_j' \neq 0\} \subseteq B$. Then use Exercise 1 (b).

Consider the following LP (Q):

$$\max \quad c^\top x + \bar{z}$$
$$\text{subject to}$$
$$Ax = b$$
$$x \geq 0,$$

where $\bar{z}$ is a constant, B is a basis of A, $A_B = I$, $c_j = 0$ for every $j \in B$ and $c_j < 0$ for every $j \notin B$. Moreover, assume that $b \geq 0$. (The previous example has such a form.) Let $\bar{x}$ be a basic solution for (Q).

(d) Show that $\bar{x}$ is an optimal solution of (Q).

(e) Show that $\bar{x}$ is the *unique* optimal solution of (Q).

 HINT: Same argument as (c).

6 (Advanced) Consider the following LP in SEF:

$$\max\{c^\top x : Ax = b, x \geq 0\}, \tag{P}$$

where A is an $m \times n$ matrix. Let $\bar{x}$ be a feasible solution to (P) and let $J = \{j : \bar{x}_j > 0\}$. Call a vector d a *good direction* for $\bar{x}$ if it satisfies the following properties:

(P1) $d_j < 0$ for some $j \in J$,

(P2) $Ad = 0$,

(P3) $d_j = 0$ for all $j \notin J$.

(a) Show that if the columns of A_J are linearly dependent, then there exists a good direction for $\bar{x}$.

 HINT: Use the definition of linear dependence to get a vector d. Then possibly replace d by $-d$.

(b) Show that if $\bar{x}$ is not basic, then there exists a good direction for $\bar{x}$.

 HINT: Use (a) and use Exercise 1.

(c) Show that if $\bar{x}$ has a good direction, then there exists a feasible solution x' of (P) such that the set $J' := \{j : x'_j > 0\}$ has fewer elements than J.

 HINT: Let $x' = x + \epsilon d$ for a suitable value $\epsilon > 0$.

(d) Show that if (P) has a feasible solution, then it has a feasible solution that is basic.

 HINT: Use (b) and (c) repeatedly.

(e) Give an algorithm that takes as input a feasible solution for (P) and returns a basic feasible solution for (P).

2.5 The simplex algorithm

2.5.1 An example with an optimal solution

Let us continue the example (2.15) which we first started in Section 2.3. At the end of the first iteration, we obtained the feasible solution $\bar{x} = (5, 0, 1, 0, 9)^\top$ which is a basic solution for the basis $B = \{1, 3, 5\}$. Using the formulae in Proposition 2.4, we can rewrite the LP so that it is in canonical form for that basis. We obtain

$$\max\{z(x) = 10 + c^{\top}x : Ax = b, x \geq 0\},$$

where

$$A = \begin{pmatrix} 1 & 1/2 & 0 & 1/2 & 0 \\ 0 & 1/2 & 1 & -1/2 & 0 \\ 0 & 3/2 & 0 & 1/2 & 1 \end{pmatrix} \qquad b = \begin{pmatrix} 5 \\ 1 \\ 9 \end{pmatrix} \qquad c = \begin{pmatrix} 0 \\ 2 \\ 0 \\ -1 \\ 0 \end{pmatrix}. \qquad (2.22)$$

Let us try to find a feasible solution x with value larger than $\bar{x}$. Recall, B and N partition the column indices of A, i.e. $N = \{2, 4\}$. Since the LP is in canonical form, $c_B = 0$. Therefore, to increase the objective function value, we must select $k \in N$ such that $c_k > 0$ and increase the component $\bar{x}_k$ of $\bar{x}$. In this case, our choice for k is $k = 2$. Therefore, we set $x_2 = t$ for some $t \geq 0$. For all $j \in N$ where $j \neq k$, we keep component x_j of x equal to zero. It means in this case that $x_4 = 0$. The matrix equation will tell us what values we need to choose for $x_B = (x_1, x_3, x_5)^{\top}$. Following the same argument as in Section 2.3, we obtain that

$$x_B = \begin{pmatrix} x_1 \\ x_3 \\ x_5 \end{pmatrix} = \begin{pmatrix} 5 \\ 1 \\ 9 \end{pmatrix} - t \begin{pmatrix} 1/2 \\ 1/2 \\ 3/2 \end{pmatrix}, \qquad (2.23)$$

which implies as $x_B \geq 0$ that

$$t \begin{pmatrix} 1/2 \\ 1/2 \\ 3/2 \end{pmatrix} \leq \begin{pmatrix} 5 \\ 1 \\ 9 \end{pmatrix}. \qquad (2.24)$$

The largest possible value for t is given by

$$t = \min \left\{ \frac{5}{1/2}, \frac{1}{1/2}, \frac{9}{3/2} \right\} = 2. \qquad (2.25)$$

Note that (2.24) can be written as $tA_k \leq b$ (where $k = 2$). Thus, t is obtained by the so-called *ratio test* : take the smallest ratio between the entry b_i and entry A_{ik} for all $A_{ik} > 0$.

Replacing $t = 2$ in (2.23) yields $(4, 2, 0, 0, 6)^{\top}$. We redefine $\bar{x}$ to be $(4, 2, 0, 0, 6)^{\top}$ and we now have $z(\bar{x}) = 14$. It can be readily checked that $\bar{x}$ is a basic solution. In fact, this is no accident as we will always obtain a basic solution proceeding this way (see Exercise 4 in Section 2.5.3). Since $\bar{x}_1, \bar{x}_2, \bar{x}_5 > 0$, the basis corresponding to $\bar{x}$ must contain each of $\{1, 2, 5\}$. As each basis of A contains exactly three basic elements, it follows that the new basis must be $\{1, 2, 5\}$. We can rewrite the LP so that it is in canonical form for that basis and repeat the same process.

To start the new iteration, it suffices to know the new basis $\{1, 2, 5\}$. We obtained $\{1, 2, 5\}$ from the old basis $\{1, 3, 5\}$ by adding element 2 and removing element 3. We will say that 2 *entered* the basis and 3 *left* the basis. Thus, it suffices at each iteration to establish which element enters and which element leaves the basis. If we set $x_k = t$ where $k \in N$, element k will enter the basis. If some basic variable x_ℓ is decreased to 0, then we can select ℓ to leave the basis. In (2.25), the minimum was attained for the second term. Thus, in (2.23) the second component of x_B will be set to zero, i.e. x_3 will be set to zero and 3 will leave the basis.

Let us proceed with the next iteration. Using the formulae in Proposition 2.4, we can rewrite the LP so that it is in canonical form for the basis $B = \{1, 2, 5\}$. We get,

$$\max\{z(x) = 14 + c^\top x : Ax = b, x \geq 0\},$$

where

$$A = \begin{pmatrix} 1 & 0 & -1 & 1 & 0 \\ 0 & 1 & 2 & -1 & 0 \\ 0 & 0 & -3 & 2 & 1 \end{pmatrix} \quad b = \begin{pmatrix} 4 \\ 2 \\ 6 \end{pmatrix} \quad c = \begin{pmatrix} 0 \\ 0 \\ -4 \\ 1 \\ 0 \end{pmatrix}. \quad (2.26)$$

Here we have $N = \{3, 4\}$. Let us first choose which element k enters the basis. We want $k \in N$ and $c_k > 0$. The only choice is $k = 4$. We compute t by taking the smallest ratio between entry b_i and entry A_{ik} (where $k = 4$) for all i where $A_{ik} > 0$, namely

$$t = \min \left\{ \frac{4}{1}, -, \frac{6}{2} \right\} = 3,$$

where "-" indicates that the corresponding entry of A_{ik} is not positive. The minimum was attained for the ratio $\frac{6}{2}$, i.e. the third row. It corresponds to the third component of x_B. As the third element of B is 5, we will have $x_5 = 0$ for the new solution. Hence, 5 will be leaving the basis and the new basis will be $\{1, 2, 4\}$.

Let us proceed with the next iteration. Using the formulae in Proposition 2.4, we can rewrite the LP so that it is in canonical form for the basis $B = \{1, 2, 4\}$. We get

$$\max \qquad z(x) = 17 + \left(0, 0, -\frac{5}{2}, 0, -\frac{1}{2} \right) x$$

subject to

$$\begin{pmatrix} 1 & 0 & 1/2 & 0 & -1/2 \\ 0 & 1 & 1/2 & 0 & 1/2 \\ 0 & 0 & -3/2 & 1 & 1/2 \end{pmatrix} x = \begin{pmatrix} 1 \\ 5 \\ 3 \end{pmatrix} \qquad (2.27)$$

$$x \geq 0.$$

The basic solution is $\bar{x} := (1, 5, 0, 3, 0)^\top$ and $z(\bar{x}) = 17$. We have $N = \{3, 5\}$. We want $k \in N$ and $c_k > 0$. However, $c_3 = -\frac{5}{2}$ and $c_5 = -\frac{1}{2}$, so there is no such choice. We claim that this occurs because the current basic solution is optimal.

Let x' be any feasible solution, then

$$z(x') = 17 + \underbrace{\left(0, 0, -\frac{5}{2}, 0, -\frac{1}{2}\right)}_{\leq 0} \underbrace{x'}_{\geq 0} \leq 17.$$

Thus, 17 is an upper bound for the value of any feasible solution. As $z(\bar{x}) = 17$, it follows that $\bar{x}$ is an optimal solution.

2.5.2 An unbounded example

Consider the following LP:

$$\max\{z(x) = c^\top x : Ax = b, x \geq 0\},$$

where

$$A = \begin{pmatrix} -2 & 4 & 1 & 0 & 1 \\ -3 & 7 & 0 & 1 & 1 \end{pmatrix} \qquad b = \begin{pmatrix} 1 \\ 3 \end{pmatrix} \qquad c = \begin{pmatrix} -1 \\ 3 \\ 0 \\ 0 \\ 1 \end{pmatrix}.$$

It is in canonical form for the basis $B = \{3, 4\}$. Then $N = \{1, 2, 5\}$. Let us choose which element k enters the basis. We want $k \in N$ and $c_k > 0$. We have choices $k = 2$ and $k = 5$. Let us select 5. We compute t by taking the smallest ratio between entry b_i and entry A_{ik} for all i, where $A_{ik} > 0$. Namely

$$\min\left\{\frac{1}{1}, \frac{3}{1}\right\}.$$

The minimum is attained for the ratio $\frac{1}{1}$, which corresponds to the first row. The first basic variable is 3. Thus, 3 is leaving the basis. Hence, the new basis is $B = \{4, 5\}$.

Using the formulae in Proposition 2.4, we can rewrite the LP so that it is in canonical form for the basis $B = \{4, 5\}$. We get

$$\max\{z(x) = 1 + c^\top x : Ax = b, x \geq 0\},$$

where

$$A = \begin{pmatrix} -1 & 3 & -1 & 1 & 0 \\ -2 & 4 & 1 & 0 & 1 \end{pmatrix} \qquad b = \begin{pmatrix} 2 \\ 1 \end{pmatrix} \qquad c = \begin{pmatrix} 1 \\ -1 \\ -1 \\ 0 \\ 0 \end{pmatrix}.$$

Here $N = \{1, 2, 3\}$. Let us choose which element k enters the basis. We want $k \in N$ and $c_k > 0$. The only possible choice is $k = 1$. We compute t by taking the smallest ratio between entry b_i and entry A_{ik} for all i, where $A_{ik} > 0$. However, as $A_k \leq 0$, this is not well defined. We claim that this occurs because the LP is unbounded.

The new feasible solution $x(t)$ is defined by setting $x_1(t) = t$ for some $t \geq 0$ and $x_2 = x_3 = 0$. The matrix equation $Ax = b$ tells us which values to choose for $x_B(t)$, namely (see argument in Section 2.3)

$$x_B(t) = \begin{pmatrix} x_4(t) \\ x_5(t) \end{pmatrix} = \begin{pmatrix} 2 \\ 1 \end{pmatrix} - t \begin{pmatrix} -1 \\ -2 \end{pmatrix}.$$

Thus, we have

$$x(t) = \begin{pmatrix} t \\ 0 \\ 0 \\ 2+t \\ 1+2t \end{pmatrix} = \underbrace{\begin{pmatrix} 0 \\ 0 \\ 0 \\ 2 \\ 1 \end{pmatrix}}_{:= \bar{x}} + t \underbrace{\begin{pmatrix} 1 \\ 0 \\ 0 \\ 1 \\ 2 \end{pmatrix}}_{:= d}.$$

Then $\bar{x}$ is feasible, $Ad = \mathbb{0}$, $d \geq \mathbb{0}$, and $c^\top d = 1 > 0$. Hence, $\bar{x}, d$ form a certificate of unboundedness.

2.5.3   Formalizing the procedure

Let us formalize the simplex procedure described in the previous sections. At each step, we have a feasible basis and we either detect unboundedness or attempt to find a new feasible basis where the associated basic solution has larger value than the basic solution for the current basis.

In light of Proposition 2.4, we can assume that the LP is in canonical form for a feasible basis B, i.e. that it is of the form

$$\max \quad z(x) = \bar{z} + c_N^\top x_N$$

subject to

$$x_B + A_N x_N = b \tag{P}$$

$$x \geq \mathbb{0},$$

where $\bar{z}$ is some real value and $b \geq \mathbb{0}$.

Let $\bar{x}$ be the basic solution for basis B, i.e. $\bar{x}_N = \mathbb{0}$ and $\bar{x}_B = b$.

Remark 2.5 *If $c_N \leq \mathbb{0}$, then $\bar{x}$ is an optimal solution to (P).*

Proof Suppose $c_N \leq \mathbb{0}$. Note that $z(\bar{x}) = \bar{z} + c_N^\top \bar{x}_N = \bar{z} + c_N^\top \mathbb{0} = \bar{z}$. For any feasible solution x', we have $x' \geq 0$. As $c_N \leq \mathbb{0}$, it implies that $c^\top x' = c_N^\top x_N' \leq 0$. It follows that $z(x') \leq \bar{z}$, i.e. $\bar{z}$ is an upper bound for (P). As $z(\bar{x}) = \bar{z}$, the result follows. $\square$

Now suppose that for some $k \in N$ we have $c_k > 0$. We define x_N', which depends on some parameter $t \geq 0$, as follows:

$$x_j' = \begin{cases} t & \text{if } j = k \\ 0 & \text{if } j \in N \setminus \{k\}. \end{cases}$$

Now we need to satisfy $x'_B + A_N x'_N = b$. Thus

$$x'_B = b - A_N x'_N = b - \sum_{j \in N} x'_j A_j = b - x'_k A_k = b - t A_k. \tag{2.28}$$

Remark 2.6 *If $A_k \leq 0$, then the LP is unbounded.*

Proof Suppose $A_k \leq 0$. Then for all $t \geq 0$, we have $x'_B = b - t A_k \geq 0$. Hence, x' is feasible. Moreover

$$z(x') = \bar{z} + c_N^\top x'_N = \bar{z} + \sum_{j \in N} c_j x'_j = \bar{z} + c_k x'_k = \bar{z} + c_k t,$$

since $c_k > 0$, $z(x')$ goes to infinity as t goes to infinity. $\square$

Thus, we may assume that $A_{ik} > 0$ for some row index i. We need to choose t so that $x'_B \geq 0$. It follows from (2.28) that

$$t A_k \leq b$$

or equivalently for every row index i for which $A_{ik} > 0$, we must have

$$t \leq \frac{b_i}{A_{ik}}.$$

Hence, the largest value t for which x' remains nonnegative is given by

$$t = \min \left\{ \frac{b_i}{A_{ik}} : A_{ik} > 0 \right\}.$$

Note that since $A_k \leq 0$ does not hold, this is well defined. Let r denote the index i where the minimum is attained in the previous equation. Then (2.28) implies that the r th entry of x'_B will be zero. Let ℓ denote the r th basic variable of B. Note that since we order the components of x_B in the same order as B, the r th component of x'_B is the basic variable x'_ℓ. It follows that $x'_\ell = 0$. Choose

$$B' = B \cup \{k\} \setminus \{\ell\}, \qquad N' = N \cup \{\ell\} \setminus \{k\}.$$

We let the reader verify that $x'_{N'} = 0$. It can be readily checked that B' is a basis (see Exercise 4 in Section 2.5.3). Then x' must be a basic solution for the basis B'.

In Algorithm 2.1 on the next page, we summarize the simplex procedure for

$$\max\{c^\top x : Ax = b, x \geq 0\}. \tag{P}$$

Note, we have argued that if the algorithm terminates, then it provides us with a correct solution. However, the algorithm as it is described need not stop. Suppose that at every step the quantity $t > 0$. Then at every step the objective function will increase. Moreover, it is clear that at no iteration will the objective function decrease or stay the same. Hence, in that case we never visit the same basis twice. As there are clearly only a finite number of bases, this would guarantee that the algorithm terminates. However, it is possible that at every iteration the quantity t equals 0. Then at the start of the next iteration we get a new basis, but the same basic solution. After a number of iterations,

---

Algorithm 2.1 Simplex algorithm

Input: Linear program (P) and feasible basis B
Output: An optimal solution of (P) or a certificate proving that (P) is unbounded.
1: Rewrite (P) so that it is in canonical form for the basis B
   Let $\bar{x}$ be the basic feasible solution for B
2: **if** $c_N \leq \mathbb{0}$ **then stop** ($\bar{x}$ is optimal) **end if**
3: Select $k \in N$ such that $c_k > 0$.
4: **if** $A_k \leq \mathbb{0}$ **then stop** ((P) is unbounded) **end if**
5: Let r be any index i where the following minimum is attained:

$$t = \min\left\{\frac{b_i}{A_{ik}} : A_{ik} > 0\right\}$$

6: Let ℓ be the rth basis element
7: Set $B := B \cup \{k\} \setminus \{\ell\}$
8: Go to step 1

---

it is possible that we revisit the same basis. If we repeat this forever, the algorithm will not terminate. This behavior is known as *cycling*.

There are a number of easy refinements to the version of the simplex algorithm we described that will guarantee termination. The easiest to state is as follows. Throughout the simplex iterations with $t = 0$, in Step 3, among all $j \in N$ with $c_j > 0$, choose $k := \min\{j \in N : c_j > 0\}$; also, in Step 5, define t as before and choose the smallest $r \in B$ with $A_{rk} > 0$ and $\frac{b_r}{A_{rk}} = t$. This rule is known as *Bland's rule*.

THEOREM 2.7   *The simplex procedure with Bland's rule terminates.*

A proof of the above theorem is outlined in Exercise 10 at the end of this section.

Exercises

1   Consider the LP problem $\max\{c^\top x : Ax = b, x \geq \mathbb{0}\}$, where

$$A = \begin{pmatrix} 1 & 2 & -2 & 0 \\ 0 & 1 & 3 & 1 \end{pmatrix} \qquad b = \begin{pmatrix} 2 \\ 5 \end{pmatrix} \qquad c = \begin{pmatrix} 0 \\ 3 \\ 1 \\ 0 \end{pmatrix}.$$

(a) Beginning with the basis $B = \{1, 4\}$, solve the problem with the simplex method. At each step, choose the entering variable and leaving variable by Bland's rule.
(b) Give a certificate of optimality or unboundedness for the problem, and verify it.

2 Consider the LP problem $\max\{c^{\mathsf{T}}x : Ax = b, x \geq 0\}$, where

$$A = \begin{pmatrix} 2 & 5 & 1 & 0 & 3 & 1 \\ 0 & 2 & 2 & -4 & 2 & -4 \\ 3 & 5 & 1 & 2 & 6 & 3 \end{pmatrix} \qquad b = \begin{pmatrix} \frac{9}{4} \\ 0 \\ 4 \end{pmatrix} \qquad c = \begin{pmatrix} 2 \\ -4 \\ 1 \\ 4 \\ 8 \\ 4 \end{pmatrix}.$$

Let $B = \{1, 3, 4\}$. You are given that

$$A_B^{-1} = \begin{pmatrix} 2 & -\frac{1}{2} & -1 \\ -3 & 1 & 2 \\ -\frac{3}{2} & \frac{1}{4} & 1 \end{pmatrix}.$$

(a) Find the basic solution x determined by B, and the canonical form of the problem corresponding to B. Is x feasible?

(b) Apply the simplex method beginning with the canonical form from part (a). Choose the entering and leaving variables by Bland's rule. Go as far as finding the entering and leaving variables on the first iteration, the new basis and the new basic solution.

3 Consider the LP problem $\max\{c^{\mathsf{T}}x : Ax = b, x \geq 0\}$, where

$$A = \begin{pmatrix} -2 & 1 & 1 & 1 & 0 & 0 \\ 1 & -1 & 0 & 0 & 1 & 0 \\ 2 & -3 & -1 & 0 & 0 & 1 \end{pmatrix} \qquad b = \begin{pmatrix} 1 \\ 2 \\ 6 \end{pmatrix} \qquad c = \begin{pmatrix} 2 \\ 1 \\ -1 \\ 0 \\ 0 \\ 0 \end{pmatrix}.$$

Notice that the problem is in canonical form with respect to the basis $B = \{4, 5, 6\}$ and that B determines a feasible basic solution x. Beginning from this canonical form, solve the problem with the simplex method. At each step, choose the entering and leaving variable by Bland's rule. At termination, give a certificate of optimality or unboundedness.

4 Suppose at some step of the simplex procedure, we have a feasible basis B and an LP

$$\max\{c^{\mathsf{T}}x : Ax = b, x \geq 0\},$$

which is in canonical form for the basis B. Following the simplex procedure, we choose an entering variable k and a leaving variable ℓ. At the next step of the simplex procedure, we will consider the basis $B' := B \cup \{k\} \setminus \{\ell\}$. Explain why the new set B' is in fact a basis.

HINT: Matrix A_B is the identity matrix. Show that the columns of $A_{B'}$ are linearly independent. Whether a set $J \subseteq \{1, \ldots, n\}$ is a basis or not does not change when we do row operations on A, or equivalently it does not change when we left-multiply A by an invertible matrix.

5   The princess' wedding ring can be made from four types of gold $1, 2, 3, 4$ with the following amounts of milligrams of impurity per gram:

Type	1	2	3	4
mg of lead	1	2	2	1
mg of cobalt	0	1	1	2
value	1	2	3	2

Set up an LP which will determine the most valuable ring that can be made containing at most 6mg of lead and at most 10mg of cobalt. Put the LP into SEF and then solve it using the simplex algorithm.

6   The simplex algorithm solves the problem

$$\max\{z = c^\top x : Ax = b, x \geq 0\}. \tag{P}$$

Consider the following LP:

$$\min\{c^\top x : Ax = b, x \geq 0\}. \tag{Q}$$

(a) Indicate what changes you need to make to the algorithm to solve (Q) instead of (P) (max was replaced by min). Do NOT convert (Q) into SEF, your algorithm should be able to deal with (Q) directly.

(b) Explain why the solution is optimal for (Q) when your algorithm claims that it is.

(c) Explain why (Q) is unbounded when your algorithm claims that it is.

7   (Advanced)

(a) Consider the following LP:

$$\max\{c^\top x : Ax = b, x \geq \ell\}. \tag{P}$$

Thus, if $\ell = 0$, then (P) is in SEF. One way to solve (P) would be for you to convert it into SEF and use the simplex. Instead come up with your own algorithm to solve (P) *directly* by modifying the simplex algorithm. Here are some guidelines:

- When $\ell = 0$, then your algorithm should be the same as the simplex algorithm,
- Given a basis B, redefine $\bar{x}$ to be basic if $A\bar{x} = b$ and for every $j \notin B$, $\bar{x}_j = \ell_j$, (when $\ell = 0$, this corresponds to the standard definition). Define B to be feasible if the corresponding basic solution $\bar{x}$ is a feasible solution of (P).
- Your algorithm should go from feasible basis to feasible basis, attempting at each step to increase the value of the corresponding basic solution. You need to indicate

how the entering and leaving variables are defined. You also need to indicate how to detect when (P) is unbounded and when your solution is optimal.

(b) Suppose you have in addition a vector u with n entries that give upper values for each of the variables, i.e. you have the following LP:

$$\max\{c^\top x : Ax = b, x \geq \ell, x \leq u\}. \tag{P$'$}$$

One way to solve (P$'$) would be for you to convert it into SEF and use the simplex algorithm. Instead come up with your own algorithm to solve (P$'$) *directly* by modifying the simplex algorithm.

HINT: Given a basis B, redefine $\bar{x}$ to be basic if $A\bar{x} = b$ and for every $j \notin B$ either $x_j = \ell_j$ or $x_j = u_j$. Suppose that you rewrite the objective function so that it is of the form

$$z = \bar{c}^\top x + \bar{z},$$

where $\bar{z}$ is some constant and $\bar{c}_j = 0$ for every $j \in B$. Then show that a feasible basic solution $\bar{x}$ (using our new definition) is optimal if the following holds for every $j \notin B$:

- if $\bar{c}_j < 0$, then $\bar{x} = \ell_j$,
- if $\bar{c}_j > 0$, then $\bar{x} = u_j$.

8 Suppose that an LP in SEF is changed by defining x_1' by $x_1 = 10x_1'$ and substituting for x_1. The new problem is equivalent to the old problem. If we now solve the new problem with the simplex method (and apply the same rule for the choice of the leaving variable), will the same choice of entering variables occur as when the old problem was solved? Discuss this question for the four entering variable rules mentioned below:

(a) *The largest coefficient rule:* When Dantzig first proposed the simplex algorithm, he suggested choosing among all $j \in N$ with $c_j > 0$, a $k \in N$ such that c_k is maximum. This rule is sometimes called *Dantzig's rule* or the *largest coefficient rule*. Based on the formula for the objective value for the new basic solution, it chooses the variable giving the largest increase in the objective function *per unit increase in the value of the entering variable*.

(b) *Bland's rule.*

(c) *The largest improvement rule.* This is the rule that chooses the entering variable to be the one that leads to the largest increase in the objective value. To choose the variable, therefore, we have to compute for each $j \in N$, for which $c_j > 0$, the amount t_j by which we will be able to increase the value of x_j, and find the maximum of $c_j t_j$ over all such j and choose the corresponding index k.

(d) *The steepest edge rule.* This rule is geometrically motivated. In moving from the current basic feasible solution to a new one, we move a certain distance in $\mathbb{R}^n$. This rule chooses the variable which gives the largest increase in the objective value *per unit distance moved*. Suppose that $k \in N$ is chosen and the new value of x_k will be t. Then the change in the value of the basic variable x_i is $-A_{ik}t$. Since there is no change in the value of the other nonbasic variables, the distance moved is

$$\sqrt{t^2 + \sum_{i \in B}(A_{ik}t)^2} = t\sqrt{1 + \sum_{i \in B}(A_{ik})^2}.$$

Since the change in the objective value is tc_k, we choose $k \in N$ with $c_k > 0$ so that

$$\frac{c_k}{\sqrt{1 + \sum_{i \in B}(A_{ik})^2}}$$

is maximized.

9   Show that the simplex method can cycle in 12 iterations with the following LP in canonical form for basis $B = \{1, 2\}$:

$$\max\{c^\top x : Ax = b, x \geq 0\},$$

where

$$A = \begin{pmatrix} 1 & 0 & 1 & 1 & 0 & 1 & -1 & 0 & -1 & -1 & 0 & -1 \\ 0 & 1 & -1 & 0 & -1 & -1 & 0 & -1 & 1 & 0 & 1 & 1 \end{pmatrix} \quad b = \begin{pmatrix} 0 \\ 0 \end{pmatrix}$$

$$c = \begin{pmatrix} 0 & 0 & 1 & 2 & 0 & 3 & -2 & 2 & -3 & 0 & -2 & -1 \end{pmatrix}^\top.$$

HINT: x_3 enters on the first iteration, and x_4 enters on the second.

10   (Advanced) Consider the simplex method applied to the following LP:

$$\max\{c^\top x : Ax = b, x \geq 0\}.$$

(a) Let $y \in \mathbb{R}^m$. Define

$$\bar{c} := c - A^\top y.$$

Prove that for every $d \in \mathbb{R}^n$, such that $Ad = 0$, we have $\bar{c}^\top d = c^\top d$.

(b) Suppose we are solving an LP problem in the above form by the simplex method with Bland's rule for choosing the entering and leaving variables. Also suppose that the method cycles on this LP problem and we encounter the bases (in order):

$$B_1, B_2, \ldots, B_t, B_{t+1} = B_1.$$

Partition the indices $\{1, 2, \ldots, n\}$ into three sets as follows:

$$J_- := \left\{ j : j \notin \bigcup_{i=1}^{t} B_i \right\},$$

$$J_+ := \left\{ j : j \in \bigcap_{i=1}^{t} B_i \right\},$$

$$J_0 := \left\{ j : j \in B_{i_1} \text{ and } j \notin B_{i_2} \text{ for some } i_1, i_2 \in \{1, 2, \ldots, t\} \right\}.$$

Let $\ell := \max\{j : j \in J_0\}$. Note that there exist $h_1, h_2 \in \{1, 2, \ldots, t\}$ such that when the current basis in the simplex method is B_{h_1}, x_ℓ leaves the basis and when the current basis is B_{h_2}, x_ℓ enters the basis. What is the value of x_i for every $i \in J_0$ in all the basic feasible solutions visited throughout the cycle? Justify your answer.

(c) When x_ℓ leaves the basis, the simplex method constructs a d vector (a direction to move along). Which d_j are zero? Which d_j are positive? Which d_j are negative? Which d_j are nonnegative? Justify your answer.

(d) Now consider the simplex iteration when x_ℓ enters the basis. Which $\bar{c}_j$ are zero? Which $\bar{c}_j$ are positive? Which $\bar{c}_j$ are nonpositive? Justify your answer.

(e) Now find a contradiction by considering $c^\top d$ and $\bar{c}^\top d$ for these two iterations. Hence, conclude that starting with a basic feasible solution, the simplex method with Bland's rule terminates after finitely many iterations.

11 (Advanced) In this exercise, we go over a classical example on which the largest coefficient rule performs $2^n - 1$ iterations. For every $n \geq 2$, consider the LP given below

$$\max \quad \sum_{j=1}^{n} 10^j x_j$$

subject to

$$
\begin{aligned}
x_1 &\leq 1, \\
10^i x_i + 2 \sum_{j=1}^{i-1} 10^j x_j &\leq 10^i, \quad (i = 2, \dots, n) \\
x &\geq 0.
\end{aligned}
$$

(a) Prove that for every $n \geq 2$, the above LP has 2^n basic solutions.

(b) Describe an optimal solution for the above family of LPs for every $n \geq 2$, and prove that your solution is the unique optimal solution for each n.

(c) Prove that the simplex method with the largest coefficient rule (see Exercise 8) on the above LP with $n = 2$, starting from the basic feasible solution $\bar{x} := 0$, visits every basic feasible solution.

(d) Using your result from the previous part and induction, prove that for every $n \geq 2$, on the above family of LP problems, the simplex method with the largest coefficient rule performs $2^n - 1$ iterations to find the optimal solution.

2.6 Finding feasible solutions

The simplex algorithm requires a feasible basis as part of its input. We will describe in this section on how to proceed when we are not given such feasible basis.

2.6.1 General scheme

Consider the following LP in SEF:

$$\max\{c^\top x : Ax = b, x \geq 0\} \tag{P}$$

and let us define the following three optimization problems:

Problem A

```
Either,
1. prove that (P) has no feasible solution, or
2. prove that (P) is unbounded, or
3. find an optimal solution to (P).
```

Problem B

```
Either,
1. prove that (P) has no feasible solution, or
2. find a feasible solution to (P).
```

Problem C

```
Given a feasible solution x̄, either,
1. prove that (P) is unbounded, or
2. find an optimal solution to (P).
```

We will see in Theorem 2.12 that for any LP, exactly one of the following holds: it is infeasible, it is unbounded, or it has an optimal solution. Thus, by solving the LP (P), we mean solving problem **A**. However, the simplex procedure (Algorithm 2.1, page 70) requires that we be given a feasible basis B as part of the input. In Exercise 6 (d), Section 2.4.2, you are asked to find an algorithm that takes as input a feasible solution for (P) and returns a basic feasible solution for (P). Together with the simplex algorithm, this allows us to solve problem **C**. The goal of this section is to show that:

PROPOSITION 2.8   *Given an algorithm to solve* **C**, *we can use it to solve problem* **A**.

In particular, you can then use any software package that implements the simplex algorithm to solve problem **A**.

The key step in proving Proposition 2.8 is to show:

PROPOSITION 2.9   *Given an algorithm to solve* **C**, *we can use it to solve problem* **B**.

The above result implies the main result of this section.

Proof of Proposition 2.8   Proposition 2.9 implies that you can solve problem **B**. If we find that (P) has no feasible solution, we can stop as we solved **A**. Otherwise, we obtain a feasible solution x̄ of (P). Using this solution x̄, we can use our algorithm to solve **C** to either deduce that (P) is unbounded or to find a feasible solution to (P). Hence, we have solved **A** as well.   □

In Appendix A, we define formally the notion of a fast algorithm. It can be readily shown (see Exercise 1, Section A.4.1), that Proposition 2.8 can in fact be strengthened to say that, given a *fast* algorithm for **C**, we can construct a *fast* algorithm for **A**. The analogous result also holds for Proposition 2.9.

In the remainder of this section, we prove Proposition 2.9. Let us first proceed on an example and suppose that (P) is the following LP with variables x_1, x_2, x_3, x_4:

$$\max \quad (1, 2, -1, 3)x$$
$$\text{subject to}$$

$$\begin{pmatrix} 1 & 5 & 2 & 1 \\ -2 & -9 & 0 & 3 \end{pmatrix} x = \begin{pmatrix} 7 \\ -13 \end{pmatrix}$$

$$x \geq 0.$$

Using an algorithm for problem **C**, we wish to find a feasible solution for (P) if it exists or show that (P) has no feasible solution. Let us first rewrite the constraints of (P) by multiplying every equation by -1, where the right-hand side is negative; we obtain

$$\begin{pmatrix} 1 & 5 & 2 & 1 \\ 2 & 9 & 0 & -3 \end{pmatrix} x = \begin{pmatrix} 7 \\ 13 \end{pmatrix}$$

$$x = (x_1, x_2, x_3, x_4)^\top \geq 0.$$

We now define the following *auxiliary* linear program:

$$\min \quad x_5 + x_6$$
$$\text{subject to}$$

$$\begin{pmatrix} 1 & 5 & 2 & 1 & 1 & 0 \\ 2 & 9 & 0 & -3 & 0 & 1 \end{pmatrix} x = \begin{pmatrix} 7 \\ 13 \end{pmatrix} \quad \text{(Q)}$$

$$x = (x_1, x_2, x_3, x_4, x_5, x_6)^\top \geq 0.$$

Variables x_5, x_6 are the *auxiliary variables*. Observe that $B = \{5, 6\}$ is a basis, and that the corresponding basic solution $\bar{x} = (0, 0, 0, 0, 7, 13)^\top$ is feasible, since all entries are nonnegative. Note that this is the case since we made sure that the right-hand side of the constraints of (Q) are all nonnegative. We can use the algorithm for problem **C** to solve (Q). Note that 0 is a lower bound for (Q), hence (Q) is not unbounded. It follows that the algorithm for **C** will find an optimal solution. In this case the optimal solution is $x' = (2, 1, 0, 0, 0, 0)^\top$. Since $x'_5 = x'_6 = 0$, it follows that $(2, 1, 0, 0)^\top$ is a feasible solution to (P). Hence, we have solved problem **B**.

Consider a second example and suppose that (P) is the following LP:

$$\max \quad (6, 1, -1)x$$
$$\text{subject to}$$

$$\begin{pmatrix} 5 & 1 & 1 \\ -1 & 1 & 2 \end{pmatrix} x = \begin{pmatrix} 1 \\ 5 \end{pmatrix}$$

$$x = (x_1, x_2, x_3)^\top \geq 0.$$

The corresponding auxiliary problem is

$$\min \quad x_4 + x_5$$

subject to

$$\left(\begin{array}{ccc|cc} 5 & 1 & 1 & 1 & 0 \\ -1 & 1 & 2 & 0 & 1 \end{array} \right) x = \left(\begin{array}{c} 1 \\ 5 \end{array} \right) \tag{Q}$$

$$x = (x_1, x_2, x_3, x_4, x_5)^\top \geq \mathbb{0}.$$

An optimal solution to (Q) is $x' = (0, 0, 1, 0, 3)^\top$, which has value 3. We claim that (P) has no feasible solution in this case. Suppose for a contradiction that there was a feasible solution $\bar{x}_1, \bar{x}_2, \bar{x}_3$ to (P). Then $\bar{x} = (\bar{x}_1, \bar{x}_2, \bar{x}_3, 0, 0)^\top$ would be a feasible solution to (Q) of value 0, contradicting the fact that x' is optimal.

Let us summarize these observations. We consider

$$\max\{c^\top x : Ax = b, x \geq \mathbb{0}\}, \tag{P}$$

where A has m rows and n columns. We may assume that $b \geq \mathbb{0}$ as we can multiply any equation by -1 without changing the problem. We construct the auxiliary problem

$$\min \quad w = x_{n+1} + \cdots + x_{n+m}$$

subject to

$$\left(A \mid I \right) \left(\begin{array}{c} x_1 \\ \vdots \\ x_{n+m} \end{array} \right) = b \tag{Q}$$

$$(x_1, \ldots, x_{n+m})^\top \geq \mathbb{0}.$$

We leave the proof of the next remark as an easy exercise (follow the argument outlined in the aforementioned examples).

Remark 2.10 *Let $x' = (x'_1, \ldots, x'_{n+m})^\top$ be an optimal solution to (Q):*

(1) *If $w = 0$, then $(x'_1, \ldots, x'_n)^\top$ is a solution to (P).*
(2) *If $w > 0$, then (P) is infeasible.*

We can now prove Proposition 2.9. Construct from (P) the auxiliary problem (Q). Find an optimal solution x' for (Q) using the algorithm for Problem **C**. Then (P) has a feasible solution if and only if $w = 0$ as indicated in the previous Remark 2.10.

2.6.2   The two phase simplex algorithm–an example

We illustrate the method presented in the previous section on an example using the simplex algorithm to solve problem **C**. The resulting algorithm is known as the *two phase simplex algorithm*. During Phase I, we look for a basic feasible solution (if one exists), and during Phase II, we find an optimal solution (if one exists), starting from the feasible basic solution obtained during Phase I. Consider

$$\text{max} \qquad \begin{pmatrix} 2 & -1 & 2 \end{pmatrix} x$$

subject to

$$\begin{pmatrix} -1 & -2 & 1 \\ 1 & -1 & 1 \end{pmatrix} x = \begin{pmatrix} -1 \\ 3 \end{pmatrix} \tag{P}$$

$$x = (x_1, x_2, x_3)^\top \geq 0.$$

Phase I

We construct the auxiliary problem

$$\text{max} \qquad \begin{pmatrix} 0 & 0 & 0 & -1 & -1 \end{pmatrix} x$$

subject to

$$\begin{pmatrix} 1 & 2 & -1 & 1 & 0 \\ 1 & -1 & 1 & 0 & 1 \end{pmatrix} x = \begin{pmatrix} 1 \\ 3 \end{pmatrix} \tag{Q}$$

$$x = (x_1, x_2, x_3, x_4, x_5)^\top \geq 0.$$

Note that the objective function is equivalent to min $x_4 + x_5$. $B = \{4, 5\}$ is a feasible basis; however, (Q) is not in canonical form for B. We could use the formulae in Proposition 2.4 to rewrite (Q) in canonical form, but a simpler approach is to add each of the two equations of (Q) to the objective function. The resulting LP is

$$\text{max} \qquad \begin{pmatrix} 2 & 1 & 0 & 0 & 0 \end{pmatrix} x - 4$$

subject to

$$\begin{pmatrix} 1 & 2 & -1 & 1 & 0 \\ 1 & -1 & 1 & 0 & 1 \end{pmatrix} x = \begin{pmatrix} 1 \\ 3 \end{pmatrix}$$

$$x = (x_1, x_2, x_3, x_4, x_5)^\top \geq 0.$$

Solving this LP using the simplex algorithm starting from $B = \{4, 5\}$, we obtain the optimal basis $B = \{1, 3\}$. The canonical form for B is

$$\text{max} \qquad \begin{pmatrix} 0 & 0 & 0 & -1 & -1 \end{pmatrix} x$$

subject to

$$\begin{pmatrix} 1 & \frac{1}{2} & 0 & \frac{1}{2} & \frac{1}{2} \\ 0 & -\frac{3}{2} & 1 & -\frac{1}{2} & \frac{1}{2} \end{pmatrix} x = \begin{pmatrix} 2 \\ 1 \end{pmatrix} \tag{Q'}$$

$$x = (x_1, x_2, x_3, x_4, x_5)^\top \geq 0.$$

The basic solution corresponding to B is $\bar{x} = (2, 0, 1, 0, 0)^\top$, which has value 0. It follows from Remark 2.10 that $(2, 0, 1)^\top$ is a feasible solution for (P). Moreover, $(2, 0, 1)^\top$ is the basic solution for basis B of (P), hence B is a feasible basis of (P). Note, it is always true that the feasible solution we construct after solving (Q) using the simplex procedure will be a basic solution of (P). It need not be the case that B be a basis of (P), however (see Exercise 2 at the end of this section).

Phase II
We can use the formulae in Proposition 2.4 to rewrite (P) in canonical form for the basis $B = \{1, 3\}$. Note that to obtain the constraints, we can use the constraints of (Q') (omitting the auxiliary variables). We obtain

$$\text{max} \quad \begin{pmatrix} 0 & 1 & 0 \end{pmatrix} x + 6$$

$$\text{subject to}$$

$$\begin{pmatrix} 1 & \frac{1}{2} & 0 \\ 0 & -\frac{3}{2} & 1 \end{pmatrix} x = \begin{pmatrix} 2 \\ 1 \end{pmatrix}$$

$$x = (x_1, x_2, x_3)^\top \geq 0.$$

Solving this LP using the simplex algorithm starting from $B = \{1, 3\}$, we obtain the optimal basis $B = \{2, 3\}$. The canonical form for B is

$$\text{max} \quad \begin{pmatrix} -2 & 0 & 0 \end{pmatrix} x + 10$$

$$\text{subject to}$$

$$\begin{pmatrix} 2 & 1 & 0 \\ 3 & 0 & 1 \end{pmatrix} x = \begin{pmatrix} 4 \\ 7 \end{pmatrix}$$

$$x \geq 0.$$

Then the basic solution $(0, 4, 7)^\top$ is an optimal solution for (P).

2.6.3   Consequences

Suppose that we are given an arbitrary LP (P) in SEF. Let us run the two phase method for (P) using Bland's rule. Theorem 2.7 implies that the two phase method will terminate. The following is now a consequence of our previous discussion.

THEOREM 2.11 (Fundamental theorem of linear programming (SEF)).
Let (P) be an LP problem in SEF. If (P) does not have an optimal solution, then (P) is either infeasible or unbounded. Moreover:

(1) *if (P) is feasible, then (P) has a basic feasible solution;*
(2) *if (P) has an optimal solution, then (P) has a basic feasible solution that is optimal.*

Since we can convert any LP problem into SEF while preserving the main property of the LP, the above theorem yields the following result for LP problems in any form:

THEOREM 2.12 (Fundamental theorem of linear programming).
Let (P) be an LP problem. Then exactly one of the following holds:

(1) *(P) is infeasible;*
(2) *(P) is unbounded;*
(3) *(P) has an optimal solution.*

Exercises

1 Consider an LP of the form $\max\{c^\top x : Ax = b, x \geq 0\}$. Use the two phase simplex method using Bland's rule, to solve the LP for each of the following cases:

(a)

$$A = \begin{pmatrix} 1 & -2 & 1 \\ -1 & 3 & -2 \end{pmatrix} \qquad b = \begin{pmatrix} 2 \\ -3 \end{pmatrix} \qquad c = \begin{pmatrix} 1 \\ 3 \\ 2 \end{pmatrix}.$$

(b)

$$A = \begin{pmatrix} 2 & 4 & 2 \\ -1 & -3 & -2 \end{pmatrix} \qquad b = \begin{pmatrix} 2 \\ 0 \end{pmatrix} \qquad c = \begin{pmatrix} 27 \\ 2 \\ -6 \end{pmatrix}.$$

(c)

$$A = \begin{pmatrix} 1 & 1 & 1 & -1 & 0 \\ 0 & 1 & 2 & 0 & 1 \end{pmatrix} \qquad b = \begin{pmatrix} 1 \\ 2 \end{pmatrix} \qquad c = \begin{pmatrix} 0 \\ 1 \\ 0 \\ -1 \\ 0 \end{pmatrix}.$$

(d)

$$A = \begin{pmatrix} -1 & 1 & 0 & 0 & -1 \\ 1 & 0 & -1 & 1 & 2 \end{pmatrix} \qquad b = \begin{pmatrix} 1 \\ 1 \end{pmatrix} \qquad c = \begin{pmatrix} 2 \\ 0 \\ 2 \\ 0 \\ 3 \end{pmatrix}.$$

(e)

$$A = \begin{pmatrix} 2 & -1 & 4 & -2 & 1 \\ -1 & 0 & -3 & 1 & -1 \end{pmatrix} \qquad b = \begin{pmatrix} 2 \\ 1 \end{pmatrix} \qquad c = \begin{pmatrix} -3 \\ -1 \\ 1 \\ 4 \\ 7 \end{pmatrix}.$$

2 Consider the following LP in SEF:

$$\max\{c^\top x : Ax = b, x \geq 0\}, \tag{P}$$

where A has m rows and n columns and $b \geq 0$. Construct the auxiliary problem

$$\min \qquad w = x_{n+1} + \ldots + x_{n+m}$$

subject to

$$\left(A \mid I \right) \begin{pmatrix} x_1 \\ \vdots \\ x_{n+m} \end{pmatrix} = b \tag{Q}$$

$$(x_1, \ldots, x_{n+m})^\top \geq 0.$$

(a) Suppose $\bar{x} := (\bar{x}_1, \ldots, \bar{x}_{n+m})$ is a basic solution of (Q), where $\bar{x}_{n+1} = \ldots, \bar{x}_{n+m} = 0$. Show that $x' := (\bar{x}_1, \ldots, \bar{x}_n)$ is a basic solution of (P).

(b) Suppose that in (a) $\bar{x}$ is a basic solution of (Q) for a basis B. Give an example that shows that B need not be a basis of (P).

(c) Show how to find a basis for the basic solution x' of (P) given in (a).

   HINT: See Exercise 1, Section 2.4.2.

2.7   Simplex via tableaus*

The simplex algorithm requires us to reformulate the problem in canonical form for every basis. In this section, we will show how to describe the computation required between two consecutive iterations in a compact way. We emphasize that this should not be viewed as a guide on how to implement the simplex algorithm. However, it is interesting from a pedagogical point of view.

2.7.1   Pivoting

Consider an $m \times n$ matrix T and let (i, j), where $i \in \{1, \ldots, m\}$ and $j \in \{1, \ldots, n\}$ such that $T_{i,j} \neq 0$. We say that matrix T' is obtained from T by *pivoting on element* (i, j) if T' is defined as follows. For every row index k

$$\text{row}_k(T') = \begin{cases} \frac{1}{T_{i,j}} \text{row}_k(T) & \text{if } k = i \\ \text{row}_k(T) - \frac{T_{k,j}}{T_{i,j}} \text{row}_i(T) & \text{if } k \neq i. \end{cases}$$

We illustrate this on an example. Consider the matrix

$$T = \begin{pmatrix} 2 & 2 & -1 & 0 & 3 \\ 0 & 3 & \boxed{2} & 3 & -5 \\ 3 & -1 & -2 & 1 & 5 \end{pmatrix}.$$

Let us compute the matrix T' obtained from T by pivoting on element $(2, 3)$. (We will use the convention that elements on which we pivot are surrounded by a square.) We get

$$T' = \begin{pmatrix} 2 & 7/2 & 0 & 3/2 & 1/2 \\ 0 & 3/2 & 1 & 3/2 & -5/2 \\ 3 & 2 & 0 & 4 & 0 \end{pmatrix}.$$

Observe, that the effect of pivoting on element (i, j) is to transform column j of the matrix T into the vector where all the entries are zero except for entry i which is equal to 1.

The students should verify that $T' = YT$, where

$$Y = \begin{pmatrix} 1 & 1/2 & 0 \\ 0 & 1/2 & 0 \\ 0 & 1 & 1 \end{pmatrix}.$$

Consider the system of equations

$$\begin{pmatrix} 2 & 2 & -1 & 0 \\ 0 & 3 & 2 & 3 \\ 3 & -1 & -2 & 1 \end{pmatrix} x = \begin{pmatrix} 3 \\ -5 \\ 5 \end{pmatrix}. \tag{2.29}$$

We can represent this system by the matrix T. Namely, T is obtained from the coefficients of the left-hand side by adding an extra column corresponding to the right-hand side. Given T', we may construct a system of equations where we do the aforementioned operations in reverse, namely the set of all but the last columns of T' corresponds to the left-hand side and the last column corresponds to the right-hand side. Then we get

$$\begin{pmatrix} 2 & 7/2 & 0 & 3/2 \\ 0 & 3/2 & 1 & 3/2 \\ 3 & 2 & 0 & 4 \end{pmatrix} x = \begin{pmatrix} 1/2 \\ -5/2 \\ 0 \end{pmatrix}. \tag{2.30}$$

Since $T' = YT$, it follows that equation (2.30) is obtained from equation (2.29) by left multiplying by the matrix Y. Observe that the matrix Y is nonsingular, hence the set of solutions for (2.30) is the same as for (2.29). Hence, we used pivoting to derive an equivalent system of equations.

We can proceed in a similar way in general. Given a system $Ax = b$, we construct a matrix T by adding to A an extra column corresponding to b, i.e. $T = (A|b)$. Let $T' = (A'|b')$ be obtained from T by pivoting. Then $T' = YT$ for some nonsingular matrix Y. It follows that $A' = YA$ and $b' = Yb$. In particular, $Ax = b$ and $A'x = b'$ have the same set of solutions. We say that T is the *augmented matrix* representing the system $Ax = b$.

2.7.2   Tableaus

To show how this discussion relates to the simplex algorithm, let us revisit the example
LP in (2.15)

$$\text{max}\qquad z = 0 + (2, 3, 0, 0, 0)x$$

$$\text{subject to}$$

$$\begin{pmatrix} 1 & 1 & 1 & 0 & 0 \\ 2 & 1 & 0 & 1 & 0 \\ -1 & 1 & 0 & 0 & 1 \end{pmatrix} x = \begin{pmatrix} 6 \\ 10 \\ 4 \end{pmatrix}$$

$$x_1, x_2, x_3, x_4, x_5 \geq 0.$$

Let us express both the equations, and the objective function as a system of equations

$$\begin{pmatrix} 1 & -2 & -3 & 0 & 0 & 0 \\ 0 & 1 & 1 & 1 & 0 & 0 \\ 0 & 2 & 1 & 0 & 1 & 0 \\ 0 & -1 & 1 & 0 & 0 & 1 \end{pmatrix} \begin{pmatrix} z \\ x_1 \\ x_2 \\ x_3 \\ x_4 \\ x_5 \end{pmatrix} = \begin{pmatrix} 0 \\ 6 \\ 10 \\ 4 \end{pmatrix}. \qquad (2.31)$$

Note, the first constraint of (2.31) states that $z - 2x_1 - 3x_2 = 0$, i.e. that $z = 2x_1 + 3x_3$,
which is the objective function. Let T^1 be the augmented matrix representing the system
(2.31), namely

$$T^1 = \begin{pmatrix} 1 & -2 & -3 & 0 & 0 & 0 & 0 \\ \hline 0 & 1 & 1 & 1 & 0 & 0 & 6 \\ 0 & \boxed{2} & 1 & 0 & 1 & 0 & 10 \\ 0 & -1 & 1 & 0 & 0 & 1 & 4 \end{pmatrix}.$$

Note, for readability the vertical bars separate the z columns and the right-hand side.
The horizontal bar separates the objective function from the equality constraints. Let us
index the column z by 0 and the constraint corresponding to the objective function by 0
as well. Thus, the first row and first column of T^1 are row and column zero. Observe, that
the LP is in canonical form for $B = \{3, 4, 5\}$ as the columns of T^1 indexed by $\{0, 3, 4, 5\}$
form an identity matrix. We say that T^1 is the *tableau representation* of the LP (2.15).
    In general, given the LP

$$\text{max}\{z = \bar{z} + c^\top x : Ax = b, x \geq 0\}, \qquad (P)$$

we construct the tableau

$$T = \begin{pmatrix} 1 & -c^\top & \bar{z} \\ \hline 0 & A & b \end{pmatrix}$$

and the LP (P) is in canonical form for basis B exactly when the columns of T formed
by columns $B \cup \{0\}$ form an identity matrix.

Let us try to solve the LP working with the tableau T^1 only. We select as an entering variable $k \in N$ such that $c_k > 0$. It means in T^1 that we are selecting a column $k \in \{1, \ldots, 5\}$, where $T^1_{0,k}$ is *smaller* than 0. We can select column 1 or 2, say select column $k = 1$. Note that b corresponds to column 6 (rows 1 to 3) of T_1. We select the row index $i \in \{1, 2, 3\}$, minimizing the ratio $T^1_{i,6}/T^1_{i,k}$, where $T^1_{i,k} > 0$, i.e. we consider

$$\min \left\{ \frac{6}{1}, \frac{10}{2}, - \right\},$$

where the minimum is attained for row $i = 2$. Let us now pivot on the element $(k, i) = (1, 2)$. We obtain the following tableau:

$$T^2 = \begin{pmatrix} 1 & 0 & -2 & 0 & 1 & 0 & 10 \\ \hline 0 & 0 & \boxed{1/2} & 1 & -1/2 & 0 & 1 \\ 0 & 1 & 1/2 & 0 & 1/2 & 0 & 5 \\ 0 & 0 & 3/2 & 0 & 1/2 & 1 & 9 \end{pmatrix}.$$

Since we pivoted on $(2, 1)$, column 1 of T^2 will have a 1 in row 2 and all other elements will be zero. Since row 2 has zeros in columns $0, 3, 5$, these columns will be unchanged in T^2. It follows that columns $0, 1, 3, 5$ will form a permutation matrix in T^2. We could permute rows 1,2 of T^2 so that columns indexed by $0, 1, 3, 5$ form an identity matrix, and therefore that T^2 represents an LP in canonical form for the basis $B = \{1, 3, 5\}$. However, reordering the rows will prove unnecessary in this procedure. The simplex procedure consists of selecting a column j, selecting a row i and pivoting on (i, j).

We state the remaining sequence of tableaus obtained to solve (2.15).

$$T^3 = \begin{pmatrix} 1 & 0 & 0 & 4 & -1 & 0 & 14 \\ \hline 0 & 0 & 1 & 2 & -1 & 0 & 2 \\ 0 & 1 & 0 & -1 & 1 & 0 & 4 \\ 0 & 0 & 0 & -3 & \boxed{2} & 1 & 6 \end{pmatrix}$$

and finally

$$T^4 = \begin{pmatrix} 1 & 0 & 0 & 5/2 & 0 & 1/2 & 17 \\ \hline 0 & 0 & 1 & 1/2 & 0 & 1/2 & 5 \\ 0 & 1 & 0 & 1/2 & 0 & 1/2 & 1 \\ 0 & 0 & 0 & -3/2 & 1 & 1/2 & 3 \end{pmatrix}.$$

The corresponding objective function is now, $z = 17 - 5/2x_3 - 1/2x_5$. Hence, the basic solution $x = (1, 5, 0, 3, 0)^\top$ is optimal.

2.8   Geometry

In this section, we introduce a number of geometric concepts and will interpret much of the material defined in the previous section through the lens of geometry. Questions that we will address include: What can we say about the shape of the set of solutions to a linear program? How are basic feasible solutions distinguished from the set of all feasible solutions? What does that say about the simplex algorithm?

2.8.1   Feasible region of LPs and polyhedra

Given an LP (P), the set of all feasible solutions of (P) is called the *feasible region*. Thus, (P) is feasible if and only if the feasible region of (P) is nonempty. In this section, we study the shape of the feasible region of an LP.

As an example, consider the following LP:

$$
\begin{aligned}
\max \quad & (c_1,\ c_2)x \\
\text{s.t.} \quad &
\end{aligned}
$$

$$
\begin{pmatrix} 1 & 1 \\ 1 & 0 \\ 0 & 1 \\ -1 & 0 \\ 0 & -1 \end{pmatrix} x \le \begin{pmatrix} 3 \\ 2 \\ 2 \\ 0 \\ 0 \end{pmatrix}.
\qquad
\begin{matrix}(1)\\(2)\\(3)\\(4)\\(5)\end{matrix}
\qquad (2.32)
$$

We represented the set of all feasible solutions to (2.32) in Figure 2.1. The set of all points $(x_1,x_2)^\top$ satisfying constraint (2) with equality correspond to line (2). The set of all points satisfying constraint (2) correspond to all points to the left of line (2). A similar argument holds for constraints (1), (3), (4), and (5). Hence, the set of all feasible solutions of (2.32) is the shaded region. Looking at examples in $\mathbb{R}^2$ as above can be somewhat misleading however. In order to get the right geometric intuition, we need to introduce, a number of definitions. Given a vector $x = (x_1,\ldots,x_n)$, the Euclidean norm

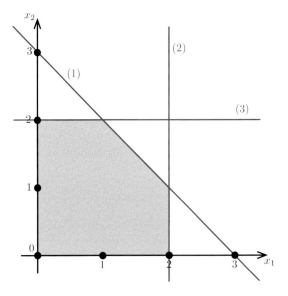

Figure 2.1 Feasible region of (2.32).

of x is defined as $\sqrt{x_1^2 + \ldots + x_n^2}$ and we will denote it by $\|x\|$. $\|x\|$ is the length of vector x and measures the distance of x from the origin $\mathbb{0}$.

Remark 2.13 *Let $a, b \in \mathbb{R}^n$. Then $a^\top b = \|a\| \|b\| \cos(\theta)$, where θ is the angle between a and b. Therefore, for every pair of nonzero vectors a, b, we have:*

- $a^\top b = 0$ *if and only if a, b are orthogonal,*
- $a^\top b > 0$ *if and only if the angle between a, b is less than 90°,*
- $a^\top b < 0$ *if and only if the angle between a, b is larger than 90°.*

Let a be a nonzero vector with n components and let $\beta \in \mathbb{R}$, we define:

(1) $H := \{x \in \mathbb{R}^n : a^\top x = \beta\}$ is a *hyperplane*, and
(2) $F := \{x \in \mathbb{R}^n : a^\top x \leq \beta\}$ is a *halfspace*.

Consider the following inequality:

$$a^\top x \leq \beta. \tag{$\star$}$$

Hence, H is the set of points satisfying constraint $(\star)$ with equality and F is the set of points satisfying constraint $(\star)$. Suppose that $\bar{x} \in H$ and let x be any other point in H. Then $a^\top \bar{x} = a^\top x = \beta$. Equivalently, $a^\top (x - \bar{x}) = 0$, i.e. a and $x - \bar{x}$ are orthogonal. This implies (1) in the following remark, we leave (2) as an exercise:

Remark 2.14 *Let $\bar{x} \in H$.*

(1) *H is the set of points x for which a and $x - \bar{x}$ are orthogonal,*
(2) *F is the set of points x for which a and $x - \bar{x}$ form an angle of at least 90°.*

We illustrate the previous remark in Figure 2.2 The line is the hyperplane H and the shaded region is the halfspace F. In $\mathbb{R}^2$, a hyperplane is a line, i.e. a one-dimensional object. What about in $\mathbb{R}^n$? Consider the hyperplane $H := \{x \in \mathbb{R}^n : a^\top x = 0\}$. Then H is a vector space and we know how to define its dimension. Recall that for any $m \times n$ matrix A, we have the relation

$$dim\{x : Ax = 0\} + rank(A) = n.$$

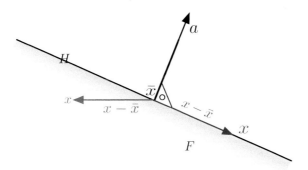

Figure 2.2

It follows that $dim\{x : a^\top x = 0\} + rank(a^\top) = n$. Since by definition $a \neq \mathbb{0}$, $rank(a^\top) = 1$, i.e. $dim(H) = dim(\{x : a^\top x = 0\}) = n - 1$. Hence, hyperplanes are $n - 1$-dimensional objects.

For any $m \times n$ matrix A and vector b, we say that $P := \{x \in \mathbb{R}^n : Ax \leq b\}$ is a *polyhedron*. Note that the set of solutions to any one of the inequalities of $Ax \leq b$ is a halfspace. Thus, equivalently, we could define a polyhedron to be the intersection of a *finite* number of halfspaces. Given an inequality $a^\top x \geq \beta$, we can rewrite it as $-a^\top x \leq -\beta$, and given an equation $a^\top x = \beta$, we can rewrite it as $a^\top x \leq \beta$ and $-a^\top x \leq -\beta$. Hence, any set of linear constraints can be rewritten as $Ax \leq b$ for some matrix A and some vector b. Thus, we proved:

PROPOSITION 2.15   *The feasible region of an LP is a polyhedron or equivalently the intersection of a finite number of halfspaces.*

In the following section, we will investigate geometric properties of polyhedra.

2.8.2   Convexity

Let $x^{(1)}, x^{(2)}$ be two points in $\mathbb{R}^n$. We define the *line through $x^{(1)}$ and $x^{(2)}$* to be the set of points

$$\{x = \lambda x^{(1)} + (1 - \lambda)x^{(2)} : \lambda \in \mathbb{R}\}.$$

Note that when $\lambda = 1$, then $x = x^{(1)}$; when $\lambda = 0$, then $x = x^{(2)}$; and when $\lambda = \frac{1}{2}$, x corresponds to the mid-point between $x^{(1)}$ and $x^{(2)}$. We define the *line segment with ends $x^{(1)}$ and $x^{(2)}$* to be the set of points

$$\{x = \lambda x^{(1)} + (1 - \lambda)x^{(2)} : 0 \leq \lambda \leq 1\}.$$

Observe that the aforementioned definitions correspond to the commonly used notions of lines and line segments. A subset C of $\mathbb{R}^n$ is said to be *convex* if for *every* pair of points $x^{(1)}$ and $x^{(2)}$ in C the line segment with ends $x^{(1)}, x^{(2)}$ is included in C.

Consider Figure 2.3. The shaded region in (i) contained in $\mathbb{R}^2$ is convex, as is the shaded region (iii) contained in $\mathbb{R}^3$ (it is a cube). The shaded regions corresponding to (ii) and (iv) are not convex. We prove this for either case by exhibiting two points $x^{(1)}, x^{(2)}$ inside the shaded region for which the line segment with ends $x^{(1)}, x^{(2)}$ is not completely included in the shaded region.

Remark 2.16 *Halfspaces are convex.*

Proof   Let H be a halfspace, i.e. $H = \{x : a^\top x \leq \beta\}$ for some nonzero vector $a \in \mathbb{R}^n$ and $\beta \in \mathbb{R}$. Let $x^{(1)}, x^{(2)} \in H$. Let $\bar{x}$ be an arbitrary point in the line segment between $x^{(1)}$ and $x^{(2)}$, i.e. $\bar{x} = \lambda x^{(1)} + (1 - \lambda)x^{(2)}$ for some $\lambda \in [0, 1]$. We need to show that $\bar{x} \in H$. We have

$$a^\top \bar{x} = a^\top \left(\lambda x^{(1)} + (1 - \lambda)x^{(2)} \right) = \underbrace{\lambda}_{\geq 0} \underbrace{a^\top x^{(1)}}_{\leq \beta} + \underbrace{(1 - \lambda)}_{\geq 0} \underbrace{a^\top x^{(2)}}_{\leq \beta} \leq \lambda\beta + (1 - \lambda)\beta = \beta.$$

Hence, $\bar{x} \in H$ as required. $\square$

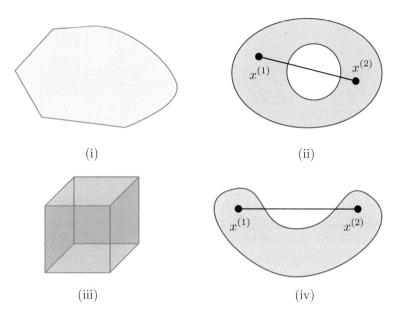

(i)   (ii)

(iii)   (iv)

Figure 2.3 Convex and nonconvex sets.

Remark 2.17 *For every $j \in J$, let C_j, denote a convex set. Then the intersection*

$$C := \bigcap \{C_j : j \in J\}$$

is convex. Note, that J can be infinite.

Proof Let $x^{(1)}$ and $x^{(2)}$ be two points that are in C. Then for every $j \in J$, $x^{(1)}, x^{(2)} \in C_j$ and since C_j is convex the line segment between $x^{(1)}$ and $x^{(2)}$ is in C_j. It follows that the line segment between $x^{(1)}$ and $x^{(2)}$ is in C. Hence, C is convex.   □

Remark 2.16 and 2.17 have the following immediate consequence:

PROPOSITION 2.18 *Polyhedra are convex.*

Note that the unit ball is an example of a convex set that is not a polyhedron. It is the intersection of an infinite number of halfspaces (hence, convex) but cannot be expressed as the intersection of a finite number of halfspaces.

2.8.3   Extreme points

We say that a point x is *properly contained* in a line segment if it is in the line segment but is distinct from its ends. Consider a convex set C and let x be a point of C. We say that x is an *extreme point* of C if no line segment that properly contains x is included in C. Equivalently:

Remark 2.19 $x \in C$ is <u>not</u> *an extreme point of C if and only if*

$$x = \lambda x^{(1)} + (1 - \lambda)x^{(2)}$$

for distinct points $x^{(1)}, x^{(2)} \in C$ and λ with $0 < \lambda < 1$.

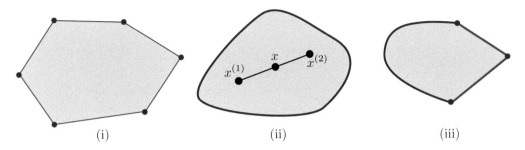

Figure 2.4 Extreme points.

Consider Figure 2.4. In each of (i), (ii), and (iii) the shaded regions represent convex sets included in $\mathbb{R}^2$. In (i), we indicate each of the six extreme points by small red circles. Note that for (ii), every point in the boundary of the shaded figure is an extreme point. This shows in particular that a convex set can have an infinite number of extreme points. In (ii), we illustrate why the point x is not extreme by exhibiting a line segment with ends $x^{(1)}, x^{(2)}$ which are contained in the shaded figure and that properly contains x. In (iii), we indicate the extreme points by red circles and a red curve. In this example, there are also an infinite number of extreme points.

Next we present a theorem that will characterize the extreme points in a polyhedron. We first need to introduce some notation and definitions. Let $Ax \leq b$ be a system of inequalities and let $\bar{x}$ denote a solution to $Ax \leq b$. We say that a constraint of $a^\top x \leq \beta$ of $Ax \leq b$ is *tight* for $\bar{x}$ if $a^\top \bar{x} = \beta$. Such constraints are also called *active* in part of the literature. We denote the set of all inequalities among $Ax \leq b$ that are tight for $\bar{x}$ by $A^= x \leq b^=$.

THEOREM 2.20 *Let $P = \{x \in \mathbb{R}^n : Ax \leq b\}$ be a polyhedron and let $\bar{x} \in P$. Let $A^= x = b^=$ be the set of tight constraints for $\bar{x}$. Then $\bar{x}$ is an extreme point of P if and only if $\mathrm{rank}(A^=) = n$.*

We will illustrate this theorem on the polyhedron P that is the feasible region of LP (2.32) (see Figure 2.1). Suppose $\bar{x} = (1, 2)^\top$. We can see in that figure that $\bar{x}$ is an extreme point. Let us verify that this is what the previous theorem also indicates. Constraints (1) and (3) are tight, hence

$$A^= = \begin{pmatrix} 1 & 1 \\ 0 & 1 \end{pmatrix}.$$

It follows that $\mathrm{rank}(A^=) = 2 = n$, hence $\bar{x}$ is indeed an extreme point. Suppose $\bar{x} = (0, 1)^\top$. We can see in that figure that $\bar{x}$ is not an extreme point. Let us verify that this is what the previous theorem also indicates. Constraint (4) is the only tight constraint, hence

$$A^= = \begin{pmatrix} -1 & 0 \end{pmatrix}.$$

It follows that $\mathrm{rank}(A^=) = 1 < n$, hence $\bar{x}$ is not an extreme point.

Proof of Theorem 2.20.   Suppose that $rank(A^=) = n$. We will show that $\bar{x}$ is an extreme point. Suppose for a contradiction this is not the case. Then there exist (see Remark 2.19) $x^{(1)}, x^{(2)} \in P$, where $x^{(1)} \neq x^{(2)}$ and λ where $0 < \lambda < 1$ for which $\bar{x} = \lambda x^{(1)} + (1-\lambda)x^{(2)}$. Thus

$$b^= = A^= \bar{x} = A^= \left(\lambda x^{(1)} + (1-\lambda)x^{(2)}\right) =$$

$$\underbrace{\lambda}_{>0} \underbrace{A^= x^{(1)}}_{\leq b^=} + \underbrace{(1-\lambda)}_{>0} \underbrace{A^= x^{(2)}}_{\leq b^=} \leq \lambda b^= + (1-\lambda)b^= = b^=.$$

Hence, we have equality throughout, which implies that $A^= x^{(1)} = A^= x^{(2)} = b^=$. As $rank(A^=) = n$, there is a unique solution to $A^= x = b^=$. Therefore, $\bar{x} = x^{(1)} = x^{(2)}$, a contradiction.

Suppose that $rank(A^=) < n$. We will show that $\bar{x}$ is not an extreme point. Since $rank(A^=) < n$, the columns of $A^=$ are linearly dependent, and hence there is a nonzero vector d such that $A^= d = 0$. Pick $\epsilon > 0$ small and define

$$x^{(1)} := \bar{x} + \epsilon d \qquad \text{and} \qquad x^{(2)} := \bar{x} - \epsilon d.$$

Hence, $\bar{x} = \frac{1}{2}x^{(1)} + \frac{1}{2}x^{(2)}$ and $x^{(1)}, x^{(2)}$ are distinct. It follows that $\bar{x}$ is in the line segment between $x^{(1)}$ and $x^{(2)}$. It remains to show that $x^{(1)}, x^{(2)} \in P$ for $\epsilon > 0$ small enough. Observe first that

$$A^= x^{(1)} = A^= (\bar{x} + \epsilon d) = \underbrace{A^= \bar{x}}_{=b^=} + \epsilon \underbrace{A^= d}_{=0} = b^=.$$

Similarly, $A^= x^{(2)} = b^=$. Let $a^\top x \leq \beta$ be any of the inequalities of $Ax \leq b$ that is not in $A^= x \leq b^=$. It follows that for $\epsilon > 0$ small enough:

$$a^\top x^{(1)} = a^\top (\bar{x} + \epsilon d) = \underbrace{a^\top \bar{x}}_{<\beta} + \epsilon a^\top d \leq \beta,$$

hence $x^{(1)} \in P$ and by the same argument, $x^{(2)} \in P$ as well.   □

Consider the following polyhedron:

$$P = \left\{ x \in \mathbb{R}^4 : \begin{pmatrix} 1 & 3 & 1 & 0 \\ 2 & 2 & 0 & 1 \end{pmatrix} x = \begin{pmatrix} 2 \\ 1 \end{pmatrix}, x \geq 0 \right\}.$$

Note that $\bar{x} = (0, 0, 2, 1)^\top$ is a basic feasible solution. We claim that $\bar{x}$ is an extreme point of P. To be able to apply Theorem 2.20, we need to rewrite P as the set of solutions to $Ax \leq b$ for some matrix A and vector b. This can be done by choosing

$$
A = \begin{pmatrix}
1 & 3 & 1 & 0 \\
2 & 2 & 0 & 1 \\
-1 & -3 & -1 & 0 \\
-2 & -2 & 0 & -1 \\
-1 & 0 & 0 & 0 \\
0 & -1 & 0 & 0 \\
0 & 0 & -1 & 0 \\
0 & 0 & 0 & -1
\end{pmatrix}
\quad \text{and} \quad
b = \begin{pmatrix}
2 \\
1 \\
-2 \\
-1 \\
0 \\
0 \\
0 \\
0.
\end{pmatrix}.
$$

Let $A^= x \leq b^=$ be the set of tight constraints for $\bar{x}$, then

$$
A^= = \begin{pmatrix}
1 & 3 & 1 & 0 \\
2 & 2 & 0 & 1 \\
-1 & -3 & -1 & 0 \\
-2 & -2 & 0 & -1 \\
-1 & 0 & 0 & 0 \\
0 & -1 & 0 & 0
\end{pmatrix}
$$

and it can be readily checked that the first two and last two rows of $A^=$ form a set of four linearly independent rows. Hence, $rank(A^=) \geq 4 = n$. This implies (as $\bar{x}$ is feasible) by Theorem 2.20 that $\bar{x}$ is an extreme point of P. Using the idea outlined in the previous example, we leave it as an exercise to prove the following theorem which relates basic feasible solutions (for problems in standard equality form) to extreme points:

THEOREM 2.21   *Let A be a matrix where the rows are linearly independent and let b be a vector. Let $P = \{x : Ax = b, x \geq 0\}$ and let $\bar{x} \in P$. Then $\bar{x}$ is an extreme point of P if and only if $\bar{x}$ is a basic feasible solution of $Ax = b$.*

2.8.4   Geometric interpretation of the simplex algorithm

Consider the following LP:

$$
\begin{aligned}
\max \quad & z = 2x_1 + 3x_2 \\
\text{s.t.} \quad &
\end{aligned}
$$

$$
\begin{array}{rlll}
2x_1 & + & x_2 & \leq 10 & \quad (1) \\
x_1 & + & x_2 & \leq 6 & \quad (2) \\
-x_1 & + & x_2 & \leq 4 & \quad (3) \\
x_1 & , & x_2 & \geq 0. &
\end{array}
\qquad \text{(P)}
$$

In Figure 2.5, we indicate the feasible region of this LP as well as a feasible solution $\bar{x} = (1, 5)^\top$. The line $z = 17$ indicates the set of all vectors for which the objective function evaluates to 17. The points above this line have objective value greater than 17 and the points below this line have value less than 17. Since there are no points in the

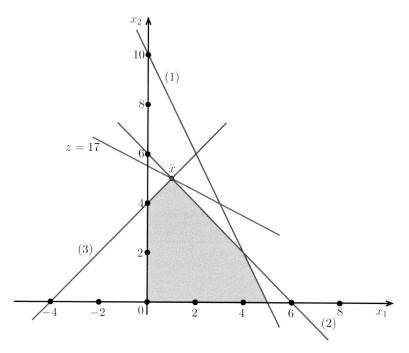

Figure 2.5 Feasible region and optimal solution.

feasible region above the line $z = 17$, it follows that $\bar{x}$ is an optimal solution. As the line $z = 17$ intersects the feasible region in only one point, namely $\bar{x}$, this also implies that $\bar{x}$ is the unique optimal solution.

Let us use the simplex algorithm to find this optimal solution $\bar{x}$. We first need to reformulate this problem in standard equality form. This can be achieved by introducing slack variables x_3, x_4, x_5 for constraints respectively (1), (2), and (3) of (P). We obtain,

$$
\begin{array}{rlllllll}
\max & z = 2x_1 & + & 3x_2 \\
\text{s.t.} \\
& 2x_1 & + & x_2 & + & x_3 & & & & & = 10 & \quad (1) \\
& x_1 & + & x_2 & & & + & x_4 & & & = 6 & \quad (2) & \quad (\hat{P}) \\
& -x_1 & + & x_2 & & & & & + & x_5 & = 4 & \quad (3) \\
& x_1 & , & x_2 & , & x_3 & , & x_4 & , & x_5 & \geq 0.
\end{array}
$$

Given any point $x = (x_1, x_2)^\top$, we define

$$
\hat{x} := \begin{pmatrix} x_1 \\ x_2 \\ 10 - 2x_1 - x_2 \\ 6 - x_1 - x_2 \\ 4 + x_1 - x_2 \end{pmatrix},
$$

i.e. the components $\hat{x}_3, \hat{x}_4, \hat{x}_5$ are defined as the value of the slack of the constraints (1), (2), and (3) respectively of (P). Thus, x is feasible for (P) if and only if $\hat{x}$ is feasible for

$(\hat{P})$. Suppose x is a feasible solution of (P), which is not an extreme point of the feasible region. Then x is properly contained in the line segment with ends $x^{(1)}, x^{(2)}$, where $x^{(1)}, x^{(2)}$ are feasible for (P), i.e. $x^{(1)} \neq x^{(2)}$ and there exists λ such that $0 < \lambda < 1$ and $x = \lambda x^{(1)} + (1 - \lambda)x^{(2)}$. It can be readily checked that $\hat{x} = \lambda \hat{x}^{(1)} + (1 - \lambda)\hat{x}^{(2)}$. Hence, $\hat{x}$ is properly contained in the line segment with ends $\hat{x}^{(1)}, \hat{x}^{(2)}$. In particular, $\hat{x}$ is not an extreme point of the feasible region of $(\hat{P})$. Conversely, if $\hat{x}$, is not an extreme point for $(\hat{P})$, then x is not an extreme point for (P). Hence:

Remark 2.22 *x is an extreme point for the feasible region of (P) if and only if $\hat{x}$ is an extreme point for the feasible region of $(\hat{P})$.*

Starting in Section 2.3, we solved the LP $(\hat{P})$. The following table summarizes the sequence of bases and basic solutions we obtained:

Iteration	Basis	$\hat{x}^{\top}$	$x^{\top}$
1	$\{3, 4, 5\}$	$(0, 0, 10, 6, 4)$	$(0, 0)$
2	$\{1, 4, 5\}$	$(5, 0, 0, 1, 9)$	$(5, 0)$
3	$\{1, 2, 5\}$	$(4, 2, 0, 0, 6)$	$(4, 2)$
4	$\{1, 2, 3\}$	$(1, 5, 3, 0, 0)$	$(1, 5)$

At each step, $\hat{x}$ is a basic solution. It follows from Theorem 2.21 that $\hat{x}$ must be an extreme point of the feasible region of $(\hat{P})$. Hence, by Remark 2.22, x must be an

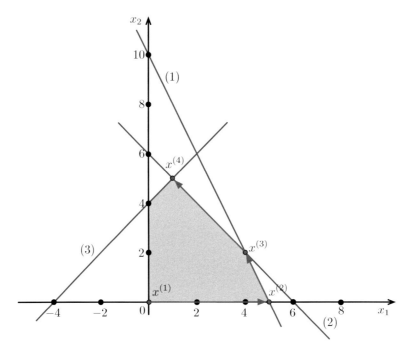

Figure 2.6 Sequence of extreme points visited by simplex.

extreme point of the feasible region of (P). We illustrate this in Figure 2.6. Each of $x^{(1)}, x^{(2)}, x^{(3)}, x^{(4)}$ is an extreme point and the simplex moves from one extreme point to another "adjacent" extreme point. In this example, at each iteration we move to a different basic feasible solution. The simplex algorithm goes from one feasible basis to another feasible basis at each iteration. It is possible however that the corresponding basic solutions for two successive bases are the same. Thus, the simplex algorithm can keep the same feasible basic solution for a number of iterations. However, when using Bland's rule for the choice of entering and leaving variables, the simplex will eventually move to a different basic solution (see Theorem 2.7).

Exercises

1 (a) Show that the set of all optimal solutions to an LP is a convex set.
(b) Deduce that an LP has either:

- no optimal solution,
- exactly one optimal solution, or
- an infinite number of optimal solutions.

2 Let F be the set of x satisfying

$$
\begin{array}{rrrrrrrrl}
x_1 & + & x_2 & + & x_3 & + & 2x_4 & = & 3 \\
x_1 & - & x_2 & - & x_3 & & & = & 1 \\
x_1 & & & - & 2x_3 & + & x_4 & = & 2 \\
x_1 & , & x_2 & , & x_3 & , & x_4 & \geq & 0.
\end{array}
$$

Find all extreme points of F. Justify your answers.

3 Let $n \geq 2$ be an integer, the n-hypercube is the polyhedron

$$
P_n = \left\{ x : 0 \leq x_j \leq 1 \, j = 1, \ldots, n \right\}.
$$

Describe all the extreme points of P_n. Justify your answer, i.e. give a proof that all the points you claim are extreme points are indeed extreme points and prove that no other point is an extreme point.

4 Let $B := \{x \in \mathbb{R}^n : \|x\| \leq 1\}$. Recall that, $\|x\| := \sqrt{x^\top x}$ corresponds to the length of x. Thus, B is the set of points in an n-dimensional space which are at distance at most 1 from the origin, i.e. B is the n-dimensional unit ball. It is intuitively clear that a ball is convex.
(a) Give an algebraic proof of the fact that B is convex.
    HINT: Use the Cauchy–Schwarz inequality, namely for every pair of vectors x, y, we have

$$
x^\top y \leq \|x\| \, \|y\|.
$$

(b) Show that z is an extreme point of B if and only if $\|z\| = 1$.

5 We say that an $n \times n$ matrix M is *positive semidefinite* if the following conditions hold:

- M is symmetric, i.e. $M^\top = M$, and
- for every vector z (of appropriate dimension), $z^\top Mz \geq 0$.

Show that the set of all $n \times n$ positive semidefinite matrices forms a convex set.

6 Let $A \subseteq \mathbb{R}^n$ be a convex set. Let M be an $n \times n$ matrix and define, $B := \{Mx : x \in A\}$, i.e. B is the set of all points in $\mathbb{R}^n$ that can be obtained by selecting a point $x \in A$ and applying the transformation $x \to Mx$.
(a) Show that B is convex.
(b) Suppose that M is invertible and consider $z \in A$, hence $Mz \in B$.
   Show that z is an extreme point of A if and only if Mz is an extreme point of B.

7 Consider three points (vectors) $x^{(1)}, x^{(2)}, x^{(3)} \in \mathbb{R}^n$. We define

$$\Delta(x^{(1)}, x^{(2)}, x^{(3)}) := \{\lambda_1 x^{(1)} + \lambda_2 x^{(2)} + \lambda_3 x^{(3)} : \lambda_1 + \lambda_2 + \lambda_3 = 1, \lambda_1, \lambda_2, \lambda_3 \geq 0\}.$$

(a) What does $\Delta(x^{(1)}, x^{(2)}, x^{(3)})$ correspond to geometrically when $n = 2$?
Let C be a subset of $\mathbb{R}^n$.
(b) Show that if for all $x^{(1)}, x^{(2)}, x^{(3)} \in C$ we have $\Delta(x^{(1)}, x^{(2)}, x^{(3)}) \subseteq C$, then C is convex.
   HINT: $x^{(1)}, x^{(2)}, x^{(3)}$ need not be distinct.
(c) Show that if C is convex, then for all $x^{(1)}, x^{(2)}, x^{(3)} \in C$ we have $\Delta(x^{(1)}, x^{(2)}, x^{(3)}) \subseteq C$.
   HINT: Every point $z \in \Delta(x^{(1)}, x^{(2)}, x^{(3)})$ is in a line segment between a point y and $x^{(3)}$ for some point y that is in a line segment between $x^{(1)}$ and $x^{(2)}$.
(ADVANCED.) We say that x is a *convex combination* of points $x^{(1)}, \ldots, x^{(k)}$ if $x = \sum_{j=1}^{k} \lambda_j x^j$ for some real numbers $\lambda_1, \ldots, \lambda_k$, where $\sum_{j=1}^{k} \lambda_j = 1$ and $\lambda_j \geq 0$ for all $j = 1, \ldots, k$. Suppose $k \geq 2$.
(d) Show that if for all $x^{(1)}, \ldots, x^{(k)} \in C$ all convex combinations of $x^{(1)}, \ldots, x^{(k)}$ are in C, then C is convex.
(e) Show that if C is convex, then for all $x^{(1)}, \ldots, x^{(k)} \in C$ all convex combinations of $x^{(1)}, \ldots, x^{(k)}$ are in C.

8 For each of the following sets, either prove that it is not a polyhedron, or give a matrix A and a vector b such that the set is the solution set to $Ax \leq b$:
(a) $\{(x_1, x_2, x_3)^\top : x_1 \geq 2 \text{ or } x_3 \geq 2\}$.
(b) $\{(x_1, x_2)^\top : x_1 \leq x_2\}$.
(c) $\{(x_1, x_2, x_3, x_4)^\top : x_1 + x_2 + x_3 = 4, x_1 + x_4 = 6, x_2 \geq 0\}$.
(d) $\{(x_1, x_2)^\top : x_1^2 + x_2^2 \leq 1\}$ (note x_1^2 is the square of component x_1.)

9 Consider the polytope P defined by the following constraints:

$$\begin{pmatrix} 1 & 0 & 1 \\ 1 & 1 & 0 \\ 0 & 1 & 1 \\ 0 & 0 & 1 \\ 0 & 1 & -1 \end{pmatrix} x \leq \begin{pmatrix} 4 \\ 4 \\ 4 \\ 3 \\ 0 \end{pmatrix}.$$

For each of the following points, determine whether it is an extreme point of P.
HINT: Consider the set of tight constraints in each case.
(a) $(2,2,2)^\top$,
(b) $(3,1,1)^\top$,
(c) $(0,1,3)^\top$,
(d) $(1,3,3)^\top$.

10 Let C be a convex set and let $\bar{x} \in C$.
(a) Show that if $\bar{x}$ is not an extreme point of C, then $C \setminus \{\bar{x}\}$ is not convex.
(b) Show that if $\bar{x}$ is an extreme point of C, then $C \setminus \{\bar{x}\}$ is convex.

11 Consider the LP $\max\{c^\top x : Ax \le b, x \ge \mathbb{0}\}$ where

$$A = \begin{pmatrix} 1 & 1 \\ 1 & 0 \end{pmatrix} \qquad b = \begin{pmatrix} 2 \\ 1 \end{pmatrix} \qquad c = \begin{pmatrix} 1 \\ 2 \end{pmatrix}.$$

(a) Convert this LP into SEF in the standard way, and call this LP2. Solve LP2 by applying the simplex algorithm. Make sure you start with the basis $\{3,4\}$. In each iteration, for the choice of the entering variable amongst all eligible variables, always choose the one with the smallest index.
(b) Give a diagram showing the set of feasible solutions of the LP, and show the order that the simplex algorithm visits its extreme points. (For each extreme point $(x_1, x_2, x_3, x_4)^\top$ of LP2 visited by the simplex algorithm, you should indicate the extreme point $(x_1, x_2)^\top$ of the original LP.)

2.9   Further reading and notes

The classical reference for the simplex algorithm is the book by Dantzig [20]. The word "simplex" is the name of a simple geometric object which generalizes a triangle (in $\mathbb{R}^2$) and a tetrahedron (in $\mathbb{R}^3$) to arbitrary dimensions. Dantzig presents in his book an attractive geometric interpretation of the algorithm (which uses simplices) to suggest that the algorithm would be efficient in practice. This is certainly worthwhile reading after this introductory course is completed.

We saw that we can replace a free variable by the difference of two new nonnegative variables. Another way of handling free variables is to find an equation which contains the free variable, isolate the free variable in the equation and use this identity elsewhere in the LP, record the variable and the equation on the side and eliminate the free variable and the equation from the original LP. This latter approach reduces the number of variables and constraints in the final LP problem and is more suitable in many situations.

As we already hinted when discussing the finite termination and Bland's rule, there are many ways of choosing entering and leaving variables (these are called pivot rules). In fact, there is a rule, based on perturbation ideas, called the lexicographic rule which only restricts the choice of the leaving variable and also ensures finite convergence of the simplex algorithm. There are some theoretical and some computational issues which can be addressed through proper choices of entering and leaving variables. See the books [13, 66] as well as the papers [9, 19, 46] and the references therein.

A basic feasible solution $\bar{x}$, determined by a basis B, is called *degenerate*, if for some $i \in B$, $\bar{x}_i = 0$. We saw that degeneracy can lead to cycling; however, cycling examples are extremely rare and therefore cycling turns out not to be a problem in practice. However, even if the simplex method is not likely to cycle in practice, it can go through a long sequence of iterations where the current degenerate basic feasible solution does not change. This is called *stalling*. In practice, stalling can be a problem, but the perturbation ideas mentioned above are helpful in mitigating stalling.

In implementing the simplex algorithm, it is extremely important that the linear algebra is done in an efficient and numerically stable way. Even though we derived various formulae involving inverses of the matrices, in practice, A_B^{-1} is almost never formed explicitly (see Chapter 24 of Chvátal [13] and the references therein). For various techniques that exploit sparsity in the data and solve linear systems of equations in a numerically stable way, in general, see Golub and Van Loan [29]. In implementing the simplex method, depending on the variant used, each iteration requires only minor modifications to the vectors and matrices computed during the preceding iterations. In our discussion in this chapter, we did not get into exploiting these aspects for the sake of simplicity of presentation; however, in successful implementations of the simplex method, exploiting this information is absolutely critical. As the hardware and software both continue to improve, the speed-up factors obtained on the software side (based on new ideas) in implementations of the simplex method have been incredible. For example, Bixby [8] reported a speed-up factor of 43500 between 1988 version of the CPLEX code (CPLEX 1.0) and the 2003 version (CPLEX 9.0) on an LP problem with approximately 1.5 million variables and 400 000 constraints. Using the same hardware, CPLEX 1.0 required 29.8 days to solve the problem, whereas CPLEX 9.0 took only 59.1 seconds.

3 Duality through examples

In this chapter, we revisit the shortest path and minimum-cost matching problems. Both were first introduced in Chapter 1, where we discussed practical example applications. We further showed that these problems can be expressed as IPs. The focus in this chapter will be on *solving* instances of the shortest path and matching problems. Our starting point will be to use the IP formulation we introduced in Section 1.5. We will show that studying the two problems through the lens of *linear programming duality* will allow us to design efficient algorithms. We develop this theory further in Chapter 4.

3.1 The shortest path problem

Recall the shortest path problem from Section 1.4.1. We are given a graph $G = (V, E)$, nonnegative lengths c_e for all edges $e \in E$, and two distinct vertices $s, t \in V$. The length $c(P)$ of a path P is the sum of the length of its edges, i.e. $\sum(c_e : e \in P)$. We wish to find among all possible st-paths one that is of minimum length.

Example 7 In the following figure, we show an instance of this problem. Each of the edges in the graph is labeled by its length. The thick black edges in the graph form an st-path $P = sa, ac, cb, bt$ of total length $3 + 1 + 2 + 1 = 7$. This st-path is of minimum length, hence is a solution to our problem.

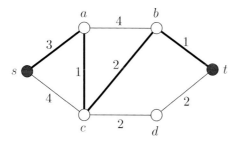

3.1.1 An intuitive lower bound

We claimed in Example 7 that the path $P = sa, ac, cb, bt$ of length 7 is a shortest st-path. How could you convince someone of that fact? Of course, we could list all possible st-paths and verify that P is indeed the shortest one. This is not a practical way of proceeding however, as we may have a huge (exponential) number of such st-paths. Our goal in this section is to find a certificate that can be used to quickly convince someone that a shortest st-path is indeed the shortest. As we will see, such a certificate is not only desirable from the user's point of view, but it also turns out to be crucial for designing the algorithm!

We will prove that the path P in Example 7 is indeed the shortest st-path. However, before we do so, it will be helpful to first consider the special case of the shortest st-path problem where all edges e have edge length $c_e = 1$. In that case, we are looking for an st-path with as few edges as possible. We refer to this case as the *cardinality case*. Consider the following graph $G = (V, E)$:

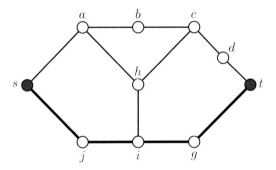

Let us show that the path $P = sj, ji, ig, gt$ is a shortest st-path of G. It has length 4. To show that it is a shortest path, it suffices to show that every st-path has length at least 4. To do this, we exhibit the following collection of st-cuts (see Figure 3.1):

$$\delta(U_1) = \{sa, sj\} \qquad\qquad \delta(U_2) = \{ab, ah, ij\}$$
$$\delta(U_3) = \{bc, hc, ig\} \qquad\qquad \delta(U_4) = \{dt, gt\},$$

where

$$U_1 = \{s\}, \quad U_2 = \{s, a, j\}, \quad U_3 = \{s, a, j, b, h, i\}, \quad \text{and} \quad U_4 = V \setminus \{t\}.$$

Observe that no two of these st-cuts share an edge. Let Q be an arbitrary st-path. We know from Remark 1.1, that every st-path and st-cut have a common edge. Hence, Q must contain an edge e_i of $\delta(U_i)$ for $i = 1, 2, 3, 4$. Since these st-cuts are disjoint, e_1, e_2, e_3, e_4 are distinct edges. In particular, Q contains at least four edges. As Q was an arbitrary st-path, every st-path contains at least four edges. It follows that $P = sj, ji, ig, gt$ is a shortest st-path of G. The collection of st-cuts $\delta(U_i)$ $(i = 1, 2, 3, 4)$ is our certificate of optimality. More generally, if a graph has k pairwise disjoint st-cuts, then every st-path has length at least k and any st-path with k edges is a shortest st-path.

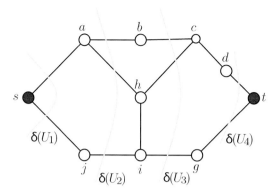

Figure 3.1

Let us now go back to the problem where the edges e of G can have arbitrary lengths $c_e \geq 0$. Rather than selecting a subset of st-cuts as in the cardinality case, we will assign to every st-cut $\delta(U)$ a nonnegative number $\bar{y}_U$ [1] that we call the *width* of the st-cut $\delta(U)$. We say that the widths $\{\bar{y}_U : U \subseteq V, s \in U, t \notin U\}$ are *feasible* if they satisfy the following:

Feasibility condition. For every edge $e \in E$, the total width of all st-cuts that contain e does not exceed the length of e. (3.1)

Example 7, continued. We assign the following widths for the graph G in Example 7:

$$\bar{y}_{U_1} = 3 \qquad\qquad \bar{y}_{U_2} = 1$$
$$\bar{y}_{U_3} = 2 \qquad\qquad \bar{y}_{U_4} = 1,$$

where

$$U_1 = \{s\}, \quad U_2 = \{s, a\}, \quad U_3 = \{s, a, c\}, \quad \text{and} \quad U_4 = \{s, a, c, b, d\}.$$

The other st-cuts are assigned a width of zero. The nonzero widths are represented in Figure 3.2. We claim that the widths are feasible. Let us check the condition for edge ab for instance. The two st-cuts that have nonzero width and that contain ab are $\delta(U_2)$ and $\delta(U_3)$ which have respectively width $\bar{y}_{U_2} = 1$ and $\bar{y}_{U_3} = 2$. Thus, the total width of all st-cuts that contain the edge ab is $1 + 2 = 3$ which does not exceed the length $c_{ab} = 4$ of ab. We leave it as an exercise to verify the feasibility condition for every other edge of the graph.

Suppose now we have a graph $G = (V, E)$ (with $s, t \in V$ and $c_e \geq 0$ for all $e \in E$) that has feasible widths. Let Q denote an arbitrary st-path. By the feasibility condition, for every edge e of Q the total width of all st-cuts using e is at most c_e. It follows that the total width of all st-cuts using *some* edge of Q is at most $\sum(c_e : e \in Q)$, i.e. the length of Q. We know however from Remark 1.1 that every st-cut of G contains some edge of Q. Hence, the total width of *all* st-cuts is at most equal to the length of Q. We summarize this result.

[1] While it might be more natural to use the notation $\bar{y}_{\delta(U)}$ to denote the width of the st-cut $\delta(U)$, the notation $\bar{y}_U$ is more compact while being unambiguous.

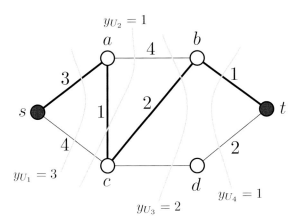

Figure 3.2

PROPOSITION 3.1 (Optimality conditions for shortest paths).   *If the widths are feasible, then the total width of all st-cuts is a lower bound on the length of any st-path. In particular, if an st-path has length equal to the total width of all st-cuts, then it is a shortest st-path.*

Example 7, continued. We can now prove that the path $P = sa, ac, cb, bt$ is a shortest st-path. We found a set of feasible widths with total width

$$\bar{y}_{U_1} + \bar{y}_{U_2} + \bar{y}_{U_3} + \bar{y}_{U_4} = 3 + 1 + 2 + 1 = 7.$$

As $c(P) = 7$, it follows from Proposition 3.1 that P is a shortest st-path.

The arguments used to prove Proposition 3.1 are fairly elementary. We will see however that this result is surprisingly powerful. Indeed, we will be able to design an efficient algorithm based on this optimality condition. Deriving good optimality conditions is a key step in designing an algorithm.

Exercises

1   Let $G = (V, E)$ be a graph and let s, t be distinct vertices of G.
(a) Show that if we have a collection of k edge disjoint st-paths, then every st-cut contains at least k edges.

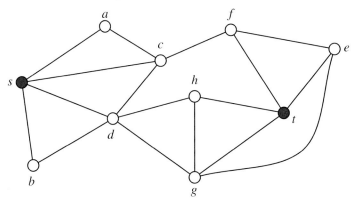

(b) For the above graph, find an *st*-cut with as few edges as possible. Use (a) to justify your answer.

Suppose now each edge *e* has a nonnegative *thickness* c_e. The thickness of an *st*-cut is the sum of the thicknesses of all the edges in the cut. Suppose we assign to each *st*-path *P* a nonnegative *width* y_P and assume that the following condition holds for every edge *e*:

> The total width of all the *st*-paths using *e* does not exceed the thickness of edge *e*.

(c) Show then that the total width of all the *st*-paths is a lower bound on the thickness of any *st*-cut.

(d) For the following graph, find an *st*-cut with minimum thickness. Edge labels indicate thickness. Use (c) to justify your answer.

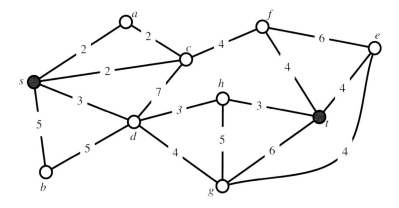

3.1.2   A general argument – weak duality

We used an *ad hoc* argument to obtain Proposition 3.1. At first glance, it is not obvious at all where the idea of assigning widths to *st*-cuts arises. We will show however that it is a natural idea when we look at the problem through the lens of duality theory. In this section, we derive a natural bound on the values a certain class of linear programs can attain. In Section 3.1.3, we show that Proposition 3.1 is a direct consequence of that result.

Example 8 Consider the LP

$$\min\{z(x) = c^\top x : Ax \geq b, x \geq 0\}, \tag{3.2}$$

where

$$A = \begin{pmatrix} 2 & 1 \\ 1 & 1 \\ -1 & 1 \end{pmatrix} \quad b = \begin{pmatrix} 20 \\ 18 \\ 8 \end{pmatrix} \quad c = \begin{pmatrix} 2 \\ 3 \end{pmatrix}.$$

It is easy to verify that the vectors $(8, 16)^\top$ and $(5, 13)^\top$ are all feasible for the LP (3.2). Their objective values are 64 and 49 respectively, and hence the first feasible solution is clearly not optimal. We will show however that $(5, 13)^\top$ is an optimal solution by proving that $z(\bar{x}) \geq 49$ for every feasible solution $\bar{x}$.

How can we find (and prove) such a lower bound? Construct a new inequality by multiplying the first constraint of $Ax \geq b$ by $y_1 \geq 0$, multiplying the second constraint of $Ax \geq b$ by $y_2 \geq 0$, multiplying the third constraint of $Ax \geq b$ by $y_3 \geq 0$, and adding the resulting three inequalities. We can write the resulting inequality in compact form as

$$(y_1, y_2, y_3) \begin{pmatrix} 2 & 1 \\ 1 & 1 \\ -1 & 1 \end{pmatrix} x \geq (y_1, y_2, y_3) \begin{pmatrix} 20 \\ 18 \\ 8 \end{pmatrix} \tag{3.3}$$

for any nonnegative vector $y = (y_1, y_2, y_3)^\top$. If we choose values $\bar{y}_1 = 0, \bar{y}_2 = 2$ and $\bar{y}_3 = 1$, we obtain the inequality

$$(1, 3)x \geq 44 \qquad \text{or equivalently} \qquad 0 \geq 44 - (1, 3)x.$$

Adding this inequality to the objective function $z(x) = (2, 3)x$ yields

$$z(x) \geq (2, 3)x + 44 - (1, 3)x = 44 + (1, 0)x.$$

Let $\bar{x}$ be any feasible solution. As $\bar{x} \geq \mathbb{0}$ and $(1, 0) \geq \mathbb{0}^\top$, we have $(1, 0)\bar{x} \geq 0$. Hence, $z(\bar{x}) \geq 44$. Thus, we have proved that no feasible solution has value smaller than 44. Note, this is not quite sufficient to prove that $(5, 13)^\top$ is optimal. It shows however that the optimum value for (3.2) is between 44 and 49. It is at most 49, as we have a feasible solution with that value, and it cannot be smaller than 44 by the previous argument.

Let us search for $y_1, y_2, y_3 \geq 0$ in a systematic way. We rewrite (3.3) as

$$0 \geq (y_1, y_2, y_3) \begin{pmatrix} 20 \\ 18 \\ 8 \end{pmatrix} - (y_1, y_2, y_3) \begin{pmatrix} 2 & 1 \\ 1 & 1 \\ -1 & 1 \end{pmatrix} x$$

and add it to the objective function $z(x) = (2, 3)x$ to obtain

$$z(x) \geq (y_1, y_2, y_3) \begin{pmatrix} 20 \\ 18 \\ 8 \end{pmatrix} + \left[(2, 3) - (y_1, y_2, y_3) \begin{pmatrix} 2 & 1 \\ 1 & 1 \\ -1 & 1 \end{pmatrix} \right] x. \tag{3.4}$$

Suppose that we pick $y_1, y_2, y_3 \geq 0$ such that

$$(2, 3) - (y_1, y_2, y_3) \begin{pmatrix} 2 & 1 \\ 1 & 1 \\ -1 & 1 \end{pmatrix} \geq \mathbb{0}^\top.$$

Then for any feasible solution $\bar{x}$, inequality (3.4), and the fact that $x \geq \mathbb{0}$ implies that

$$z(\bar{x}) \geq (y_1, y_2, y_3) \begin{pmatrix} 20 \\ 18 \\ 8 \end{pmatrix}.$$

For a minimization problem, the larger the lower bound the better. Thus, the best possible lower bound for (3.2) we can achieve using the above argument is given by the optimal value to the following LP:

$$\max \quad (y_1, y_2, y_3) \begin{pmatrix} 20 \\ 18 \\ 8 \end{pmatrix}$$

subject to

$$(2, 3) - (y_1, y_2, y_3) \begin{pmatrix} 2 & 1 \\ 1 & 1 \\ -1 & 1 \end{pmatrix} \geq 0^\top$$

$$y_1, y_2, y_3 \geq 0,$$

which we can rewrite as

$$\max \quad (20, 18, 8)\, y$$

subject to

$$\begin{pmatrix} 2 & 1 \\ 1 & 1 \\ -1 & 1 \end{pmatrix}^\top y \leq \begin{pmatrix} 2 \\ 3 \end{pmatrix} \tag{3.5}$$

$$y \geq 0.$$

Solving this LP gives

$$\bar{y}_1 = 0, \bar{y}_2 = \frac{5}{2}, \text{ and } \bar{y}_3 = \frac{1}{2},$$

and this solution has objective value 49. Since solution $(5, 13)^\top$ has value 49, it is an optimal solution of (3.2).

Let us generalize the previous argument and consider the following LP:

$$\min\{c^\top x : Ax \geq b, x \geq 0\}. \tag{3.6}$$

We first choose a vector $y \geq 0$ and create a new inequality

$$y^\top Ax \geq y^\top b.$$

This last inequality is obtained from $Ax \geq b$ by multiplying the first inequality by $y_1 \geq 0$, the second by $y_2 \geq 0$, the third by $y_3 \geq 0$, etc., and by adding all of the resulting inequalities together. This inequality can be rewritten as

$$0 \geq y^\top b - y^\top Ax,$$

which holds for every feasible solution $\bar{x}$ of (3.6). Thus, adding this inequality to the objective function $z(x) = c^\top x$ yields

$$z(x) \geq y^\top b + c^\top x - y^\top Ax = y^\top b + (c^\top - y^\top A)x. \tag{3.7}$$

Suppose that because of the choice of y, $c^\top - y^\top A \geq \mathbb{0}^\top$. Let $\bar{x}$ be any feasible solution. As $\bar{x} \geq \mathbb{0}$, we have that $(c^\top - y^\top A)\bar{x} \geq 0$. It then follows by (3.7) that $z(\bar{x}) \geq y^\top b$. Thus, we have shown that for all $y \geq \mathbb{0}$ such that $c^\top - y^\top A \geq \mathbb{0}^\top$ the value $y^\top b$ is a lower bound on the value of the objective function. Finally, note that the condition $c^\top - y^\top A \geq \mathbb{0}^\top$ is equivalent to $y^\top A \leq c^\top$, i.e. to $A^\top y \leq c$.

The best lower bound we can get in this way is the optimal value of

$$\max\{b^\top y : A^\top y \leq c, y \geq \mathbb{0}\}. \tag{3.8}$$

Let us summarize our result and provide a more direct proof.

THEOREM 3.2 (Weak duality–special form).   *Consider the following pair of LPs:*

$$\min\{c^\top x : Ax \geq b, x \geq \mathbb{0}\}, \tag{P}$$
$$\max\{b^\top y : A^\top y \leq c, y \geq \mathbb{0}\}. \tag{D}$$

Let $\bar{x}$ be a feasible solution for (P) and $\bar{y}$ be a feasible solution for (D). Then $c^\top \bar{x} \geq b^\top \bar{y}$. Moreover, if equality holds, then $\bar{x}$ is an optimal solution for (P).

In the previous theorem, we define (D) to be the *dual* of (P).

Proof of Theorem 3.2   Let $\bar{x}$ be a feasible solution of (P) and let $\bar{y}$ be a feasible solution (D). Then

$$b^\top \bar{y} = \bar{y}^\top b \leq \bar{y}^\top (A\bar{x}) = (\bar{y}^\top A)\bar{x} = (A^\top \bar{y})^\top \bar{x} \leq c^\top \bar{x}.$$

The first inequality follows from the fact that $\bar{y} \geq \mathbb{0}$ and that $A\bar{x} \geq b$. The second inequality follows from the fact that $A^\top \bar{y} \leq c$ and that $\bar{x} \geq \mathbb{0}$. Finally, as $b^\top \bar{y}$ is a lower bound on (P), if $c^\top \bar{x} = b^\top \bar{y}$, it follows that $\bar{x}$ is optimal for (P). $\square$

3.1.3   Revisiting the intuitive lower bound

Suppose we are given a graph $G = (V, E)$ with distinct vertices s, t and edge lengths $c_e \geq 0$ for every $e \in E$. The following is an integer programming formulation for the shortest st-path problem (see Section 1.5.2):

$$\min \qquad \sum (c_e x_e : e \in E)$$

subject to

$$\sum (x_e : e \in \delta(U)) \geq 1 \qquad (U \subseteq V, s \in U, t \notin U) \tag{3.9}$$
$$x_e \geq 0 \qquad\qquad (e \in E)$$
$$x_e \text{ integer} \qquad\qquad (e \in E).$$

Example 9 In the following figure, we show a simple instance of the shortest path problem. Each of the edges in the graph is labeled by its length.

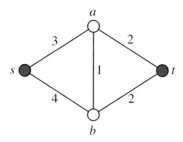

Let $x := (x_{sa}, x_{sb}, x_{ab}, x_{at}, x_{bt})^\top$. Then (3.9) specializes to the following:

$$\min \qquad (3, 4, 1, 2, 2)x$$
$$\text{subject to}$$

$$
\begin{array}{c}
\begin{array}{ccccc} sa & sb & ab & at & bt \end{array} \\
\begin{array}{c}
\{s\} \\
\{s, a\} \\
\{s, b\} \\
\{s, a, b\}
\end{array}
\left(
\begin{array}{ccccc}
1 & 1 & 0 & 0 & 0 \\
0 & 1 & 1 & 1 & 0 \\
1 & 0 & 1 & 0 & 1 \\
0 & 0 & 0 & 1 & 1
\end{array}
\right) x \geq \mathbb{1}
\end{array}
\qquad (3.10)
$$

$$x \geq 0 \qquad \text{integer.}$$

Each of the first four constraints corresponds to one of the four distinct st-cuts $\delta(U)$. For instance, the second constraint, corresponding to $\delta(\{s, a\}) = \{sb, ab, at\}$, states that $x_{sb} + x_{ab} + x_{at} \geq 1$.

For an integer program (IP), we call the linear program, obtained by removing the condition that some variables have to take integer values, the *linear programming relaxation* of (IP), or *LP relaxation* for short. Consider an instance of the shortest path problem $G = (V, E)$ with $s, t \in V$ and $c_e \geq 0$ for all $e \in E$. Suppose Q is some arbitrary st-path of G. Then we can construct a feasible solution $\bar{x}$ to the integer program (3.9) as follows:

$$\bar{x}_e = \begin{cases} 1 & \text{if } e \text{ is an edge of } Q \\ 0 & \text{otherwise.} \end{cases} \qquad (3.11)$$

Moreover, the length $c(Q)$ of the st-path Q is equal to the value of $\bar{x}$ for (3.9). It follows in particular that $c(Q)$ is greater than or equal to the optimal value of (3.9). Let (P) denote the LP relaxation of (3.9). Since the constraints of (P) are a subset of the constraints of (3.9) and since (3.9) is a minimization problem, the optimal value of (3.9) is greater than or equal to the optimal value of (P). Let (D) be the dual of (P). We know from weak duality (Theorem 3.2) that the optimal value of (P) is greater than or equal to the value of any feasible solution $\bar{y}$ of (D). Hence, we have proved the following result:

Remark 3.3 *If (D) is the dual of the LP relaxation of (3.9), then the value of any feasible solution of (D) is a lower bound on the length of any st-path.*

Example 9, continued. Let us compute the dual of the LP relaxation of (3.10), as defined in Theorem 3.2. Observe that by taking the dual, we interchange the role of the variables and the constraints. We now have one variable y_U for each st-cut $\delta(U)$ and one constraint for each edge of G. Hence, let us define $y := \left(y_{\{s\}}, y_{\{s,a\}}, y_{\{s,b\}}, y_{\{s,a,b\}}\right)^{\top}$. The dual is given by

$$\max \qquad \mathbb{1}^{\top} y$$

$$\text{subject to}$$

$$
\begin{array}{c}
 \quad \{s\}\,\{s,a\}\,\{s,b\}\,\{s,a,b\} \\
\begin{array}{c} sa \\ sb \\ ab \\ at \\ bt \end{array}
\begin{pmatrix}
1 & 0 & 1 & 0 \\
1 & 1 & 0 & 0 \\
0 & 1 & 1 & 0 \\
0 & 1 & 0 & 1 \\
0 & 0 & 1 & 1
\end{pmatrix}
y \le
\begin{pmatrix}
3 \\ 4 \\ 1 \\ 2 \\ 2
\end{pmatrix}
\end{array}
\qquad (3.12)
$$

$$y \ge \mathbb{0},$$

where $\mathbb{1}$ denotes a column vector of 1s. Note, a feasible solution to (3.12) assigns some nonnegative width y_U to every st-cut $\delta(U)$. The constraint for edge sb states that $y_{\{s\}} + y_{\{s,a\}} \le 4$. Observe that $\delta(\{s\}) = \{sa, sb\}$ and $\delta(\{s, a\}) = \{at, ab, sb\}$ are the two st-cuts of G that contain edge sb. Finally, 4 is the length of the edge sb. Hence, the constraint $y_{\{s\}} + y_{\{s,a\}} \le 4$ says that the total width of all st-cuts of G that contain edge sb does not exceed the length of sb. The corresponding condition holds for every other edge. Hence, if y is feasible for (3.12), the widths y satisfy the feasibility conditions (3.1). The objective function $\mathbb{1}^{\top} y$ calculates the total width of all st-cuts. Hence, it follows from Remark 3.3 that if the widths are feasible, then the total width of all st-cuts is a lower bound on the length of any st-path. This was precisely the statement of Proposition 3.1.

Consider now a general instance of the shortest path problem. We are given a graph $G = (V, E)$, vertices $s, t \in V$, and lengths $c_e \ge 0$ for all $e \in E$. We can rewrite the LP relaxation of (3.9) as

$$\min\{c^{\top} x \ : \ Ax \ge \mathbb{1}, x \ge \mathbb{0}\}, \qquad (3.13)$$

where c is the vector of edge lengths, and the matrix A is defined as follows:

(1) the rows of A are indexed by sets $U \subseteq V$ with $s \in U, t \notin U$,
(2) columns are indexed by edges $e \in E$, and

(3) for every row U and every column e

$$A[U, e] = \begin{cases} 1 & \text{if } e \text{ is an edge in } \delta(U) \\ 0 & \text{otherwise.} \end{cases}$$

Remark 3.4

(1) *In row U of A, entries with a 1 correspond to the edges in $\delta(U)$.*
(2) *In column e of A, entries with a 1 correspond to the st-cuts containing edge e.*

The dual of (3.13), as defined in Theorem 3.2, is given by

$$\max\{\mathbb{1}^\top y \; : \; A^\top y \le c, y \ge 0\}. \tag{3.14}$$

Let us try to understand this dual. There is a variable y_U for every st-cut $\delta(U)$ and a constraint for every edge $e \in E$. Consider the constraint for edge $e \in E$. The right-hand side of this constraint is the length c_e of the edge, and the left-hand side corresponds to column e of A. Remark 3.4(2) implies that the left-hand side of this constraint is the sum of the variables y_U over all st-cuts $\delta(U)$ that contain e. We can therefore rewrite (3.14) as follows:

max $\quad \sum\left(y_U \; : \; \delta(U) \text{ is an } st\text{-cut}\right)$

subject to

$$\sum\left(y_U \; : \; \delta(U) \text{ is an } st\text{-cut containing } e\right) \le c_e \quad (e \in E)$$
$$y_U \ge 0 \qquad\qquad\qquad\qquad\qquad (\delta(U) \text{ is an } st\text{-cut}). \tag{3.15}$$

A feasible solution $\bar{y}$ to (3.15) assigns nonnegative width $\bar{y}_U$ to every st-cut $\delta(U)$. The constraint for each edge e states that the total width of all st-cuts of G that contain edge e does not exceed the length of e. Hence, if $\bar{y}$ is feasible for (3.15), the widths $\bar{y}$ satisfy the feasibility conditions (3.1). The objective function $\mathbb{1}^\top y$ calculates the total width of all st-cuts. Hence, as in the previous example, it follows from Remark 3.3 that if the widths are feasible, then the total width of all st-cuts is a lower bound on the length of any st-path. Hence, we now have an alternate proof for Proposition 3.1.

While the derivation of Proposition 3.1 using duality may, at first glance, seem more technical than the *ad hoc* argument we had in Section 3.1.1 – notice that our derivation was completely mechanical. We formulated the shortest path problem as an integer program, wrote the dual of its LP relaxation, and used weak duality to obtain bounds on the possible values of our original optimization problem. After generalizing the notion of duals in Chapter 4, we will be able to apply the aforementioned strategy to arbitrary optimization problems that can be formulated as integer programs. A word of caution, however: the quality of the bounds obtained through this procedure depend on the problem, and on the LP relaxation used. In the shortest path example discussed here, the bound produced was sufficient to prove optimality of a certain primal solution. This may not always be possible as we will see in Chapter 6.

Exercises

1 A *vertex-cover* of a graph $G = (V, E)$ is a set $\mathscr{S}$ of vertices of G such that each edge of G is incident with at least one vertex of $\mathscr{S}$. The following IP finds a vertex-cover of minimum cardinality (see Exercise 2 in Section 1.5.2):

$$\text{min} \qquad \sum (x_e : e \in E)$$

subject to

$$x_i + x_j \geq 1 \qquad \text{(for all } ij \in E)$$
$$x \geq 0.$$

Denote by (P) the LP relaxation of this IP.
(a) Find the dual (D) of (P).
(b) Show that the largest size (number of edges) of a matching in G is a lower bound on the size of a minimum vertex-cover of G.
       HINT: Use part (a) and Theorem 3.2.
(c) Give an example where all matchings have size strictly less than the optimal solutions of (P) and of (D) and where these are strictly less than the size of the minimum vertex-cover.

2 The following graph has vertices of two types, a set $H = \{1, 2, 3, 4, 5, 6, 7\}$ of hubs, indicated by filled circles, and a set $C = \{a, b, c, d, e, f\}$ of connectors, indicated by squares. A subset S of the hubs is dominant if each connector in C has an edge to at

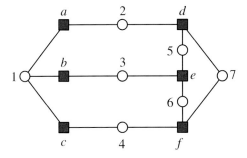

least one hub in S. For instance, $S = \{1, 3, 7\}$ is dominant because a, b, and c have edges to 1, d and f have edges to 7, and e has an edge to 3.
(a) Formulate as an IP the problem of finding a dominant set $S \subseteq H$ which is a small as possible.
       HINT: Assign a binary variable to each hub and a constraint for each connector.
Denote by (P) the LP relaxation of the IP given in (a):
(b) State the dual (D) of (P).
(c) Find a solution of (D) of value greater than 2.
(d) Using Theorem 3.2, prove that every dominant set of hubs contains at least three hubs.

3.1.4   An algorithm

We will present an algorithm to solve the shortest path problem based on the optimality conditions given in Proposition 3.1. Before we do so, however, we require a number of definitions.

We will need to generalize our definition of a graph $G = (V, E)$ by allowing both edges and *arcs*. Recall that an edge uv is an unordered pair of vertices. We call an *ordered* pair uv of vertices an *arc*, and denote it by $\overrightarrow{uv}$. Vertex u is the  *tail* of $\overrightarrow{uv}$ and vertex v is the *head* of $\overrightarrow{uv}$. We represent an arc as an arrow going from the tail of the arc to the head of the arc. A *directed st-path* is a sequence of arcs

$$\overrightarrow{v_1 v_2}, \overrightarrow{v_2 v_3}, \ldots, \overrightarrow{v_{k-2} v_{k-1}}, \overrightarrow{v_{k-1} v_k}$$

such that $v_1 = s$, $v_k = t$ and $v_i \neq v_j$ for all $i \neq j$. In other words, a directed st-path is an st-path in the graph obtained by ignoring the orientation, with the additional property that for any two consecutive arcs, the head of the first arc is the tail of the second arc. The following figure gives an example of a graph with both edges and arcs. The thick arcs form a directed st-path.

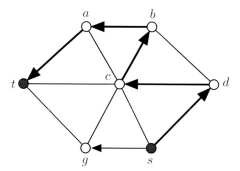

Consider now a general instance of the shortest path problem. We are given a graph $G = (V, E)$, vertices $s, t \in V$, and lengths $c_e \geq 0$ for all $e \in E$. Suppose that y is a feasible solution to (3.15). In other words, every st-cut $\delta(U)$ has a nonnegative width $y_U \geq 0$ and the feasibility condition (3.1) holds, namely for every edge e the total width of all st-cuts using e is smaller than or equal to the length of e. We define the *slack* of edge e for y to be the quantity

$$\text{slack}_y(e) := c_e - \sum\left(y_U \; : \; \delta(U) \text{ is an } st\text{-cut containing } e\right).$$

In other words, $\text{slack}_y(e)$ is the length of e minus the total width of all st-cuts using e.

We are now ready to describe our algorithm.[2]

Algorithm 3.2 Shortest path

Input: Graph $G = (V, E)$, costs $c_e \geq 0$ for all $e \in E$, $s, t \in V$, where $s \neq t$.
Output: A shortest st-path P.
  1: $y_W := 0$ for all st-cuts $\delta(W)$. Set $U := \{s\}$
  2: **while** $t \notin U$ **do**
  3:    Let ab be an edge in $\delta(U)$ of smallest slack for y where $a \in U$, $b \notin U$
  4:    $y_U := \text{slack}_y(ab)$
  5:    $U := U \cup \{b\}$
  6:    change edge ab into an arc $\overrightarrow{ab}$
  7: **end while**
  8: **return** A directed st-path P.

Example 10 Consider the shortest path problem described in Figure 3.3 (i).
We start with $y = \mathbb{0}$ and $U = \{s\}$. We compute the slack of the edges in $\delta(\{s\})$

$$\text{slack}_y(sa) = 6, \quad \text{slack}_y(sb) = 2, \quad \text{slack}_y(sc) = 4,$$

and therefore sb is the edge in $\delta(\{s\})$ of smallest slack for y. We let $y_{\{s\}} = 2$, we set $U := \{s, b\}$ and change edge sb into an arc $\overrightarrow{sb}$ (see Figure 3.3 (ii)).
Since $t \notin U = \{s, b\}$, we compute the slack of the edges in $\delta(\{s, b\})$

$$\text{slack}_y(sa) = 6 - y_{\{s\}} = 6 - 2 = 4$$
$$\text{slack}_y(sc) = 4 - y_{\{s\}} = 4 - 2 = 2$$
$$\text{slack}_y(bc) = 1$$
$$\text{slack}_y(bt) = 5,$$

and therefore bc is the edge in $\delta(\{s, b\})$ with smallest slack for y. We let $y_{\{s,b\}} = 1$, we set $U = \{s, b, c\}$, and change edge bc into an arc $\overrightarrow{bc}$ (see Figure 3.3 (iii)).
Since $t \notin U = \{s, b, c\}$, we compute the slack of the edges in $\delta(\{s, b, c\})$

$$\text{slack}_y(sa) = 6 - y_{\{s\}} - y_{\{s,b\}} = 6 - 2 - 1 = 3$$
$$\text{slack}_y(ca) = 1$$
$$\text{slack}_y(ct) = 2$$
$$\text{slack}_y(bt) = 5 - y_{\{s,b\}} = 5 - 1 = 4,$$

and therefore ca is the edge in $\delta(\{s, b, c\})$, with smallest slack for y. We let $y_{\{s,b,c\}} = 1$, we set $U = \{s, b, c, a\}$, and change edge ca into an arc $\overrightarrow{ca}$ (see Figure 3.3 (iv)).

---

[2] We will assume that at least one st-path exists in the graph G.

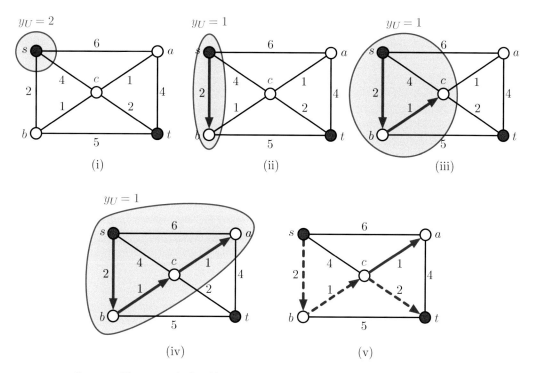

Figure 3.3 Shortest path algorithm – an example.

Since $t \notin U = \{s, b, c, a\}$, we compute the slack of the edges in $\delta(\{s, b, c, a\})$

$$\text{slack}_y(at) = 4$$
$$\text{slack}_y(ct) = 2 - y_{\{s,b,c\}} = 2 - 1 = 1$$
$$\text{slack}_y(bt) = 5 - y_{\{s,b\}} - y_{\{s,b,c\}} = 5 - 1 - 1 = 3,$$

and therefore ct is the edge in $\delta(\{s, b, c, a\})$ with smallest slack for y. We let $y_{\{s,b,c,a\}} = 1$, we set $U = \{s, b, c, a, t\}$, and change edge ct into an arc $\overrightarrow{ct}$ (see Figure 3.3 (v)). Note, now that $t \in U$ and there exists a directed st-path P, namely $P = \overrightarrow{sb}, \overrightarrow{bc}, \overrightarrow{ct}$. It can be readily checked that entries of y are feasible widths. Moreover, the total width of all st-cuts is equal to $2 + 1 + 1 + 1 = 5$ and the length $c(P)$ of P is equal to $2 + 1 + 2 = 5$. It follows from Proposition 3.1 that P is a shortest st-path.

We will prove that this simple algorithm is guaranteed to always find a shortest st-path in the next section. Suppose after running the algorithm and finding a shortest path P we define variables $\bar{x}$ as in (3.11). Then $\bar{x}$ is a feasible solution to the linear program (3.9) and $\bar{y}$ is a feasible solution to the dual linear program (3.15). Moreover, we will see that the algorithm will guarantee that the total width of all st-cuts is equal to the length of the shortest st-path. In other words, that the value of $\bar{x}$ in (3.9) is equal to the value of $\bar{y}$ in (3.15). It follows from weak duality (Theorem 3.2) that $\bar{x}$ is an optimal solution to (3.9). Hence, correctness of the algorithm will imply the following result:

THEOREM 3.5   *If $c_e \geq 0$ for all $e \in E$, then the linear program (3.9) has an integral optimal solution; i.e. it has an optimal solution all of whose variables have integer values.*

Note that our shortest path algorithm preserves at each step a feasible solution to the dual LP relaxation of (3.15). This is an example of a *primal–dual algorithm*. We will see another example of such an algorithm in Section 5.1.

Exercises

1   For each of the following two graphs, find the shortest path between s and t using the algorithm described in this section. In the figures below, each edge is labeled by its length. Make sure to describe for each step of the algorithm which edge becomes an arc, which vertex is added to the set U, and which st-cut is assigned a positive width. At the end of the procedure, give the shortest st-path and certify that it is indeed a shortest st-path by exhibiting feasible widths.

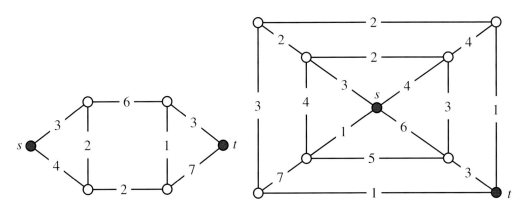

3.1.5   Correctness of the algorithm

Consider a graph $G = (V, E)$ with distinct vertices $s, t \in V$ and lengths $c_e \geq 0$ for all $e \in E$. Suppose that $\bar{y}$ is a feasible solution to (3.15). We say that an edge (resp. arc) uv is an equality edge (resp. arc) if its slack for $\bar{y}$ is zero, i.e. if $\text{slack}_{\bar{y}}(uv) = 0$. We say that a cut $\delta(U)$ is *active* for $\bar{y}$ if $\bar{y}_U > 0$.
   We shall first refine our optimality condition given in Proposition 3.1.

PROPOSITION 3.6   *Let $\bar{y}$ be feasible widths and let P be an st-path. Then P is a shortest st-path if both of the following conditions hold:*

(1) *all edges of P are equality edges for $\bar{y}$,*
(2) *all active cuts for $\bar{y}$ contain exactly one edge of P.*

Proof   Suppose that P is an st-path that satisfies both (1) and (2) for feasible widths $\bar{y}$. For every edge e of P, denote by $\mathscr{C}_e$ the set of all active st-cuts that contain edge e.

To prove that P is a shortest path, it suffices (because of Proposition 3.1) to verify that the following relations hold:

$$\text{Total width of } st\text{-cuts} = \sum \left(y_U \; : \; \delta(U) \text{ is an active } st\text{-cut} \right)$$

$$\overset{(a)}{=} \sum_{e \in P} \left[\sum \left(y_U \; : \; \delta(U) \in \mathscr{C}_e \right) \right]$$

$$\overset{(b)}{=} \sum_{e \in P} c_e = \text{length of } P.$$

Finally, note that (a) holds because of (2) and that (b) holds because of (1). □

Note that the algorithm is guaranteed to terminate after at most $|V|$ iterations since at every step one vertex is added to the set U. Hence, to show that the algorithm is correct it will suffice to verify the following result:

PROPOSITION 3.7 (Correctness of shortest path algorithm).
In STEP 8, *a directed st-path P exists and it is a shortest st-path.*

Proof. Throughout the execution, we maintain the following properties:

(I1) entries of y are feasible widths,
(I2) all arcs are equality arcs for y,
(I3) there is no active cut $\delta(W)$ for y and arc uv with $u \notin W$ and $v \in W$,
(I4) for every $u \in U$, where $u \neq s$, there exists a directed su-path,
(I5) all arcs have both ends in U.

Above, U refers to the current set U as defined in step 5 of Algorithm 3.2.

CLAIM   *If (I1)–(I4) hold in* STEP 8, *then a directed st-path P exists and it is a shortest st-path.*

Proof of claim: Since the loop completed, $t \in U$ and by (I4) there exists a directed st-path P.

We will show that P is a shortest st-path. (I1) states that y is feasible. (I2) implies that all arcs of P are equality arcs. Because of Proposition 3.6, it suffices to verify that active cuts for y contain exactly one edge of P. Suppose for a contradiction there is an active cut $\delta(W)$ that contains at least two edges of P (it contains at least one edge because of Remark 1.1). Denote by f the second arc of P, starting from s, that is in $\delta(W)$. Then (see figure), the tail of f is not in W but the head of f is in W, contradicting (I3).   ◇

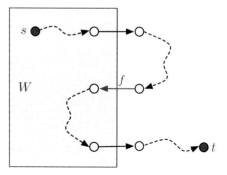

We leave it as an exercise to verify that (I1)–(I5) hold at the end of STEP 1. Assume that (I1)–(I5) hold at the beginning of the LOOP, we will show that (I1)–(I5) hold at the end of the LOOP. Together with the Claim this will complete the proof. In the following argument, U and y represent the quantities at the start of the LOOP and U' and y' the end of the LOOP. The only difference between y and y' is that $y_U = 0$ and that $y'_U =$ slack$_y(ab)$. Hence, to verify (I1) it suffices to consider $f \in \delta(U)$. Then

$$\text{slack}_{y'}(f) = c_f - \sum \left(y'_U : \delta(U) \text{ is an } st\text{-cut containing } f \right)$$

$$= c_f - \sum \left(y_U : \delta(U) \text{ is an } st\text{-cut containing } f \right) - y'_U \qquad (\star)$$

$$= \text{slack}_y(f) - \text{slack}_y(ab).$$

By the choice of ab in Step 3, slack$_y(f) \geq$ slack$_y(ab)$. Hence, $(\star)$, implies that slack$_{y'}$ $(f) \geq 0$. Thus, the feasibility condition (3.1) holds for y', i.e. (I1) holds at the end of the LOOP. By $(\star)$, we have slack$_{y'}(ab) = 0$. Since ab is the only new arc between the start and the end of the LOOP, (I2) holds at the end of the LOOP. The only cut that is active in y' but not y is $\delta(U)$. By (I5), all arcs different from ab have both ends in U. Arc ab has tail in U and head outside U. It follows that (I3) holds at the end of the LOOP. By (I4), there exists an sa-directed path Q. Moreover, by (I5) all arcs of Q have both tail and head in U. Hence, b is distinct from all vertices in Q, and adding arc ab at the end of Q gives a directed sb-path. It follows that (I4) holds at the end of the LOOP. Finally, since the only new arc is ab and $a, b \in U'$, (I5) holds at the end of the LOOP.   □

Exercises

1 Let $G = (V, E)$ be a graph with vertices s and t and suppose there exists at least one st-path in G. The *cardinality case* of the shortest path problem is the problem of finding a shortest path with as few edges as possible.
(a) Simplify the shortest path algorithm in Section 3.1.4 to deal with the cardinality case. You should strive to make the resulting algorithm as simple as possible.
(b) Simplify Proposition 3.6 for the cardinality case. Try to make the resulting statement as simple as possible.
(c) Simplify Proposition 3.7 for the cardinality case. Try to make the proof of correctness as simple as possible.

2 Let $G = (V, E)$ be a graph with distinct vertices s and t and nonnegative edge weights c.
(a) Show that if G has no st-path, then the LP (3.15) is unbounded.
(b) Show that (3.9) has an optimal solution if and only if G has an st-path.
       HINT: Use Theorems 3.2 and 2.11.

3.2   Minimum cost perfect matching in bipartite graphs

Recall the minimum cost perfect matching problem from Section 1.4.2. We are given a graph $G = (V, E)$ and costs c_e for all edges $e \in E$. (Note, we allow the costs to be negative.) A perfect matching M is a subset of the edges with the property that for every vertex v exactly one edge of M is incident to v. The cost $c(M)$ of a perfect matching is defined as the sum of the costs of the edges of M, i.e. as $\sum(c_e : e \in M)$. We wish to find among all possible perfect matchings one that is of minimum cost.

Example 11 In the following figure, we show an instance of this problem. Each of the edges in the graph is labeled by its cost. The thick edges in the graph form a perfect matching $M = \{ag, hb, cd\}$ of total cost $3 + 2 + 1 = 6$. This perfect matching is of minimum cost, and hence is a solution to our problem.

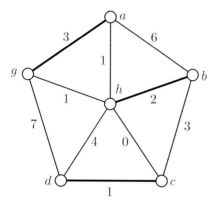

3.2.1   An intuitive lower bound

We claimed in the previous example that the matching $M = \{ag, hb, cd\}$ of cost 6 is a minimum cost perfect matching. How could you convince someone of that fact? Of course, we could list all possible perfect matchings and verify that M is indeed the one of minimum cost. This is not a practical way of proceeding however, as we may have a huge (exponential) number of perfect matchings. Our goal in this section is to find a certificate that can be used to quickly convince someone that a perfect matching is indeed a minimum cost perfect matching. Once again, we will see that this kind of a certificate will also be of crucial importance for the design of an algorithm.

Example 11, continued. Suppose that for every edge incident to vertex b we decrease the cost of all edges incident to vertex b by the value 3 (see Figure 3.4(i)). Since every perfect matching has exactly one edge that is incident to vertex b, this will decrease the cost of *every* perfect matching by exactly 3. In particular, if a matching M is a minimum cost perfect matching with these new costs, then it must be a minimum cost perfect matching with the original costs.

In Figure 3.4 (ii), we pick values for every vertex in the graph and repeat the same argument for each vertex. For instance, if we choose value 3 for vertex b and value 2 for vertex a, the new cost of edge ab is given by $6 - 3 - 2 = 1$. We indicate in Figure 3.4 (ii) the new cost for every edge. Observe now, that for the graph given in Figure 3.4 (ii), every edge has nonnegative cost. This implies in particular, that every perfect matching

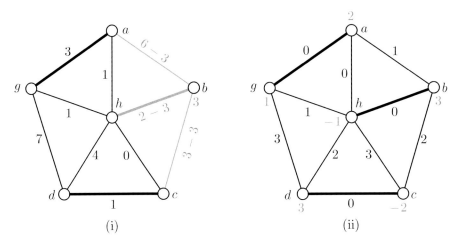

Figure 3.4 Reduced costs.

has nonnegative cost. However, as the matching $M = \{ag, hb, cd\}$ has cost 0, M must be a minimum cost perfect matching with these new costs. But then M must be a minimum cost perfect matching with the original costs.

Let us try to extend these ideas to an arbitrary graph $G = (V, E)$ with edge costs c_e for all $e \in E$. Let us assign to every vertex u, a number y_u that we call the *potential* of the vertex u. The *reduced cost* of the edge uv is defined as

$$\bar{c}_{uv} := c_{uv} - y_u - y_v.$$

Let M be a perfect matching. Since by definition M has exactly one edge that is incident to every vertex u, the difference between the cost of M with costs c and costs $\bar{c}$ is given by the sum of all the potentials y_u over every vertex u, i.e. by $\sum(y_u : u \in V)$. Since this quantity is a constant (for a fixed y), it follows in particular that, if M is a minimum cost perfect matching with the reduced costs $\bar{c}$, then it must be a minimum cost perfect matching with the original costs c. If we selected the potentials y so that every reduced cost is nonnegative, then every perfect matching would have nonnegative costs. If in addition all edges in the perfect matching M have zero reduced costs, then M must be a minimum cost perfect matching with respect to the reduced costs $\bar{c}$. But then it means that M is a minimum cost perfect matching with respect to the original costs c.

We formalize this result by introducing a few definitions. Let us say that an edge uv is an *equality edge* with respect to some potentials y if its reduced cost $\bar{c}_{uv} = c_{uv} - y_u - y_v = 0$. We say that potentials y are *feasible* if they satisfy the following:

Feasibility condition. For every edge $e \in E$, $\bar{c}_{uv} = c_{uv} - y_u - y_v \geq 0$.   (3.16)

We have thus proved the following:

PROPOSITION 3.8 (Optimality condition for perfect matchings).   *If the potentials y are feasible and all edges of M are equality edges with respect to y, then M is a minimum cost perfect matching.*

The arguments used to prove Proposition 3.8 are fairly elementary. We will see how-ever that this result is surprisingly powerful. Indeed, we will be able to design an efficient algorithm to find a minimum cost perfect matching for the class of bipartite graphs, based on this optimality condition. Deriving good optimal conditions is a key step in designing an algorithm.

3.2.2 A general argument–weak duality

We used an *ad hoc* argument to obtain Proposition 3.8. At first glance, it is not obvious at all where the idea of assigning potentials to vertices arises from. We will show however that the result stated in Proposition 3.8 is an almost immediate consequence of looking at the matching problem through the lens of duality theory.

Example 12 Consider the LP

$$\min\{z(x) = c^\top x : Ax = b, x \geq 0\},\tag{3.17}$$

where

$$A = \begin{pmatrix} -2 & 1 & -1 & 0 \\ 9 & 2 & 1 & 1 \\ 5 & 1 & 0 & 3 \end{pmatrix} \qquad b = \begin{pmatrix} -2 \\ 7 \\ 7 \end{pmatrix} \qquad c = \begin{pmatrix} 4 \\ 3 \\ -2 \\ 3 \end{pmatrix}.$$

It is easy to verify that the vectors $(1/2, 0, 1, 3/2)^\top$ and $(0, 1, 3, 2)^\top$ are both feasible for the linear program (3.17). Their objective values are $9/2$ and 3, respectively, and hence the first feasible solution is clearly not optimal. We will show however that $(0, 1, 3, 2)^\top$ is an optimal solution by proving that $z(\bar{x}) \geq 3$ for every feasible solution $\bar{x}$.

How can we find (and prove) such a lower bound? As before, recall from Chapter 2 that, for any vector $y = (y_1, y_2, y_3)^\top \in \mathbb{R}^3$, the equation

$$(y_1, y_2, y_3) \begin{pmatrix} -2 & 1 & -1 & 0 \\ 9 & 2 & 1 & 1 \\ 5 & 1 & 0 & 3 \end{pmatrix} x = (y_1, y_2, y_3) \begin{pmatrix} -2 \\ 7 \\ 7 \end{pmatrix}\tag{3.18}$$

holds for every feasible solution of (3.17). In particular, if we choose values, $\bar{y}_1 = 1$, $\bar{y}_2 = -1$ and $\bar{y}_3 = 1$, we obtain the equality

$$(-6, 0, -2, 2)x = -2 \quad \text{or equivalently} \quad 0 = -2 - (-6, 0, -2, 2)x.$$

Adding this equality to the objective function $z(x) = (4, 3, -2, 3)x$ yields

$$z(x) = (4, 3, -2, 3)x - (-6, 0, -2, 2)x - 2 = (10, 3, 0, 1)x - 2.$$

Let $\bar{x}$ be any feasible solution. As $\bar{x} \geq 0$, and $(10, 3, 0, 1) \geq 0^\top$ we have $(10, 3, 0, 1)\bar{x} \geq 0$. Hence, $z(\bar{x}) \geq -2$. Thus, we have proved that no feasible solution has value smaller than -2. Note, this is not quite sufficient to prove that $(0, 1, 3, 2)^\top$ is optimal. It shows however that the optimum value for (3.17) is between -2 and 3. It is at most 3 as we have a feasible solution with that value, and it cannot be smaller than -2 by the previous argument.

Let us search for y_1, y_2, y_3 in a systematic way. We rewrite (3.18) as

$$0 = (y_1, y_2, y_3) \begin{pmatrix} -2 \\ 7 \\ 7 \end{pmatrix} - (y_1, y_2, y_3) \begin{pmatrix} -2 & 1 & -1 & 0 \\ 9 & 2 & 1 & 1 \\ 5 & 1 & 0 & 3 \end{pmatrix} x$$

and add it to the objective function $z(x) = (4, 3, -2, 3)x$ to obtain

$$z(x) = (y_1, y_2, y_3) \begin{pmatrix} -2 \\ 7 \\ 7 \end{pmatrix} + \left[(4, 3, -2, 3) - (y_1, y_2, y_3) \begin{pmatrix} -2 & 1 & -1 & 0 \\ 9 & 2 & 1 & 1 \\ 5 & 1 & 0 & 3 \end{pmatrix} \right] x.$$

$$(3.19)$$

Suppose that we pick, y_1, y_2, y_3 such that

$$(4, 3, -2, 3) - (y_1, y_2, y_3) \begin{pmatrix} -2 & 1 & -1 & 0 \\ 9 & 2 & 1 & 1 \\ 5 & 1 & 0 & 3 \end{pmatrix} \geq 0^\top.$$

Then for any feasible solution $\bar{x}$, inequality (3.19), and the fact that $\bar{x} \geq 0$ implies that

$$z(\bar{x}) \geq (y_1, y_2, y_3) \begin{pmatrix} -2 \\ 7 \\ 7 \end{pmatrix}.$$

For a minimization problem, the larger the lower bound the better. Thus, the best possible upper bound for (3.17) we can achieve using the above argument is given by the optimal value to the following LP:

$$\max \quad (y_1, y_2, y_3) \begin{pmatrix} -2 \\ 7 \\ 7 \end{pmatrix}$$

subject to

$$(4, 3, -2, 3) - (y_1, y_2, y_3) \begin{pmatrix} -2 & 1 & -1 & 0 \\ 9 & 2 & 1 & 1 \\ 5 & 1 & 0 & 3 \end{pmatrix} \geq 0^\top$$

$$y_1, y_2, y_3 \geq 0,$$

which we can rewrite as

$$\max \quad (-2, 7, 7)y$$

subject to

$$\begin{pmatrix} -2 & 1 & -1 & 0 \\ 9 & 2 & 1 & 1 \\ 5 & 1 & 0 & 3 \end{pmatrix}^\top y \leq \begin{pmatrix} 4 \\ 3 \\ -2 \\ 3 \end{pmatrix} \qquad (3.20)$$

$$y \geq 0.$$

Solving this LP gives

$$\bar{y}_1 = 2, \bar{y}_2 = 0, \text{ and } \bar{y}_3 = 1,$$

and this solution has objective value 3. Thus, 3 is lower bound for (3.17). In particular, since the feasible solution $(0, 1, 3, 2)^\top$ of (3.17) has value 3, it is an optimal solution.

Let us generalize the previous argument and consider the following LP:

$$\min\{c^\top x : Ax = b, x \geq 0\}. \tag{3.21}$$

We first choose a vector y and create a new equality

$$y^\top Ax = y^\top b.$$

This last equality is obtained from $Ax = b$ by multiplying the first equality by y_1, the second by y_2, the third by y_3, etc, and by adding all of the resulting equalities together. This equality can be rewritten as

$$0 = y^\top b - y^\top Ax,$$

which holds for every feasible solution $\bar{x}$ of (3.21). Thus, adding this equality to the objective function $z(x) = c^\top x$ yields

$$z(x) = y^\top b + c^\top x - y^\top Ax = y^\top b + (c^\top - y^\top A)x. \tag{3.22}$$

Suppose that because of the choice of y, $c^\top - y^\top A \geq 0^\top$. Let $\bar{x}$ be any feasible solution. As $\bar{x} \geq 0$, we have that $(c^\top - y^\top A)\bar{x} \geq 0$. It then follows by (3.22) that $z(\bar{x}) \geq y^\top b$. Thus, we have shown that for all y such that $c^\top - y^\top A \geq 0^\top$ the value $y^\top b$ is a lower bound on the value of the objective function. Finally, note that the condition $c^\top - y^\top A \geq 0^\top$ is equivalent to $y^\top A \leq c^\top$, i.e. to $A^\top y \leq c$.

The best lower bound we can get in this way is the optimal value of

$$\max\{b^\top y : A^\top y \leq c\}, \tag{3.23}$$

where we note that the variables y are *free* (i.e. the variables are unrestricted). Let us summarize our result and provide a more direct proof.

THEOREM 3.9 (Weak duality–special form). *Consider the following pair of LPs:*

$$\min\{c^\top x : Ax = b, x \geq 0\}, \tag{P}$$
$$\max\{b^\top y : A^\top y \leq c\}. \tag{D}$$

Let $\bar{x}$ be a feasible solution for (P) and $\bar{y}$ be a feasible solution for (D). Then $c^\top \bar{x} \geq b^\top \bar{y}$. Moreover, if equality holds, then $\bar{x}$ is an optimal solution for (P).

In the previous theorem, we define (D) to be the *dual* of (P).

Proof of Theorem 3.9 Let $\bar{x}$ be a feasible solution of (P) and let $\bar{y}$ be a feasible solution (D). Then

$$b^\top \bar{y} = \bar{y}^\top b = \bar{y}^\top (A\bar{x}) = (\bar{y}^\top A)\bar{x} = (A^\top \bar{y})^\top \bar{x} \leq c^\top \bar{x}.$$

The inequality follows from the fact that $A^\top \bar{y} \le c$ and that $\bar{x} \ge 0$. Finally, as $b^\top \bar{y}$, is a lower bound on (P), if $c^\top \bar{x} = b^\top \bar{y}$, it follows that $\bar{x}$ is optimal for (P).   □

3.2.3   Revisiting the intuitive lower bound

Suppose we are given a graph $G = (V, E)$ with edge costs c_e (possibly negative) for every $e \in E$. The following is an integer programming formulation for the minimum cost perfect matching problem (see Section 1.5.2)

$$\text{min} \qquad \sum \left(c_e x_e : e \in E \right)$$

subject to

$$\sum \left(x_e : e \in \delta(v) \right) = 1 \qquad (v \in V) \qquad (3.24)$$
$$x_e \ge 0 \qquad\qquad\qquad (e \in E)$$
$$x_e \text{ integer} \qquad\qquad (e \in E).$$

Example 13 In the following figure, we show a simple instance of the perfect matching problem. Each of the edges in the graph is labeled by its cost.

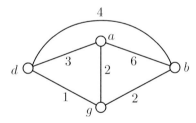

Let $x := (x_{ab}, x_{bg}, x_{dg}, x_{ad}, x_{ag}, x_{bd})^\top$. Then (3.24) specializes to the following in this case:

$$\text{min} \qquad (6, 2, 1, 3, 2, 4)x$$

subject to

$$
\begin{array}{c}
\quad\quad ab \;\; bg \;\; dg \;\; ad \;\; ag \;\; bd \\
\begin{array}{c} a \\ b \\ d \\ g \end{array}
\left(
\begin{array}{cccccc}
1 & 0 & 0 & 1 & 1 & 0 \\
1 & 1 & 0 & 0 & 0 & 1 \\
0 & 0 & 1 & 1 & 0 & 1 \\
0 & 1 & 1 & 0 & 1 & 0
\end{array}
\right) x = 1
\end{array}
\qquad (3.25)
$$
$$x \ge 0 \quad \text{integer.}$$

Each of the first four constraints corresponds to one of the four vertices of G. For instance, the second constraint, corresponding to vertex b, states that $x_{ab} + x_{bg} + x_{bd} = 1$, i.e. that we should select exactly one of the edges incident to b.

Recall that for an integer program (IP) we call the linear program, obtained by removing the condition that some variables have to take integer values, the *linear programming relaxation* of (IP) or *LP relaxation* for short. Consider an instance of the minimum cost perfect matching problem $G = (V, E)$ with c_e for all $e \in E$. Suppose M is some arbitrary perfect matching of G. Then we can construct a feasible solution $\bar{x}$ to the integer program (3.24) as follows:

$$
\bar{x}_e = \begin{cases} 1 & \text{if } e \text{ is an edge of } M \\ 0 & \text{otherwise.} \end{cases} \tag{3.26}
$$

Moreover, the cost $c(M)$ of the perfect matching M is equal to the value of $\bar{x}$ for (3.24). It follows in particular that $c(M)$ is greater than or equal to the optimal value of (3.24). Let (P) denote the LP relaxation of (3.24). Since the constraints of (P) are a subset of the constraints of (3.24) and since (3.24) is a minimization problem, the optimal value of (3.24) is greater than or equal to the optimal value of (P). Let (D) be the dual of (P). We know from the weak duality theorem (3.9) that the optimal value of (P) is greater than or equal to the value of any feasible solution $\bar{y}$ of (D). Hence, we have proved the following result:

Remark 3.10 *If (D) is the dual of the LP relaxation of (3.24), then the value of any feasible solution of (D) is a lower bound on the cost of any perfect matching.*

Example 9, continued. Let us compute the dual of the LP relaxation of (3.25), as defined in Theorem 3.9. Observe that by taking the dual we interchange the role of the variables and the constraints. We now have one variable y_v for every vertex v and one constraint for each edge of G. Hence, let us define $y := (y_a, y_b, y_d, y_g)$. The dual is given by

$$
\max \quad \mathbb{1}^T y
$$

subject to

$$
\begin{array}{c}
 \\ ab \\ bg \\ dg \\ ad \\ ag \\ bd
\end{array}
\begin{array}{cccc}
a & b & d & g \\
\end{array}
\left(\begin{array}{cccc}
1 & 1 & 0 & 0 \\
0 & 1 & 0 & 1 \\
0 & 0 & 1 & 1 \\
1 & 0 & 1 & 0 \\
1 & 0 & 0 & 1 \\
0 & 1 & 1 & 0
\end{array} \right) y \leq \begin{pmatrix} 6 \\ 2 \\ 1 \\ 3 \\ 2 \\ 4 \end{pmatrix}. \tag{3.27}
$$

A feasible solution to (3.27) assigns some potential y_v to every vertex v. The constraint for edge ab states that $y_a + y_b \leq 6 = c_{ab}$ or equivalently that the reduced cost $c_{ab} - y_a - y_b$ of edge ab is nonnegative. Similarly, for every other edge the corresponding constraint states that the reduced cost is nonnegative. Hence, if $\bar{y}$ is feasible for (3.27), then the potentials $\bar{y}$ satisfy the feasibility conditions (3.16). For instance, $\bar{y}_a = 2$, $\bar{y}_b = 2$, $\bar{y}_d = 1$ and $\bar{y}_g = 0$ are feasible potentials. Observe that

$$c_{ad} = 3 = 2 + 1 = \bar{y}_a + \bar{y}_d \quad \text{and} \quad c_{bg} = 2 = 2 + 0 = \bar{y}_b + \bar{y}_g,$$

i.e. ad and bc are equality edges with respect to $\bar{y}$. It follows that

$$\mathbb{1}^\top \bar{y} = (\bar{y}_a + \bar{y}_d) + (\bar{y}_b + \bar{y}_g) = c_{ad} + c_{bg} = c(M).$$

Hence, the value of $\bar{y}$ in (3.27) is equal to the cost of the perfect matching $M = \{ad, bc\}$. It follows by Remark 3.10 that M is a minimum cost perfect matching. Hence, in this example we see that if the potentials $\bar{y}$ are feasible and every edge of M is an equality edge, then M is a perfect matching as predicted by Proposition 3.8.

Consider now a general instance of the perfect matching problem. We are given a graph $G = (V, E)$ and costs c_e for all $e \in E$. We can rewrite the LP relaxation of (3.24) as

$$\min\{c^\top x \; : \; Ax = \mathbb{1}, x \geq 0\}, \tag{3.28}$$

where c is the vector of edge costs, and the matrix A is defined as follows:

(1) the rows of A are indexed by vertices $v \in V$,
(2) columns are indexed by edges $e \in E$, and
(3) for every row U and every column e

$$A[v, e] = \begin{cases} 1 & \text{if } v \text{ is an endpoint of } e \\ 0 & \text{otherwise.} \end{cases}$$

Remark 3.11

(1) *In row v of A, entries with a 1 correspond to the edges incident to v.*
(2) *In column e of A, entries with a 1 correspond to the endpoints of e.*

The dual of (3.28), as defined in Theorem 3.9, is given by

$$\max\{\mathbb{1}^\top y \; : \; A^\top y \leq c\}. \tag{3.29}$$

Let us try and understand this dual. There is a variable y_v for every vertex v and a constraint for every edge $e \in E$. Consider the constraint for edge $e \in E$. The right-hand side is the cost c_e of the edge, and the left-hand side corresponds to column $e = uv$ of A. Remark 3.11(2) implies that the left-hand side of this constraint is $y_u + y_v$. We can therefore rewrite (3.29) as follows:

$$\begin{aligned} \max \quad & \sum (y_v : v \in V) \\ \text{subject to} \quad & \\ & y_u + y_v \leq c_{uv} \quad (uv \in E). \end{aligned} \tag{3.30}$$

A feasible solution $\bar{y}$ to (3.30) assigns some potential y_v to every vertex v. The constraint for each edge uv states that $y_u + y_v \leq c_{uv}$ or equivalently that the reduced cost $c_{uv} - y_u - y_v$ of edge uv is nonnegative. Hence, if $\bar{y}$ is feasible for (3.30), then the potentials satisfy the Feasibility Conditions (3.16). Let M be a perfect matching and suppose that every edge $uv \in M$ is an equality edge, i.e. $\bar{y}_u + \bar{y}_v = c_{uv}$. Then

$$\mathbb{1}^\top \bar{y} = \sum_{v \in V} \bar{y}_v = \sum_{uv \in M} (\bar{y}_u + \bar{y}_v) = \sum_{uv} c_{uv} = c(M),$$

where the second equality follows from the fact that M is a perfect matching, i.e. that every vertex of G is the end of exactly one edge of M. Hence, the value of $\bar{y}$ in (3.30) is equal to the cost of the perfect matching M. It follows by Remark 3.10 that M is a minimum cost perfect matching. Hence, we see that if the potentials $\bar{y}$ are feasible and every edge of M is an equality edge, then M is a perfect matching. Hence, we now have an alternate proof for Proposition 3.8.

While the derivation of Proposition 3.8 using duality may, at first glance, seem more technical than the *ad hoc* argument we had in Section 3.2.1, notice that our derivation was completely mechanical. We formulated the minimum cost perfect matching problem as an integer program, wrote the dual of its LP relaxation, and used weak duality to obtain bounds on the possible values of our original optimization problem. After generalizing the notion of duals in Chapter 4, we will be able to apply the aforementioned strategy for arbitrary optimization problems that can be formulated as integer programs. Once again a word of caution at this point. While the dual bound obtained through Proposition 3.8 was sufficient to prove the optimality of a certain perfect matching, this may not always work. The success of the mechanical procedure outlined above depends heavily on the quality of the underlying LP relaxation as well as the problem.

3.2.4 An algorithm

In this section, we give an algorithm to find a minimum cost perfect matching in a bipartite graph only. While there exist efficient algorithms for solving the minimum cost perfect matching problem for general graphs, these algorithms are more involved and beyond the scope of this book (the interested reader is referred to Cook et al [16] for details). One of the key reasons is that for general graphs we cannot always certify that a minimum cost perfect matching is indeed a minimum cost perfect matching using Proposition 3.8.

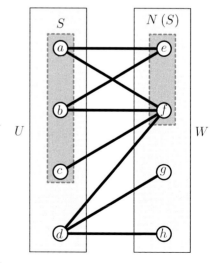

We first need to characterize which bipartite graphs have a perfect matching. Recall from Section 1.4.2 that a graph $G = (V, E)$ is bipartite with bipartition U, W, if V is the disjoint union of U and W and every edge has one end in U and one end in W. A necessary condition for a graph G with bipartition U, W to have a perfect matching is that $|U| = |W|$. However, this is not sufficient as the example in the figure illustrates. Suppose M is an arbitrary matching of this graph. Observe that all edges that have an endpoint in the set $\{a, b, c\}$ have their other endpoint in the set $\{e, f\}$. Since M is a matching, at most two edges of M can have an endpoint in $\{e, f\}$. Hence, at most two of the vertices $\{a, b, c\}$ can be an endpoint of some edge of M. In particular, M cannot be a perfect matching.

Consider a graph $G = (V, E)$ and a subset of vertices $S \subseteq V$. The *set of neighbors* of S is the set of all vertices outside S that are joined by an edge to some vertex of S, i.e. it is the set $\{r \in V \setminus S : sr \in E \text{ and } s \in S\}$. We denote the set of neighbors of S in graph G by $N_G(S)$. For instance, in the previous example the set of neighbors of $\{a, b, c\}$ is the set $\{e, f\}$, the neighbors of $\{a, b, c\}$ are the vertices e and f. Consider a bipartite graph $G = (V, E)$ with a bipartition U, W. Suppose that there exists a set $S \subseteq U$ such that $|S| > |N_G(S)|$. Let M be an arbitrary matching of G. Observe that all edges of M that have an endpoint in S have an endpoint in $N_G(S)$. Since M is a matching, at most $|N_G(S)|$ edges of M can have an endpoint in $N_G(S)$. Hence, at most $|N_G(S)|$ of the vertices in S can be an endpoint of some edge of M. As $|S| > |N_G(S)|$, M cannot be a perfect matching. We call a set $S \subseteq U$ such that $|S| > |N_G(S)|$ a *deficient set*. Thus, we have argued that if G contains a deficient set, then G has no perfect matching. The following result states that the converse is true:

THEOREM 3.12 (Hall's theorem). *Let $G = (V, E)$ be a bipartite graph with bipartition U, W, where $|U| = |W|$. Then there exists a perfect matching M in G if and only if G has no deficient set $S \subseteq U$. Moreover, there exists an efficient (polynomial-time)* [3] *algorithm that given G will either find a perfect matching M or find a deficient set $S \subseteq U$.*

We postpone the proof of the previous result and the description of the associated algorithm until Section 3.2.6.

We are now ready to start the description of the matching algorithm. We consider a bipartite graph $G = (V, E)$ with bipartition U, W, where $|U| = |W|$. In addition, we are given edge cost c_e for every edge $e \in E$. At each step, we have a feasible solution $\bar{y}$ to the LP (3.30). To get an initial feasible dual solution, we let α denote the value of the minimum cost edge and set $\bar{y}_v := \frac{1}{2}\alpha$ for all $v \in V$. (This is clearly feasible as for every edge $uv \in E$, $\bar{y}_u + \bar{y}_v = \alpha \le c_{uv}$.)

Given a dual feasible solution $\bar{y}$, we construct a graph H as follows: H has the same set of vertices as G, and the set of edges of H consist of all edges of G that are equality edges with respect to $\bar{y}$. We know from Theorem 3.12 that we can either find:

(a) a perfect matching M in H, or
(b) a deficient set $S \subseteq U$ of H, i.e. $S \subseteq U$ such that $|N_H(S)| < |S|$.

In case (a), since all edges of M are equality edges, Proposition 3.8 implies that M is a minimum cost perfect matching of G, in which case we can stop the algorithm. Thus,

---

[3] Polynomial-time algorithms are defined in Appendix A.

we may assume that (b) occurs. We will use the set S to define a new feasible solution y' to (3.30) with a larger objective value than $\bar{y}$ as follows. For every vertex v

$$y'_v := \begin{cases} \bar{y}_v + \epsilon & \text{for } v \in S \\ \bar{y}_v - \epsilon & \text{for } v \in N_H(S) \\ \bar{y}_v & \text{otherwise.} \end{cases}$$

We wish to choose $\epsilon \geq 0$ as large as possible such that y' is feasible for (3.30), i.e. for every edge uv we need to satisfy $c_{uv} - y'_u - y'_v \geq 0$. As y is feasible for (3.30), $c_{uv} - \bar{y}_u - \bar{y}_v \geq 0$. Consider an edge uv, we may assume $u \in U$ and $v \in W$. There are four possible cases for such an edge (see figure):

Case 1. $u \notin S, v \notin N_H(S)$.
Case 2. $u \notin S, v \in N_H(S)$.
Case 3. $u \in S, v \in N_H(S)$.
Case 4. $u \in S, v \notin N_H(S)$.

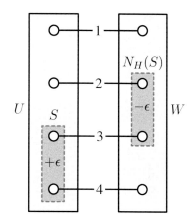

Let us investigate each case:

Case 1. $c_{uv} - y'_u - y'_v = c_{uv} - \bar{y}_u - \bar{y}_v \geq 0$.
Case 2. $c_{uv} - y'_u - y'_v = c_{uv} - \bar{y}_u - (\bar{y}_v - \epsilon) \geq 0$.
Case 3. $c_{uv} - y'_u - y'_v = c_{uv} - (\bar{y}_u + \epsilon) - (\bar{y}_v - \epsilon) \geq 0$.

Thus, we only need to worry about Case 4. Note, we may assume that there is such an edge, for otherwise S is a deficient set for G and we may stop as G has no perfect matching. We want, $0 \leq c_{uv} - y'_u - y'_v = c_{uv} - (\bar{y}_u + \epsilon) - \bar{y}_v$, i.e. $\epsilon \leq c_{uv} - \bar{y}_u - \bar{y}_v$. Note that since uv is not an edge of H, it is not an equality edge. Hence, $c_{uv} - \bar{y}_u - \bar{y}_v > 0$. Thus, we can choose $\epsilon > 0$ as follows:

$$\epsilon = \min\{c_{uv} - \bar{y}_u - \bar{y}_v : uv \in E, u \in S, v \notin N_H(S)\}. \tag{3.31}$$

Note, that $\mathbb{1}^\top y' - \mathbb{1}^\top \bar{y} = \epsilon(|S| - |N_H(S)|) > 0$. Hence, the new dual solution y' has a higher (better) value than $\bar{y}$. Algorithm 3.3 on the next page is a formal description of our algorithm.

Example 14 Consider the minimum cost perfect matching problem described in Figure 3.5(i) (left) where the edges are labeled by the costs. Since the minimum cost is 2 we initially set potentials $\bar{y}_a = \bar{y}_b = \bar{y}_c = \bar{y}_d = \bar{y}_e = \bar{y}_f = 1$. In Figure 3.5(i) (right), we indicate the graph H and the deficient set S. The edges of G with one endpoint in S and the other endpoint not in $N_H(S)$ are ae, be, bf, hence

$$\epsilon = \min\{c_{ae} - \bar{y}_a - \bar{y}_e, c_{be} - \bar{y}_b - \bar{y}_e, c_{bf} - \bar{y}_b - \bar{y}_f\}$$
$$= \min\{4 - 1 - 1, 5 - 1 - 1, 6 - 1 - 1\} = 2.$$

The new potentials are given in Figure 3.5 (ii) (left). In Figure 3.5 (ii) (right), we indicate the graph H and the deficient set S. The edges of G with one endpoint in S and the other endpoint not in $N_H(S)$ are bf, cf, hence

$$\epsilon = \min\{c_{bf} - \bar{y}_b - \bar{y}_f, c_{cf} - \bar{y}_c - \bar{y}_f\}$$
$$\min\{6 - 3 - 1, 3 - 1 - 1\} = 1.$$

The new potentials are given in Figure 3.5 (iii) (left). In Figure 3.5 (iii) (right), we indicate the graph H. This graph has no deficient set, but it has a perfect matching $M = \{ae, bd, cf\}$. This matching is a minimum cost perfect matching of G.

Algorithm 3.3 Minimum cost perfect matching in bipartite graphs

---

Input: Graph $G = (V, E)$ with bipartition U, W where $|U| = |W|$ and costs c.
Output: A minimum cost perfect matching M or a deficient set S.

1: $\bar{y}_v := \frac{1}{2}\min\{c_e : e \in E\}$, for all $v \in V$
2: **loop**
3:    Construct graph H with vertices V and edges $\{uv \in E : c_{uv} = \bar{y}_u + \bar{y}_v\}$
4:    **if** H has perfect matching M **then**
5:       **stop** (M is a minimum cost perfect matching of G)
6:    **end if**
7:    Let $S \subseteq U$ be a deficient set for H
8:    **if** all edges of G with an endpoint in S have an endpoint in $N_H(S)$ **then**
9:       **stop** (S is a deficient set of G)
10:   **end if**
11:   $\epsilon := \min\{c_{uv} - \bar{y}_u - \bar{y}_v : uv \in E, u \in S, v \notin N_H(S)\}$
12:   $\bar{y}_v := \begin{cases} \bar{y}_v + \epsilon & \text{for } v \in S \\ \bar{y}_v - \epsilon & \text{for } v \in N_H(S) \\ \bar{y}_v & \text{otherwise} \end{cases}$
13: **end loop**

We will prove, in the next section, that this algorithm terminates and hence is guaranteed to always find a minimum cost perfect matching. Suppose that after running the algorithm and finding a minimum cost matching M we define variables $\bar{x}$ as in (3.26). Then $\bar{x}$ is a feasible solution to the LP (3.24) and $\bar{y}$ is a feasible solution to the dual of the LP relaxation (3.30). Moreover, as all edges of M are equality edges the total sum of the potentials will be equal to the cost of the matching M. In other words, the value of $\bar{x}$ in (3.24) is equal to the value of $\bar{y}$ in (3.30). It follows from weak duality (Theorem 3.9), that $\bar{x}$ is an optimal solution to (3.24). Hence, correctness of the algorithm will imply the following result:

THEOREM 3.13   *If G is a bipartite graph and there exists a perfect matching, then the LP relaxation of (3.24) has an integral optimal solution; i.e. there is an optimal solution all of whose entries are integers.*

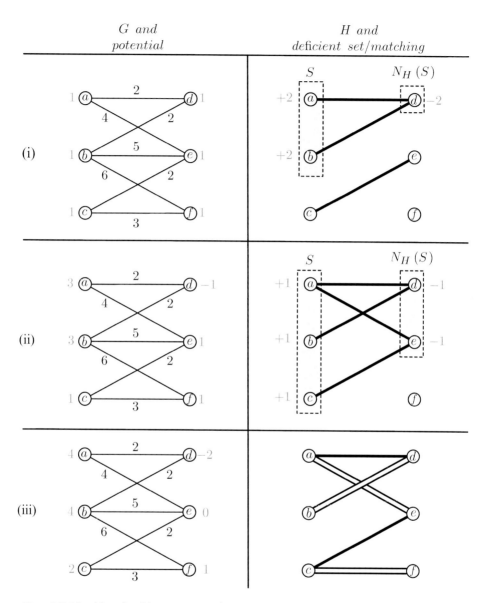

Figure 3.5 Matching algorithm – an example.

Note that our matching algorithm preserves at each step a feasible solution to the dual linear program (3.30). This is an example of a *primal–dual algorithm*. We will see another example in Section 5.1.

3.2.5   Correctness of the algorithm

It follows immediately from Proposition 3.8 that if the algorithm terminates with a perfect matching, then that matching is a minimum cost perfect matching. Moreover, if the algorithm terminates with a deficient set, then Theorem 3.12 implies that G does not have a perfect matching. Hence, to show that the algorithm is correct it suffices to prove that the algorithm terminates. The proof is very simple if we assume that the original graph has a perfect matching. Since we use as a subroutine an algorithm that checks for the existence of a perfect matching, this is not a real limitation. Moreover, in the next section we will see an efficient implementation of the minimum cost perfect matching algorithm and a proof of termination that does not rely on the assumption that the original graph has a perfect matching.

PROPOSITION 3.14   *Algorithm 3.3 terminates if G has a perfect matching and c is a rational vector.*

Proof   Let us first assume that c_e is an even integer for all $e \in E$. We claim that, for all $uv \in$, the reduced cost $\bar{c}_{uv} := c_{uv} - \bar{y}_u - \bar{y}_v$ is an integer at all times during the execution of the algorithm. This is clearly true at the beginning since $\bar{y}_u$ is set to half the cost of the minimum cost edge – an integer by our assumption. Let us show that integrality is maintained. In Step 7 of the algorithm, we choose a deficient set $S \subseteq U$. We then determine ϵ according to (3.31), add it to $\bar{y}_v$ for all $v \in S$, and subtract it from $\bar{y}_v$ for $v \in N_H(S)$. Inductively, we know that $\bar{c}_e$ is an integer for all $e \in E$, and hence ϵ is a positive integer. Thus, $\bar{y}_v$ changes by an integer amount in the current iteration of the algorithm.

Denote by ℓ the value $\sum_{v \in V} \bar{y}_v$ after the initialization Step 1. By hypothesis, G has a perfect matching M. Then

$$\sum_{v \in V} \bar{y}_v = \sum_{uv \in M} (\bar{y}_u + \bar{y}_v) \leq \sum_{uv \in M} c_{uv} = c(M),$$

where the first equality follows from the fact that every vertex is the endpoint of exactly one edge of M, and the inequality follows from the fact that throughout the algorithm $\bar{y}_u + \bar{y}_v \leq c_{uv}$. Since ϵ is a positive integer in every iteration, and since $|S| \geq |N_G(S)| + 1$ for deficient sets S, $\sum_{u \in V} \bar{y}_u$ must increase by at least 1 in each iteration of the algorithm. It follows that the number of iterations is bounded by $c(M) - \ell$. In particular, the algorithm terminates.

Suppose now that c is an arbitrary rational vector. For some large enough integer ω, ωc_e is in an even integer for all $e \in E$. It can be readily checked that $\epsilon \geq \frac{1}{\omega}$ and that the number of iterations is bounded by $\omega(c(M) - \ell)$. In particular, the algorithm terminates.   □

We conclude with a brief discussion of the iteration bound obtained in Proposition 3.14. The earmark of an *efficient* algorithm is that its running time – the number of elementary steps that it performs – is bounded by a polynomial in the *size* of the input (see Appendix A). Roughly speaking, the size of the input is the number of bits needed to store it. In the case of the matching problem, this is a linear function of $|V|$, $|E|$, and $\sum_e \log(c_e + 1)$. Proposition 3.14 bounds the number of iterations by a function of $c(E)$, and this can be an *exponential* function of $\sum_e \log(c_e + 1)$ in the worst case. Hence, the analysis provided does not show that the minimum cost perfect matching algorithm is an efficient algorithm. Efficient implementations of this algorithm are however known, and we provide some details in the next section.

Exercises

1 C&O transportation company has three trucks with different capabilities (affectionately called Rooter, Scooter, and Tooter) and three moving jobs in different cities (Kitchener, London, and Missisauga). The following table shows the profit gained (in $1000s) if a given truck is used for the job in a given city. Your task is to find an allocation of the trucks to the three different cities to maximize profit.

	K	L	M
R	12	9	9
S	12	4	3
T	12	3	4

(a) Formulate this as an IP problem.
(b) Formulate it as a maximum matching problem.
(c) Solve the maximum matching problem using the minimum matching algorithm.

2 Consider a bipartite graph with vertices $U = \{a, b, c, d\}$ and $W = \{g, h, i, j\}$ and all edges between U and W. You also have a weight for each edge of G as indicated in the following table:

	g	h	i	j
a	2	7	1	2
b	3	4	3	2
c	6	5	5	5
d	2	6	2	3

For instance, edge ag has weight 2. Find a maximum weight perfect matching using the matching algorithm.

3 Let $G = (V, E)$ be a graph with edge weights c. Consider the following three problems:
(P1) find a perfect matching of minimum weight,
(P2) find a perfect matching of maximum weight,
(P3) find a matching of maximum weight.

(a) Show that if you have an algorithm to solve (P1), then you can use it to solve (P2).
(b) Show that if you have an algorithm to solve (P2), then you can use it to solve (P1).
Suppose that in addition all the weights are nonnegative.
(c) Show that if you have an algorithm to solve (P2), then you can use it to solve (P3).
(d) Show that if you have an algorithm to solve (P3), then you can use it to solve (P2).

4 Let $G = (V, E)$ be a bipartite graph. Consider the following LP (P):

$$\max \quad 0$$

subject to

$$\sum (x_e : e \in \delta(i)) = 1 \qquad (i \in V)$$

$$x_e \geq 0 \qquad\qquad (e \in E).$$

(a) Find the dual (D) of (P).
(b) Show that if G has a deficient set, then (D) is unbounded.
   HINT: Look at how the variables y are updated in the matching algorithm.
(c) Deduce from (b) and Theorem 3.9 that if G has a deficient set, then it has no perfect matching.

5 Let $G = (V, E)$ be a graph. A *vertex-cover* of G is a subset S of V with the property that for every edge ij, we have that either $i \in S$ or $j \in S$ (or possibly both).
(a) Show that for every matching M and every vertex-cover S, we have $|S| \geq |M|$. In particular

$$\max\{|M| : M \text{ is a matching of } G\} \leq \min\{|S| : S \text{ is vertex-cover of } G\}. \qquad (\star)$$

   HINT: How many vertices do you need to cover just the edges of M?
(b) Show that there are graphs for which the inequality $(\star)$ is strict.

6 Suppose $G = (V, E)$ is a bipartite graph with partition U, W. Suppose also that $|U| = |W|$. Let H be the complete bipartite graph with partition U and W (the edges of H are *all* pairs ij where $i \in U$ and $j \in W$). Assign weights to the edges of H as follows: for every edge ij

$$w_{ij} = \begin{cases} 1 & \text{if } ij \text{ is an edge of } G \\ 0 & \text{otherwise.} \end{cases}$$

Recall that y is feasible if for every edge ij of H, $y_i + y_j \leq w_{ij}$.
(a) Show that there exists a feasible $0, 1$ vector y (i.e. all entries of y are 0 or 1).
(b) Show that if y is integer at the beginning of an iteration of the matching algorithm, then y will be integer at the beginning of the next iteration.
(c) Show that there exists a feasible integer vector y and a matching M of G such that

$$\sum_{i \in V} y_i = |M|.$$

   HINT: Look at what you have when the matching algorithm terminates.
(d) Show that there exists a feasible $0, 1$ vector y and a matching M of G such that

$$\sum_{i \in V} y_i \leq |M|.$$

   HINT: Add (resp. substract) the same value α to all $y_i \in W$ (resp. $y_i \in U$).
(e) Show that there exists a vertex-cover S of G and a matching M of G such that $|M| = |S|$. In particular, $(\star)$ holds with equality for bipartite graphs.

3.2.6 Finding perfect matchings in bipartite graphs*

Hall's theorem (Theorem 3.12) states that in a bipartite graph G with bipartition U, W, where $|U| = |W|$, there is an efficient algorithm that finds a perfect matching in G if one exists, and otherwise returns a deficient set $S \subseteq U$; i.e. a set S that has fewer than $|S|$ neighbors. In this section, we show how this can be accomplished.

We start by introducing a few more matching related terms and definitions. Consider a (possibly empty) matching $M \subseteq E$; we say that a path P in G is M-*alternating* if its edges alternate between matching and non-matching edges, or equivalently if $P \setminus M$ is a matching. Examples of alternating paths are given in Figure 3.6. Given a matching M, a vertex v is M-*covered* if M has an edge that is incident to v, otherwise v is M-*exposed*. In Figure 3.7(i) vertices $v_1, v_2, v_4, v_6, v_7, v_8$ are M-covered and vertices v_3, v_5 are M-exposed. In Figure 3.7(ii), all vertices are M-covered, or equivalently M is a perfect matching. An alternating path P is M-*augmenting* if its endpoints are both M-exposed. In Figure 3.7(i), the path $P_1 = v_5v_1, v_1v_6, v_6v_3$ is M-augmenting as both v_5, v_3 are M-exposed, but the path $P_2 = v_1v_6, v_6v_2, v_2v_7, v_7v_4, v_4v_8, v_8v_3$ is M-alternating but not M-augmenting as v_1 is M-covered. Observe that the matching M' in Figure 3.7(ii) is obtained from the matching M in Figure 3.7(i), by removing $v_1, v_6 \in P_1$ from M and adding edges $v_5v_1, v_6v_3 \in P_1$. In fact, given an M-augmenting path, we can always construct a new matching M', where $|M'| = |M| + 1$. We formalize this operation next. For two sets A and B, we let $A \triangle B$ denote the set of those elements in $A \cup B$ that are not

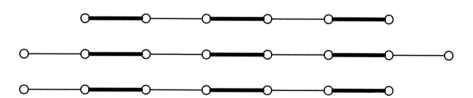

Figure 3.6 M-alternating paths (edges in M are bold).

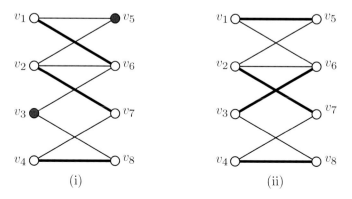

Figure 3.7 M-exposed and M-covered vertices (edges in M are bold).

in both A and B; i.e. $A \triangle B = (A \setminus B) \cup (B \setminus A)$. A matching is a *maximum matching* if there is no matching with more edges.

We leave the following remark as an easy exercise:

Remark 3.15 *Let M be matching of a graph G and let P be an M-augmenting path. Then $M' := M \triangle P$ is a matching and $|M'| = |M| + 1$. In particular, M is not a maximum matching.*

For instance, in Figure 3.7(i) we had $M = \{v_1 v_6, v_2 v_7, v_4 v_8\}$, an augmenting path $P_1 = v_5 v_1, v_1 v_6, v_6 v_3$ and a matching M' in (ii) where

$$M' = \{v_1 v_6, v_2 v_7, v_4 v_8\} \triangle \{v_5 v_1, v_1 v_6, v_6 v_3\} = \{v_5 v_1, v_2 v_7, v_6 v_3, v_4 v_8\}.$$

In fact, the converse of Remark 3.15 holds (but we will not need it), namely that, if a matching is not maximum, then there must exist an augmenting path. Thus, a plausible strategy for finding a perfect matching is as follows: start with the empty matching $M = \emptyset$. As long as there exists an M-augmenting path P, replace M by $M \triangle P$. If the final matching is not perfect, then no perfect matching exists. Two problems remain however: how do we find an augmenting path, and if no augmenting path exist, how do we find a deficient set? The key to addressing both of these problems is the concept of M-alternating trees, to be defined next.

Let us start with a few definitions. A *cycle* is a sequence of edges

$$C = v_1 v_2, v_2 v_3, \dots, v_{k-1} v_k,$$

where $k \geq 4$ and where $v_1, \dots, v_{k-1}$ are distinct vertices and $v_k = v_1$. In other words, a cycle is what we obtain if we identify the endpoints of a path. For instance, in Figure 3.8(i) the bold edges form a cycle of the graph. We say that a graph is *connected* if there exists a path between every pair of distinct vertices. For instance, in Figure 3.8(ii) the graph is not connected as there is no path with endpoints u and v. A graph is a *tree* if it is connected but has no cycle. In Figure 3.8, the graph in (iii) is a tree, but (i) is not a tree as it has a cycle, and (ii) is not a tree as it is not connected.

If G is a tree, then for any pair of distinct vertices u and v there exists at least one path between u and v (as G is connected). Moreover, it is easy to check that if there existed two distinct paths between u and v, we would have a cycle, thus:

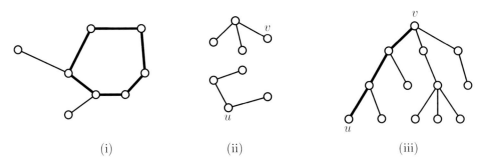

(i)   (ii)   (iii)

Figure 3.8 Cycles and trees.

Remark 3.16 *In a tree, there exists a unique path between every pair of distinct ver-tices.*

Given a tree T with distinct vertices u, v, we denote by T_{uv} the unique path in Remark 3.16. For instance, in Figure 3.8(iii) the path T_{uv} is denoted in bold. Given a graph G, we say that T is a *tree of G* if T is a tree and T is obtained from G by removing edges and removing vertices. For instance, in Figure 3.9 the graph in (ii) is a tree of (i). Given a tree $T = (V, E)$, we call a vertex $u \in V$ a *leaf* if it is incident to exactly one edge. For instance, u is a leaf of the tree in Figure 3.9(ii). We are now ready for our key definition. Let $G = (V, E)$ be a graph with a matching M and let $r \in V$. We say that a tree T of G is an *M-alternating tree rooted at r* if:

- r is a vertex of T that is M-exposed in G,
- all vertices of T distinct from r are M-covered,
- for every vertex $u \neq r$ of T, the unique ru-path T_{ru} is M-alternating.

The graph in Figure 3.9(ii) is an M-alternating tree rooted at r. For instance, for vertex u the path T_{ru} indicated by dashed lines is an M-alternating path. Given an M-alternating tree T rooted at r, we partition the vertices of T into sets $A(T)$ and $B(T)$ where $r \in B(T)$ and for every vertex $u \neq r$ of T, $u \in B(T)$ if and only if the path T_{ru} has an even number of edges. In Figure 3.9(ii), vertices $B(T)$ correspond to squares and vertices $A(T)$ to circles.

We can now state the algorithm that proves Theorem 3.12. (See Algorithm 3.4 on next page). We denote by $V(T)$ the set of vertices of T and by $E(T)$ the set of edges of T. Note, in this algorithm we will view paths as sets of edges. At any time during its execution, the algorithm will maintain a matching M, and an M-alternating tree T. Initially, the matching M is set to the empty set.

We illustrate the algorithm in Figure 3.10. Suppose that $U = \{b, d, g\}$ and that $W = \{a, c, f\}$. Suppose that after a number of iterations we just increased the size of the matching in Step 7 to obtain the matching $M = \{ab, fd\}$; see Figure 3.10 (i). As M is not a

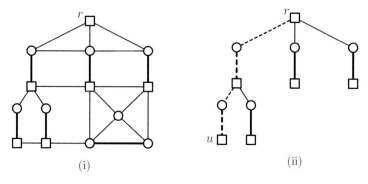

(i)   (ii)

Figure 3.9 *M*-alternating tree (edges in *M* are bold).

Algorithm 3.4 Perfect matching

Input: Bipartite graph $H = (V, E)$ with bipartition U, W where $|U| = |W| \geq 1$.
Output: A perfect matching M, or a deficient set $B \subseteq U$.
  1: $M := \emptyset$
  2: $T := (\{r\}, \emptyset)$ where $r \in U$ is any M-exposed vertex
  3: **loop**
  4:    **if** $\exists$ edge uv where $u \in B(T)$ and $v \notin V(T)$ **then**
  5:      **if** v is M-exposed **then**
  6:        $P := T_{ru} \cup \{uv\}$
  7:        $M := M \triangle P$
  8:        **if** M is a perfect matching **then stop end if**
  9:        $T := (\{r\}, \emptyset)$ where $r \in U$ is any M-exposed vertex
10:      **else**
11:        Let $w \in V$ where $vw \in M$
12:        $T := (V(T) \cup \{v, w\}, E(T) \cup \{uv, vw\})$
13:      **end if**
14:    **else**
15:      **stop** $B(T) \subseteq U$ is a deficient set
16:    **end if**
17: **end loop**

---

perfect matching, and since $|U| = |W|$, there must exist a vertex in U that is M-exposed. In our case, $g \in U$ is M-exposed and we define our tree T to consists of vertex g (with no edges), i.e. $T = (\{g\}, \emptyset)$. It is trivially an M-alternating tree rooted at g. At the next iteration, we look in Step 4 for an edge incident to a vertex of $B(T) = \{g\}$ where its other endpoint is not in T. For instance, gf is such an edge. We then observe that f is M-covered because of edge $fd \in M$. We now increase the tree by adding edges gf and fd, i.e. $T = (\{g, f, d\}, \{gf, fd\})$ (see Figure 3.10 (ii)). It remains an M-alternating tree rooted at g. At the next iteration, we look for an edge incident to $B(T) = \{g, d\}$ where its other endpoint is not in T. For instance, ga is such an edge. We then observe that a is M-covered because of edge $ab \in M$. We now increase the tree by adding edges ga and ab, i.e. $T = (\{g, f, d, a, b\}, \{gf, fd, ga, ab\})$ (see Figure 3.10 (iii)). It remains an M-alternating tree rooted at g. At the next iteration, we look for an edge incident to $B(T) = \{g, h, t\}$, where its other endpoint is not in T. Edge dc is such an edge, (see Figure 3.10 (iv)). We now observe that c is M-exposed. Since $T_{gd} = \{gf, fd\}$ is an M-alternating path and since $dc \notin M$, it follows that $P := T_{gd} \cup \{dc\} = \{gf, fd, dc\}$ is an M-augmenting path (indicated by dashed lines in Figure 3.10 (iv)). We replace the current matching M by $M \triangle P = \{gf, ab, dc\}$. As this matching is perfect, we stop.

Consider another example in Figure 3.11(i). Suppose that after a number of iteration we have matching $M = \{ac, fh, bd, gi\}$ and the M-alternating tree $T = (\{j, f, h, i, g\}, \{jf, fh, ji, ig\})$ rooted at j (see Figure 3.11 (ii)). We look for an edge incident to $B(T) = \{j, h, g\}$

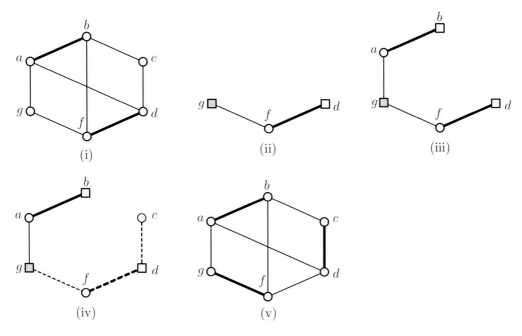

Figure 3.10 An example with perfect matching (edges in M are bold).

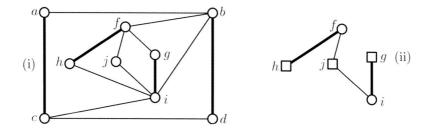

Figure 3.11 An example without perfect matching (edges in M are bold).

where its other endpoint is not in T. Since there is no such edge, the algorithm claims that $B(T)$ is a deficient set, and we can stop as there is no perfect matching.

It remains to prove that the algorithm is correct and that it runs in polynomial-time. At any time during its execution, the algorithm will maintain a matching M, and an M-alternating tree T. Observe that:

(P1) $|M|$ never decreases.
(P2) If $|M|$ does not increase in an iteration, then either $|V(T)|$ increases or the algorithm terminates with a deficient set.

PROPOSITION 3.17   *The perfect matching algorithm (Algorithm 3.4) terminates in at most $\frac{1}{2}n^2$ iterations, where n is the number of vertices.*

Proof Let us group those iterations during which the matching M has size i into the *ith phase*. Clearly, by property (P1), the iterations of Phase i are consecutive. Moreover, every iteration within a phase grows the alternating tree by (P2), and therefore, a phase can have at most n iterations (trees have at most $n - 1$ vertices). Finally, there are clearly no more than $n/2$ phases (matchings have at most $n/2$ edges) and this completes the argument.   □

Let us show that M remains a matching at each step of the algorithm. Initially, $M = \emptyset$, which is trivially a matching. The set M is only updated in Step 7 of the algorithm. Because of Remark 3.15, we only need to verify that $P = T_{ru} \cup \{uv\}$ is an M-augmenting path in Step 6. Since T is an M-alternating tree, T_{ru} is an M-alternating path and r is M-exposed. Moreover, since v is M-exposed, $T_{ru} \cup \{uv\}$ is an M-augmenting path as required.

Let us show that T remains an M-alternating tree at each step of the algorithm. In Step 2, there exists a vertex $r \in U$ that is M-exposed as $|U| \geq 1$. In Step 9, there exists a vertex $r \in U$ that is M-exposed as M is not a perfect matching and as $|U| = |W|$. Trivially, trees consisting of a single M-exposed vertex are M-alternating. Thus, T defined in Step 2 and Step 9 is M-alternating. Suppose that T is an M-alternating tree rooted at r. We need to show that $T' = (V(T) \cup \{u, v\}, E(T) \cup \{uv, vw\})$ (Step 12) is an M-alternating tree rooted at r. Because of Step 4, $v \notin V(T)$. We claim $w \notin V(T)$. Clearly, $w \neq r$ as w is M-covered. Moreover, $w \notin V(T) \backslash \{r\}$ as all vertices of T distinct from r are M-covered by edges of $M \cap E(T)$ but $vw \notin E(T)$. Thus, T' has no cycle. Since T' is clearly connected, T' is a tree. Finally, note that T'_{rv} and T'_{rw} are M-alternating. Thus, T' is an M-alternating tree rooted at r as required.

It only remains to show in Step 15 that $B(T)$ is indeed a deficient set contained in U. Thus, the next result will complete the proof of correctness:

PROPOSITION 3.18   *Let $G = (V, E)$ be a bipartite graph with partition U, W. Suppose that T is an M-alternating tree rooted at $r \in U$ and that every neighbor of a vertex of $B(T)$ is a vertex of T. Then $B(T)$ is a deficient set contained in U.*

Proof First observe that $B(T) \subseteq U$, for if $u \in B(T)$, then the path T_{ru} has an even number of edges. As $r \in U$ and since the graph is bipartite, we must have $u \in U$ as well. Let v be a neighbor of $B(T)$, i.e. $uv \in E$. By hypothesis, $v \in V(T)$. Suppose for a contradiction that $v \in B(T)$. Then T_{ru}, T_{rv} are both paths with an even number of edges. It can be readily checked that $C = T_{ru} \triangle T_{rv} \cup \{uv\}$ is a cycle that contains an odd number of edges. But a bipartite graph cannot have such a cycle (we leave this as an exercise). It follows that $N(B(T)) \subseteq A(T)$. Finally, observe that $|B(T)| = |A(T)| + 1$, as every vertex of $B(T)$ distinct from the root r is paired by the matching with exactly one vertex of $A(T)$. Hence, $B(T)$ is a deficient set.   □

The attentive reader may have noticed that the condition that G is bipartite is only used in the last proposition, i.e. to guarantee a correct outcome in Step 15.

An efficient implementation of the minimum cost perfect matching algorithm

The Perfect matching Algorithm 3.4 could now simply be used as a "black box" in Step 7 of the minimum cost perfect matching Algorithm 3.3. We will see, however, that a more careful implementation gives a polynomial-time algorithm.

By the *subroutine*, we mean the algorithm that takes as input a graph H, a matching M, and an M-alternating tree rooted at r, and applies Step 4 through Step 16 of Algorithm 3.4. At the end of the subroutine, we have either:

(a) increased $|M|$, or
(b) increased $|V(T)|$, or
(c) found a deficient set $B(T)$ of H.

Moreover, M remains a matching and T remains an M-alternating tree rooted at r. Combining our subroutine and the minimum cost perfect algorithm (Algorithm 3.3), we obtain a fast algorithm for finding a minimum cost perfect matching in a bipartite graph described in Algorithm 3.5.

Algorithm 3.5 Minimum cost perfect matching in bipartite graphs (fast version)

Input: Bipartite graph $H = (V, E)$ with bipartition U, W where $|U| = |W| \geq 1$.
Output: A minimum cost perfect matching M or a deficient set S.
  1: $M := \emptyset$
  2: $T := (\{r\}, \emptyset)$ where $r \in U$ is any M-exposed vertex
  3: $\bar{y}_v := \frac{1}{2} \min\{c_e : e \in E\}$, for all $v \in V$
  4: **loop**
  5:    Construct graph H with vertices V and edges $\{uv \in E : c_{uv} = \bar{y}_u + \bar{y}_v\}$
  6:    Invoke the subroutine with H, M, and T.
  7:    **if** outcome (a) of subroutine occurs **then**
  8:      **if** M is a perfect matching of H **then**
  9:        **stop** (M is a minimum cost perfect matching of G)
10:      **end if**
11:    **else if** outcome (c) of subroutine occurs **then**
12:      Let $S := B(T)$
13:      **if** all edges of G with an endpoint in S have an endpoint in $N_H(S)$ **then**
14:        **stop** (G has no perfect matching)
15:      **end if**
16:      $\epsilon = \min\{c_{uv} - \bar{y}_u - \bar{y}_v : u \in S, v \notin V(T)\}$
17:      $\bar{y}_v := \begin{cases} \bar{y}_v + \epsilon & \text{for } v \in S \\ \bar{y}_v - \epsilon & \text{for } v \in N_H(S) \\ \bar{y}_v & \text{otherwise} \end{cases}$
18:    **end if**
19: **end loop**

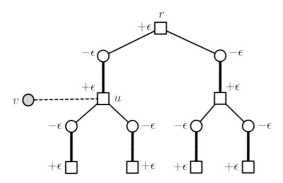

Figure 3.12 *M*-alternating tree (edges in *M* are bold), dashed edge *u, v* is nontree edge.

It remains to prove that the algorithm is correct and that it runs in polynomial-time. At any time during its execution, the algorithm will maintain a matching *M* and an *M*-alternating tree *T*. We will show that the algorithm has the following important properties:

(P1) $|M|$ never decreases.
(P2) If neither $|M|$ nor $|V(T)|$ increase, then either the algorithm terminates with a deficient set or one of $|M|$ or $|V(T)|$ will increase in the next iteration.

Proceeding similarly as in the proof of Proposition 3.17, we obtain:

PROPOSITION 3.19   *The minimum cost perfect matching algorithm (Algorithm 3.5) terminates in at most n^2 iterations where n is the number of vertices.*

Property (P1) trivially follows from the description of the algorithm. Let us prove that property (P2) is preserved. Suppose that during one of the iterations neither $|M|$ nor $|V(T)|$ increases. We must have outcome (c) for the subroutine, i.e. $S := B(T)$ is a deficient set of *H*. If *S* is a deficient set of *G*, then we stop as *G* has no perfect matching. Otherwise, by Proposition 3.18 there exists an edge *uv* with $u \in B(T)$ and $v \notin V(T)$. We illustrate this in Figure 3.12. Because of the choice of ϵ in Step 16, one of these edges, say *uv*, will have zero reduced cost in the next iteration. Then in the next iteration, *uv* is an edge of *H* and *uv* can be added to the tree in the subroutine. If *v* is *M*-exposed, the subroutine will find a larger matching. Otherwise, there exists an edge $vw \in M$ and the subroutine will find a larger *M*-alternating tree. Thus, property (P2) holds.

3.3   Further reading and notes

The shortest path algorithm presented in Section 3.1 is equivalent to the well-known algorithm by Dijkstra [22]. This algorithm has several very efficient implementations using sophisticated heap data structures, e.g., see the work by Fredman and Tarjan [26].
    The algorithm presented for the minimum cost perfect matching problem in bipartite graphs is sometimes known as the Hungarian Algorithm. Efficient algorithms for this

problem exist even in general graphs. We refer the reader to Cook et al [16] for more information.

The algorithms presented in Sections 3.1 and 3.2.3 are examples of so-called *primal–dual* algorithms. The primal–dual technique has more recently also been used successfully for literally hundreds of NP-hard optimization problems. A good introduction to this topic which is beyond the scope of these notes can be found in the books of Vazirani [68], and Williamson and Shmoys [69].

4 Duality theory

In Chapter 3, we introduced the notion of a dual of a linear program, and demonstrated how this can be used to design algorithms to solve optimization problems, such as the shortest path problem and the minimum-cost perfect matching problem in bipartite graphs. In this chapter, we shall generalize these results, and develop a general theoretical framework.

4.1 Weak duality

In Section 3.1.2, we defined the dual of a linear program

$$\min\{c^\top x : Ax \geq b, x \geq \mathbb{0}\} \tag{4.1}$$

to be the linear program

$$\max\{b^\top y : A^\top y \leq c, y \geq \mathbb{0}\}. \tag{4.2}$$

Similarly, in Section 3.2.2 we defined the dual of a linear program

$$\min\{c^\top x : Ax = b, x \geq \mathbb{0}\} \tag{4.3}$$

to be the linear program

$$\max\{b^\top y : A^\top y \leq c, y \text{ free}\}. \tag{4.4}$$

Our goal in this section is to generalize these definitions, as well as the corresponding weak duality theorems (Theorems 3.2 and 3.9).

Consider the table on the next page describing a pair of linear programs, the notation Ax ? b (resp. $A^\top y$? c) indicates that constraints of ($\mathrm{P_{max}}$) and ($\mathrm{P_{min}}$) are arbitrary linear constraints, i.e. the relation between the left-hand side and right-hand side of the constraint is either $\leq$, $=$, or $\geq$. The notation x ? $\mathbb{0}$ and y ? $\mathbb{0}$ indicates that the variables of ($\mathrm{P_{max}}$) and ($\mathrm{P_{min}}$) are either nonnegative, free, or nonpositive. We pair constraint i of ($\mathrm{P_{max}}$), i.e. $\mathrm{row}_i(A)^\top x$? b_i, with the variable y_i of ($\mathrm{P_{min}}$) and we pair constraint j of ($\mathrm{P_{min}}$), i.e. $\mathrm{col}_j(A)y$? c_j, with the variable x_j of ($\mathrm{P_{max}}$). The table indicates the relation between a constraint of ($\mathrm{P_{max}}$) and the corresponding variable of ($\mathrm{P_{min}}$). For instance, if $\mathrm{row}_i(A)^\top x \leq b_i$ in ($\mathrm{P_{max}}$), then we have $y_i \geq 0$ in ($\mathrm{P_{min}}$). The table also indicates the

Table 4.1 Primal–dual pairs

(P_{max})				(P_{min})	
		$\leq$ constraint	≥ 0 variable		
max	$c^\top x$	$=$ constraint	free variable	min	$b^\top y$
subject to		$\geq$ constraint	≤ 0 variable	subject to	
	$Ax\ ?\ b$	≥ 0 variable	$\geq$ constraint		$A^\top y\ ?\ c$
	$x\ ?\ 0$	free variable	$=$ constraint		$y\ ?\ 0$
		≤ 0 variable	$\leq$ constraint		

relation between a constraint of (P_{min}) and the corresponding variable of (P_{max}). For instance, if $\text{col}_j(A)y \leq c_j$ in (P_{min}), then we have $x_j \leq 0$ in (P_{max}).

Consider a pair of linear programs (P_{max}) and (P_{min}) satisfying the relations given in Table 4.1. We then define (P_{min}) to be the *dual* of (P_{max}) and we define (P_{max}) to be the *dual* of (P_{min}). We call the pair (P_{max}) and (P_{min}) a *primal–dual pair*. The reader should verify that these definitions imply in particular that the dual of the dual of a linear program (P) is the original linear program (P).

Example 15 Let us apply these definitions to try to find the dual of the linear program (4.1). This linear program can be rewritten as

$$\min\{\hat{b}^\top \hat{y} : \hat{A}^\top \hat{y} \geq \hat{c}, \hat{y} \geq 0\}, \tag{P}$$

where $\hat{b} = c$, $\hat{A} = A^\top$, $\hat{c} = b$ and $\hat{y} = x$. Since (P) is a minimization problem, it plays the role of (P_{min}) in Table 4.1. It follows that the dual (D) of (P) is of the form (P_{max}) in Table 4.1

$$\max\{\hat{c}^\top \hat{x} : \hat{A}\hat{x}\ ?\ \hat{b}, \hat{x}\ ?\ 0\}. \tag{D}$$

Since:

- we have $\hat{A}^\top \hat{y} \geq \hat{c}$ in (P), we have $\hat{x} \geq 0$ in (D)
- $\hat{y} \geq 0$ in (P), we have $\hat{A}\hat{x} \leq \hat{b}$ in (D).

Hence, the dual (D) of (P) is given by

$$\max\{\hat{c}^\top \hat{x} : \hat{A}\hat{x} \leq \hat{b}, \hat{x} \geq 0\},$$

which is equal to (4.2) if we define y as being equal to $\hat{x}$. Similarly, (4.4) is the dual of (4.3). Hence, our new definition of dual is consistent with the definitions given in Sections 3.1 and 3.2.

Example 16 Consider the following linear program:

$$\text{max} \qquad (12, 26, 20)x$$

subject to

$$\begin{pmatrix} 1 & 2 & 1 \\ 4 & 6 & 5 \\ 2 & -1 & -3 \end{pmatrix} x \begin{array}{c} \geq \\ \leq \\ = \end{array} \begin{pmatrix} -2 \\ 2 \\ 13 \end{pmatrix} \begin{array}{c} (1) \\ (2) \\ (3) \end{array} \qquad (P)$$

$$x_1 \geq 0, \ x_2 \text{ free}, \ x_3 \geq 0.$$

Suppose we wish to find the dual of (P). Since (P) is a maximization problem, it plays the role of ($\mathbb{P}_{max}$) in Table 4.1. It follows that the dual (D) of (P) is of the form ($\mathbb{P}_{min}$) in Table 4.1

$$\text{min} \qquad (-2, 2, 13)y$$

subject to

$$\begin{pmatrix} 1 & 4 & 2 \\ 2 & 6 & -1 \\ 1 & 5 & -3 \end{pmatrix} y \ ? \begin{pmatrix} 12 \\ 26 \\ 20 \end{pmatrix} \begin{array}{c} (1) \\ (2) \\ (3) \end{array} \qquad (D)$$

$$y_1 \ ? \ 0, \ y_2 \ ? \ 0, \ y_3 \ ? \ 0.$$

Since:

- constraint (1) of (P) is of type "$\geq$", we have $y_1 \leq 0$ in (D),
- constraint (2) of (P) is of type "$\leq$", we have $y_2 \geq 0$ in (D),
- constraint (3) of (P) is of type "$=$", we have y_3 is free in (D),
- $x_1 \geq 0$ in (P), we have that constraint (1) of (D) is of type "$\geq$",
- x_2 is free in (P), we have that constraint (2) of (D) is of type "$=$",
- $x_3 \geq 0$ in (P), we have that constraint (3) of (D) is of type "$\geq$".

Hence, the dual (D) of (P) is given by

$$\text{min} \qquad (-2, 2, 13)y$$

subject to

$$\begin{pmatrix} 1 & 4 & 2 \\ 2 & 6 & -1 \\ 1 & 5 & -3 \end{pmatrix} y \begin{array}{c} \geq \\ = \\ \geq \end{array} \begin{pmatrix} 12 \\ 26 \\ 20 \end{pmatrix} \begin{array}{c} (1) \\ (2) \\ (3) \end{array} \qquad (D)$$

$$y_1 \leq 0, \ y_2 \geq 0, \ y_3 \text{ free}.$$

We proved in Theorem 3.2 that if $\bar{x}$ is a feasible solution to (4.1) and $\bar{y}$ is a feasible solution to (4.2), then $c^\top \bar{x} \geq b^\top \bar{y}$. Similarly, we proved in Theorem 3.9 that if $\bar{x}$ is a feasible solution to (4.3) and $\bar{y}$ is a feasible solution to (4.4), then $c^\top \bar{x} \geq b^\top \bar{y}$. The following is a common generalization to both of these results.

THEOREM 4.1 (Weak duality theorem). *Let (P) and (D) be a primal–dual pair where (P) is a maximization problem and (D) a minimization problem. Let $\bar{x}$ and $\bar{y}$ be feasible solutions for (P), and (D), respectively:*

1. *Then $c^{\top}\bar{x} \le b^{\top}\bar{y}$.*
2. *If $c^{\top}\bar{x} = b^{\top}\bar{y}$, then $\bar{x}$ is an optimal solution to (P) and $\bar{y}$ is an optimal solution to (D).*

Observe, that (2) in the previous theorem follows immediately from (1). This is because as $\bar{y}$ is feasible, (1) implies that for every feasible solution x of (P), $c^{\top}x \le b^{\top}\bar{y}$, i.e. $b^{\top}\bar{y}$ is an upper bound of (P). But then $\bar{x}$ is an optimal solution to (P) since its value is equal to its upper bound. The argument to prove that $\bar{y}$ is an optimal solution to (D) is similar.

Example 16, continued. Consider the primal–dual pair given by (P) and (D). Let

$$\bar{x} = (5, -3, 0)^{\top} \qquad \text{and} \qquad \bar{y} = (0, 4, -2)^{\top}.$$

It can be readily checked that $\bar{x}$ is feasible for (P) and $\bar{y}$ is feasible for (D). Moreover, $(-2, 2, 13)\bar{y} = (12, 26, 20)\bar{x} = -18$. It follows from Theorem 4.1(2) that $\bar{x}$ is an optimal solution to (P) and that $\bar{y}$ is an optimal solution to (D).

Proof of Theorem 4.1   Let (P) be an arbitrary linear program where the goal is to maximize the objective function. Then (P) can be expressed as

$$
\begin{aligned}
\max \quad & c^{\top}x \\
\text{subject to} \quad & \\
& \text{row}_i(A)x \le b_i && (i \in R_1) \\
& \text{row}_i(A)x \ge b_i && (i \in R_2) \\
& \text{row}_i(A)x = b_i && (i \in R_3) \\
& x_j \ge 0 && (j \in C_1) \\
& x_j \le 0 && (j \in C_2) \\
& x_j \text{ free} && (j \in C_3)
\end{aligned}
\tag{4.5}
$$

for some partition R_1, R_2, R_3 of the row indices, and some partition C_1, C_2, C_3 of the column indices. Its dual (D) is given by (see Table 4.1)

$$
\begin{aligned}
\min \quad & b^{\top}y \\
\text{subject to} \quad & \\
& \text{col}_j(A)^{\top}y \ge c_j && (j \in C_1) \\
& \text{col}_j(A)^{\top}y \le c_j && (j \in C_2) \\
& \text{col}_j(A)^{\top}y = c_j && (j \in C_3) \\
& y_i \ge 0 && (i \in R_1) \\
& y_i \le 0 && (i \in R_2) \\
& y_i \text{ free} && (i \in R_3).
\end{aligned}
\tag{4.6}
$$

After adding slack variables s, we can rewrite (4.5) as

$$\begin{aligned}
\max \quad & c^\top x \\
\text{subject to} \quad & \\
& Ax + s = b \\
& s_i \geq 0 && (i \in R_1) \\
& s_i \leq 0 && (i \in R_2) \\
& s_i = 0 && (i \in R_3) \\
& x_j \geq 0 && (j \in C_1) \\
& x_j \leq 0 && (j \in C_2) \\
& x_j \text{ free} && (j \in C_3).
\end{aligned} \tag{4.7}$$

After adding slack variables w, we can rewrite (4.6) as

$$\begin{aligned}
\min \quad & b^\top y \\
\text{subject to} \quad & \\
& A^\top y + w = c \\
& w_j \leq 0 && (j \in C_1) \\
& w_j \geq 0 && (j \in C_2) \\
& w_j = 0 && (j \in C_3) \\
& y_i \geq 0 && (i \in R_1) \\
& y_i \leq 0 && (i \in R_2) \\
& y_i \text{ free} && (i \in R_3).
\end{aligned} \tag{4.8}$$

Let $\bar{x}$ be a feasible solution to (4.5) and let $\bar{y}$ be a feasible solution to (4.6). Then for $\bar{s} := b - A\bar{x}$, the vectors $\bar{x}, \bar{s}$ are a solution to (4.7). Moreover, for $\bar{w} := c - A^\top \bar{y}$, the vectors $\bar{y}, \bar{w}$ are a solution to (4.8). Since $A\bar{x} + \bar{s} = b$, we have

$$\bar{y}^\top b = \bar{y}^\top (A\bar{x} + \bar{s}) = (\bar{y}^\top A)\bar{x} + \bar{y}^\top \bar{s} \overset{\star}{=} (c - \bar{w})^\top \bar{x} + \bar{y}^\top \bar{s} = c^\top \bar{x} - \bar{w}^\top \bar{x} + \bar{y}^\top \bar{s},$$

where equality $(\star)$ follows from the fact that $A^\top \bar{y} + \bar{w} = c$ or equivalently that $\bar{y}^\top A = (c - \bar{w})^\top$. Hence, to prove that $b^\top \bar{y} \geq c^\top \bar{x}$ it suffices to show that

$$\bar{w}^\top \bar{x} = \sum_{\substack{j \in C_1 \\ }} \underbrace{\bar{w}_j}_{\leq 0} \underbrace{\bar{x}_j}_{\geq 0} + \sum_{\substack{j \in C_2 \\ }} \underbrace{\bar{w}_j}_{\geq 0} \underbrace{\bar{x}_j}_{\leq 0} + \sum_{\substack{j \in C_3 \\ }} \underbrace{\bar{w}_j \bar{x}_j}_{=0} \leq 0 \qquad \text{and}$$

$$\bar{y}^\top \bar{s} = \sum_{\substack{i \in R_1 \\ }} \underbrace{\bar{s}_i}_{\geq 0} \underbrace{\bar{y}_i}_{\geq 0} + \sum_{\substack{i \in R_2 \\ }} \underbrace{\bar{s}_i}_{\leq 0} \underbrace{\bar{y}_i}_{\leq 0} + \sum_{\substack{i \in R_3 \\ }} \underbrace{\bar{s}_i \bar{y}_i}_{=0} \geq 0. \qquad \square$$

We close this section by noting the following consequences of the weak duality theorem and the fundamental theorem of linear programming:

COROLLARY 4.2   *Let (P) and (D) be a primal–dual pair. Then:*

(1) *if (P) is unbounded, then (D) is infeasible,*
(2) *if (D) is unbounded, then (P) is infeasible,*
(3) *if (P) and (D) are both feasible, then (P) and (D) both have optimal solutions.*

Proof We may assume that (P) is a maximization problem with objective function $c^\top x$ and that (D) is a minimization problem with objective function $b^\top y$. (1) Suppose that (D) has a feasible solution $\bar{y}$. Then by the weak duality (Theorem 4.1), $b^\top \bar{y}$ is an upper bound for (P). In particular, (P) is not unbounded. (2) Suppose that (P) has a feasible solution $\bar{x}$. Then by the weak duality (Theorem 4.1), $c^\top \bar{x}$ is a lower bound for (D). In particular, (D) is not unbounded. (3) By (1), we know that (P) is not unbounded and by (2) we know that (D) is not unbounded. Since by the fundamental theorem of linear programming (see Theorem 2.11), (P) is either unbounded, infeasible, or has an optimal solution, (P) must have an optimal solution. By a similar argument, (D) has an optimal solution. □

It is often easy when given a primal–dual pair to verify that both linear programs are feasible. In that case, statement (3) in the previous result implies immediately that both linear programs have an optimal solution.

Exercises

1 Find the dual of each of the following LPs:

(a)

$$\max \quad (1,3,1)x$$
$$\text{subject to}$$
$$\begin{pmatrix} 1 & 2 & 7 \\ 0 & 1 & -1 \\ 9 & 0 & 0 \\ 1 & -1 & 1 \end{pmatrix} x \begin{matrix} \le \\ = \\ \le \\ = \end{matrix} \begin{pmatrix} -3 \\ 9 \\ 5 \\ 0 \end{pmatrix}$$
$$x_1, x_2 \ge 0, x_3 \text{ free.}$$

(b)

$$\min \quad (5,6)x$$
$$\text{subject to}$$
$$\begin{pmatrix} 3 & -1 \\ 2 & 4 \\ 3 & 2 \\ -1 & 2 \end{pmatrix} x \begin{matrix} \ge \\ = \\ \le \\ = \end{matrix} \begin{pmatrix} 8 \\ -2 \\ 2 \\ -3 \end{pmatrix}$$
$$x_1 \ge 0, x_2 \text{ free.}$$

2 Find the dual of each of the following LPs:

(a)

$$\min \quad c^\top x + d^\top u$$
$$\text{subject to}$$
$$Ax + Du = b$$
$$x \ge 0,$$

where A and D are matrices, c, d, and b are vectors, and x and u are vectors of variables.

(b)

$$\max \qquad c^\top x$$

$$\text{subject to}$$

$$\begin{aligned} Ax &\ge b \\ Dx - Iu &\le d \end{aligned}$$

$$x \le 0, \ u \ge 0,$$

where A and D are matrices, I is the identity matrix, c, b, and d are vectors, and x and u are vectors of variables.

3 Let $a_1, \ldots, a_n$ be n distinct real numbers. Consider the LP problem

$$\max\{x : x \le a_i, \text{ for } i = 1, \ldots, n\}. \tag{P}$$

(a) Prove that the optimal value of (P) is the minimum of $a_1, \ldots, a_n$.
(b) Find the dual (D) of (P).
(c) Explain in words what (D) is doing?

4 Give an example of an infeasible LP problem whose dual is also infeasible. You need to prove algebraically that your example and its dual are infeasible.

5 (Advanced) A directed graph (or *digraph*) $\vec{G}$ is a pair (V, E), where V is a finite set and $\vec{E}$ is a set of *ordered* pairs of elements of V. Members of V are called vertices and members of $\vec{E}$ are called *arcs*. Directed graphs can be represented in a drawing where vertices correspond to points and arcs ij are indicated by an arrow going from i to j. For instance, the directed graph with $V = \{1, 2, 3, 4, 5, 6\}$ and $\vec{E} = \{a, b, c, e, f, g, h, k, l\}$ where $a = \vec{12}, b = \vec{13}, c = \vec{43}, e = \vec{35}, f = \vec{54}, g = \vec{25}, h = \vec{52}, k = \vec{65}, l = \vec{46}$ can be represented as in Figure 4.1. A directed cycle is a sequence of arcs of the form $\vec{i_1 i_2}, \vec{i_2 i_3}, \ldots, \vec{i_{k-1} i_k}, \vec{i_k i_1}$ where $i_1, i_2, \ldots, i_k$ denote distinct vertices. For instance, c, e, f is a directed cycle and so is g, h. Given an arc $\vec{ij}$, vertex i is the *tail* of $\vec{ij}$ and vertex j is the *head* of $\vec{ij}$. Given a vertex v, the set of all arcs with tail v is denoted $\delta^+(v)$

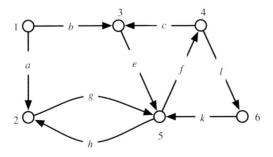

Figure 4.1

and the set of all arcs with head v is denoted $\delta^-(v)$. For instance, $\delta^+(5) = \{f, h\}$ and $\delta^-(5) = \{e, g, k\}$.

(a) A *topological ordering* of the vertices of a digraph $\vec{G} = (V, \vec{E})$ is an assignment of value y_i to each vertex i such that for every arc ij, $y_i \geq y_j + 1$. Show that if there exists a topological ordering then there is no directed cycle.

Let $\vec{G} = (V, \vec{E})$ be a directed graph and consider the following linear program (P):

$$\max \quad \sum (x_e : e \in \vec{E})$$

subject to

$$\sum \left(x_e : e \in \delta^+(v)\right) - \sum \left(x_e : e \in \delta^-(v)\right) = 0 \qquad \text{(for all } v \in V)$$

$$x_e \geq 0 \qquad \text{(for all } e \in \vec{E}).$$

For instance, if $\vec{G}$ is the digraph represented in Figure 4.1, the equality constraint for vertex 5 says that

$$x_f + x_h - x_e - x_g - x_k = 0.$$

(b) Show that (P) is feasible.
(c) Show that if $\vec{G}$ has a directed cycle, then (P) is unbounded.
(d) Write the dual (D) of (P).

HINT: You will have one constraint for each arc ij and the constraints will be of the form

$$\pm y_i \pm y_j \,?\, 1.$$

(e) Show that a topological ordering y_i ($\forall i \in V$) of $\vec{G}$ is a feasible solution to (D).
(f) Find an alternate proof to the statement you proved in (a) by using (e) and weak duality.
(g) Show that if (P) has a feasible solution of value greater than 0, then $\vec{G}$ has a directed cycle.
(h) Using (g) prove that if $\vec{G}$ has no directed cycle, then $\vec{G}$ must have a topological ordering. (Use Table 4.2 in the next section.)

6 Consider a graph G. A *dyad* of G is a pair of edges that share exactly one vertex. A *packing of dyads* is a set $\mathscr{S}$ of dyads with the property that no two dyads in $\mathscr{S}$ share a common vertex. In the following example, the dyads are represented by bold edges:

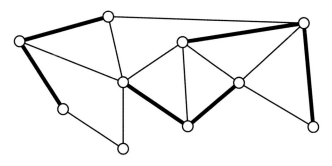

(a) Show that the following IP finds a packing of dyads of maximum cardinality:

$$\max \qquad \sum (x_D : D \text{ is a dyad})$$

subject to

$$\sum (x_D : \text{dyad } D \text{ contains } v) \leq 1 \qquad (v \text{ vertex})$$

$$x_D \geq 0, x_D \quad \text{integer} \qquad\qquad (D \text{ is a dyad}).$$

(b) Denote by (P) the linear program obtained from (IP) by relaxing the conditions that x be integer. Find the dual (D) of (P).

(c) Suppose that G is bipartite and let A, B be a partition of V such that all edges are going between A and B. Find a solution for (D) of value $\min\{|A|, |B|\}$.

(d) Using (c), show that packing of dyads have cardinality at most $\min\{|A|, |B|\}$.

4.2   Strong duality

Let (P) and (D) be a primal–dual pair where (P) is a maximization problem and (D) is a minimization problem. We know from Theorem 4.1 that for any feasible solution $\bar{x}$ of (P) and any feasible solution $\bar{y}$ of (D), the value of $\bar{x}$ in (P) is at at most the value of $\bar{y}$ in (D). It is natural to ask whether we can find a pair of feasible solutions for (P) and (D) that have the same optimal value. The next theorem states that this is indeed always the case.

THEOREM 4.3 (Strong duality).   *Let (P) and (D) be a primal–dual pair. If there exists an optimal solution $\bar{x}$ of (P), then there exists an optimal solution $\bar{y}$ of (D). Moreover, the value of $\bar{x}$ in (P) equals the value of $\bar{y}$ in (D).*

We will only prove this theorem for the special case when (P) is in SEF. It is routine however to derive the result for arbitrary linear program by first converting the linear program to an equivalent linear program in SEF.

Proof of Theorem 4.3 for SEF.   If (P) is in SEF, the primal–dual pair (P) and (D) is of the form

$$\max\{c^\top x : Ax = b, x \geq 0\} \qquad\qquad (\text{P})$$

$$\min\{b^\top y : A^\top y \geq c\}. \qquad\qquad (\text{D})$$

We know from Theorem 2.7 that the two-phase simplex procedure terminates. Since (P) has an optimal solution, the simplex will terminate with an optimal basis B. Let us rewrite (P) in canonical form for B (see Proposition 2.4)

$$\max \qquad z = \bar{y}^\top b + \bar{c}^\top x$$

subject to

$$x_B + A_B^{-1} A_N x_N = A_B^{-1} b \qquad\qquad (\text{P}')$$

$$x \geq 0,$$

where

$$\bar{y} = A_B^{-\top} c_B \qquad \text{and} \qquad \bar{c}^\top = c^\top - \bar{y}^\top A.$$

Let $\bar{x}$ be the basic solution for B, i.e.

$$\bar{x}_B = A_B^{-1} b \qquad \text{and} \qquad \bar{x}_N = \mathbb{0}.$$

(P') has the property that for any feasible solution the values in (P) and (P') are the same. (P') also has the property that $\bar{c}_B = \mathbb{0}$. Hence

$$z = c^\top \bar{x} = \bar{y}^\top b + \bar{c}^\top \bar{x} = b^\top \bar{y} + \underbrace{\bar{c}_N^\top}_{=\mathbb{0}} \bar{x}_N + \underbrace{\bar{c}_B^\top}_{=\mathbb{0}} \bar{x}_B = b^\top \bar{y}.$$

Since the simplex procedure stopped, we must have $\bar{c} \leq \mathbb{0}$ i.e.

$$c^\top - \bar{y}^\top A \leq \mathbb{0} \qquad \text{or equivalently} \qquad A^\top \bar{y} \geq c.$$

It follows that $\bar{y}$ is feasible for (D). Since $c^\top \bar{x} = b^\top \bar{y}$, we know from weak duality (Theorem 4.1) that $\bar{x}$ is an optimal solution to (P) and $\bar{y}$ is an optimal solution to (D). □

Combining the previous result together with Corollary 4.2(3), we obtain.

THEOREM 4.4 (Strong duality–feasibility version).   *Let (P) and (D) be a primal–dual pair. If (P) and (D) are both feasible, then there exists an optimal solution $\bar{x}$ of (P) and an optimal solution $\bar{y}$ of (D) and the objective value of $\bar{x}$ in (P) equals the objective value of $\bar{y}$ in (D).*

We can now classify the possibilities (infeasible, optimal, unbounded) for a primal–dual pair of linear programs. On the face of it, the three possibilities for each LP would lead to nine possibilities for the primal–dual pair, but certain of these cannot occur. Table 4.2 describes the set of all possible outcomes.
Corollary 4.2(1) proves entries ②, ⑤ and ⑧. Corollary 4.2(2) proves entries ④, ⑤ and ⑥. Strong duality (Theorem 4.3) proves entries ①, ②, ③, ④, ⑦. We can construct an example to show that entry ⑨ is possible. Here we take advantage of the fact

Table 4.2 Outcomes for primal–dual pairs

Dual	Primal					
	Optimal solution		Unbounded		Infeasible	
Optimal solution	possible	①	impossible	②	impossible	③
Unbounded	impossible	④	impossible	⑤	possible	⑥
Infeasible	impossible	⑦	possible	⑧	possible	⑨

that infeasibility depends only on the constraints, and not on the objective function. Therefore, the problem

$$\max \quad c_1 x_1 + c_2 x_2$$

subject to

$$\begin{pmatrix} 1 & -1 \\ -1 & 1 \end{pmatrix} \begin{pmatrix} x_1 \\ x_2 \end{pmatrix} \le \begin{pmatrix} -2 \\ 1 \end{pmatrix}$$

$$x \ge 0$$

is infeasible, regardless of the values of c_1 and c_2. To give an example for ⑨, we need to only choose c_1 and c_2 such that the dual is infeasible. We leave this as an exercise.

We know from weak duality that in order to prove that a vector $\bar{x}$ is an optimal solution to a linear program (P), it suffices to find a vector $\bar{y}$ satisfying the following three properties:

(A) $\bar{x}$ is feasible for (P);
(B) $\bar{y}$ is feasible for the dual (D) of (P);
(C) the value of $\bar{x}$ in (P) is equal to the value of $\bar{y}$ in (D).

In the simplex method, we consider (P) of the form $\max\{c^\top x : Ax = b, x \ge 0\}$. At each iteration, we have a feasible basis B. The vector $\bar{x}$ is the basic feasible solution for B. Hence, property (A) holds. We define $\bar{y}$ as $A_B^{-T} c_B$, which implies (see the proof of Theorem 4.3) that (C) holds. The algorithm terminates when (B) is satisfied. Thus, the simplex method keeps two out of the three properties (A), (B), (C) satisfied at each step, and terminates when the third property holds. There are other algorithms that follow the same scheme. The dual simplex procedures maintain properties (B), (C) at each iteration and terminate when (A) holds. Interior-point methods (see Section 7.8) maintain properties (A), (B) at each iteration and terminate when (C) holds.

Exercises

1   Let (P) denote the LP problem $\max\{c^\top x : Ax \le b, x \ge 0\}$, and let (D) be the dual of (P). The goal of this exercise is to prove the strong duality theorem for (P) and (D). (In the book we proved strong duality only for problems in SEF.)
(a) Convert (P) into an equivalent problem (P') in SEF.
(b) Find the dual (D') of (P').
(c) Suppose that (P) has an optimal solution x. Construct an optimal solution x' of (P').
(d) Using Theorem 4.3 for problems in SEF, deduce that there exists an optimal solution y' to (D') where the value of x' for (P') is equal to the value of y' for (D').

2   Consider the following linear program (P), $\max\{c^\top x : Ax \le b, x \ge 0\}$. Let (D) denote the dual of (P), and suppose that (P) has an optimal solution $\bar{x}$ and (D) has an optimal solution $\bar{y}$.

(a) Let (P2) denote the LP resulting by multiplying the first inequality of (P) by 2, and let (D2) denote the dual of (P2). Find optimal solutions of (P2) and (D2) in terms of $\bar{x}$ and $\bar{y}$. Justify your answers.

(b) Let (P3) denote the LP obtained from (P) by replacing the constraint $Ax \leq b$ by $MAx \leq Mb$ where M is a square non-singular matrix of suitable dimension. Let (D3) denote the dual of (P3). Find optimal solutions of (P3) and (D3) in terms of $\bar{x}$ and $\bar{y}$. Justify your answers.

3   Consider the following linear program

$$\max\{c^\mathsf{T}x : Ax = b, x \geq 0\}. \tag{P}$$

Assume that (P) has a feasible solution. The goal of this exercise is to use duality to show that (P) is unbounded if and only if there exists r such that

$$r \geq 0, \qquad Ar = 0, \qquad \text{and} \qquad c^\mathsf{T}r > 0. \tag{$\star$}$$

(a) Show that if there exists r satisfying ($\star$), then (P) is unbounded.
(b) Find the dual (D) of (P).
Consider the following linear program

$$\max\{c^\mathsf{T}r : Ar = 0, r \geq 0\}. \tag{P$'$}$$

Suppose that (P) is unbounded:
(a) Find the dual (D$'$) of (P$'$).
(b) Show that (D$'$) is infeasible.
(c) Show that (P$'$) is feasible.
(d) Show that (P$'$) is unbounded.
          HINT: Consider the possible pairs of outcomes for (P$'$) and (D$'$).
(e) Deduce from f) that there exists a feasible solution $\bar{r}$ to (P$'$) that satisfies ($\star$).

4.3   A geometric characterization of optimality

In this section, we will provide a geometric characterization for when a feasible solution to a linear program of the form

$$\max\{c^\mathsf{T}x : Ax \leq b\}$$

is also an optimal solution. Observe that every linear program can be transformed into a linear program of the aforementioned form. As an intermediate step, we will provide an alternate statement of the strong duality theorem by carefully analyzing when the weak duality relation holds with equality. This will lead to the notion of *complementary slackness*.

4.3.1   Complementary slackness

Consider the following primal–dual pair of linear programs (see Table 4.1)

$$\max\{c^\top x : Ax \le b\} \tag{4.9}$$

$$\min\{b^\top y : A^\top y = c, y \ge 0\}. \tag{4.10}$$

Let us revisit the proof of weak duality (Theorem 4.1) for this particular case. After adding slack variables s, we can rewrite (4.9) as

$$\max\{c^\top x : Ax + s = b, s \ge 0\}. \tag{4.11}$$

Let $\bar{x}$ be a feasible solution to (4.9) and let $\bar{y}$ be a feasible solution to (4.10). Then for $\bar{s} = b - A\bar{x}$, the vectors $\bar{x}, \bar{s}$ are a solution to (4.11). Since $A\bar{x} + \bar{s} = b$, we have

$$b^\top \bar{y} = \bar{y}^\top b = \bar{y}^\top (A\bar{x} + \bar{s}) = (\bar{y}^\top A)\bar{x} + \bar{y}^\top \bar{s} = c^\top \bar{x} + \bar{y}^\top \bar{s},$$

where the last equality follows from the fact that $\bar{y}$ is feasible for (4.10). We know from strong duality (Theorem 4.3) that $\bar{x}$ and $\bar{y}$ are *both* optimal for (4.9) and (4.10) respectively, if and only if $c^\top \bar{x} = b^\top \bar{y}$, or equivalently $\bar{y}^\top \bar{s} = 0$. Let m denote the number of constraints of (4.9), then

$$\bar{y}^\top \bar{s} = \sum_{i=1}^{m} \bar{s}_i \bar{y}_i. \tag{$\star$}$$

As $\bar{s}_i \ge 0$ and $\bar{y}_i \ge 0$, every term in the sum ($\star$) is nonnegative. It follows $\bar{y}^\top \bar{s} = 0$ if and only if for every $i \in \{1, \ldots, m\}$, $\bar{s}_i = 0$ or $\bar{y}_i = 0$. Note, that $\bar{s}_i = 0$ means that constraint i of (4.9) holds with equality for $\bar{x}$; we will sometimes also say that this constraint is *tight* for $\bar{x}$. The following result summarizes these observations:

THEOREM 4.5 (Complementary slackness – special case).   *Let $\bar{x}$ be a feasible solution to (4.9) and let $\bar{y}$ be a feasible solution to (4.10). Then $\bar{x}$ is an optimal solution to (4.9) and $\bar{y}$ is an optimal solution to (4.10) if and only if for every row index i of A, constraint i of (4.9) is tight for $\bar{x}$ or the corresponding dual variable $\bar{y}_i = 0$.*

Example 17 Consider the following primal–dual pair:

$$\max \quad (5, 3, 5)\, x$$

subject to

$$\begin{pmatrix} 1 & 2 & -1 \\ 3 & 1 & 2 \\ -1 & 1 & 1 \end{pmatrix} x \le \begin{pmatrix} 2 \\ 4 \\ -1 \end{pmatrix} \tag{P}$$

and

$$\min \quad (2, 4, -1)y$$

subject to

$$\begin{pmatrix} 1 & 3 & -1 \\ 2 & 1 & 1 \\ -1 & 2 & 1 \end{pmatrix} y = \begin{pmatrix} 5 \\ 3 \\ 5 \end{pmatrix} \tag{D}$$

$$y \geq 0.$$

It can be readily checked that $\bar{x} = (1, -1, 1)^\top$ is a feasible solution to (P) and that $\bar{y} = (0, 2, 1)^\top$ is a feasible solution to (D). Theorem 4.5 states that to show $\bar{x}$ is an optimal solution to (P) and $\bar{y}$ is an an optimal solution to (D), it suffices to check the following conditions:

- $\boxed{\bar{y}_1 = 0}$ OR $(1\ 2\ -1)\bar{x} = 2;$

- $\bar{y}_2 = 0$ OR $\boxed{(3\ 1\ 2)\bar{x} = 4}$

- $\bar{y}_3 = 0$ OR $\boxed{(-1\ 1\ 1)\bar{x} = -1}$

We indicate by a $\square$ the part of the OR condition that is satisfied. In our particular case, all OR conditions are satisfied. It follows that $\bar{x}$ is optimal for (P) and that $\bar{y}$ is optimal for (D). Consider now the feasible solution $x' = (2, -2, 0)^\top$ of (P). The conditions of Theorem 4.5 do not hold for the pair x' and $\bar{y}$, since $(-1, 1, 1)x' < -1$ and $\bar{y}_3 = 1 > 0$. Observe that in this case Theorem 4.5 only allows us to deduce that either x' is not optimal for (P), or $\bar{y}$ is not optimal for (D). The theorem does not give any indication as to which outcome occurs.

Theorem 4.5 generalizes to arbitrary linear programs. Consider a primal–dual pair $(\text{P}_{\max})$ and $(\text{P}_{\min})$ as given in Table 4.1. Let $\bar{x}$ be a feasible solution to $(\text{P}_{\max})$ and let $\bar{y}$ be a feasible solution to $(\text{P}_{\min})$. We say that $\bar{x}, \bar{y}$ satisfy the *complementary slackness* (CS) conditions if:

- for every variable x_j of $(\text{P}_{\max})$,
  $\bar{x}_j = 0$ or the corresponding constraint j of $(\text{P}_{\min})$ is satisfied with equality,
- for every variable y_i of $(\text{P}_{\min})$,
  $\bar{y}_i = 0$ or the corresponding constraint i of $(\text{P}_{\max})$ is satisfied with equality.

Note, that if a variable x_j is free, then the corresponding constraint j of $(\text{P}_{\min})$ is an equality constraint, hence the condition is trivially satisfied. Similarly, if y_i is free, then the corresponding constraint i of $(\text{P}_{\max})$ is an equality constraint. Thus, we only need to state the CS conditions for non-free variables of $(\text{P}_{\max})$ and $(\text{P}_{\min})$.

We leave as an exercise the following generalization of Theorem 4.5:

THEOREM 4.6 (Complementary slackness theorem). *Let (P) and (D) be an arbitrary primal–dual pair. Let $\bar{x}$ be a feasible solution to (P) and let $\bar{y}$ be a feasible solution to*

(D). Then $\bar{x}$ is an optimal solution to (P) and $\bar{y}$ is an optimal solution to (D) if and only if the complementary slackness conditions hold.

Example 18 Consider the following linear program:

$$\max \quad (12, 26, 20)x$$

subject to

$$\begin{pmatrix} 1 & 2 & 1 \\ 4 & 6 & 5 \\ 2 & -1 & -3 \end{pmatrix} x \begin{matrix} \leq \\ \leq \\ = \end{matrix} \begin{pmatrix} -1 \\ 2 \\ 13 \end{pmatrix} \tag{P}$$

$$x_1 \geq 0, \; x_3 \geq 0.$$

Its dual is given by

$$\min \quad (-1, 2, 13)y$$

subject to

$$\begin{pmatrix} 1 & 4 & 2 \\ 2 & 6 & -1 \\ 1 & 5 & -3 \end{pmatrix} \begin{matrix} \geq \\ = \\ \geq \end{matrix} \begin{pmatrix} 12 \\ 26 \\ 20 \end{pmatrix} \tag{D}$$

$$y_1 \geq 0, \; y_2 \geq 0.$$

It can be readily checked that $\bar{x} = (5, -3, 0)^\top$ is a feasible solution to (P) and that $\bar{y} = (0, 4, -2)^\top$ is a feasible solution to (D). Theorem 4.6 states that to show $\bar{x}$ is an optimal solution to (P) and $\bar{y}$ is an optimal solution to (D), it suffices to check the following conditions:

- $\bar{x}_1 = 0$ OR $\boxed{\bar{y}_1 + 4\bar{y}_2 + 2\bar{y}_3 = 12}$;

- $\boxed{\bar{x}_3 = 0}$ OR $\bar{y}_1 + 5\bar{y}_2 - 3\bar{y}_3 = 20$;

- $\boxed{\bar{y}_1 = 0}$ OR $\boxed{\bar{x}_1 + 2\bar{x}_2 + \bar{x}_3 = -1}$;

- $\bar{y}_2 = 0$ OR $\boxed{4\bar{x}_1 + 6\bar{x}_2 + 5\bar{x}_3 = 2}$.

We indicate by a $\square$ the part of the OR condition that is satisfied. In our particular case, all OR conditions are satisfied. It follows that $\bar{x}$ is optimal for (P) and that $\bar{y}$ is optimal for (D). Observe that in the third condition, both OR alternatives are satisfied. This does not contradict the theorem as the complimentary slackness conditions require that at least one of the alternative holds, but they do not preclude both alternatives to simultaneously hold.

Exercises

1 For each of (a) and (b), do the following:

- Write the dual (D) of (P).
- Write the complementary slackness (CS) conditions for (P) and (D).
- Use weak duality to prove that $\bar{x}$ is optimal for (P) and $\bar{y}$ is optimal for (D).
- Use CS to prove that $\bar{x}$ is optimal for (P) and $\bar{y}$ is optimal for (D).

(a)

$$\max \quad (-2, -1, 0)x$$
$$\text{subject to}$$
$$\begin{pmatrix} 1 & 3 & 2 \\ -1 & 4 & 2 \end{pmatrix} x \begin{array}{c} \geq \\ \leq \end{array} \begin{pmatrix} 5 \\ 7 \end{pmatrix} \qquad \text{(P)}$$
$$x_1 \leq 0, \quad x_2 \geq 0$$

and

$$\bar{x} = (-1, 0, 3)^\top \qquad \bar{y} = (-1, 1)^\top.$$

(b)

$$\min \quad (-5, -7, -5)x$$
$$\text{subject to}$$
$$\begin{pmatrix} 1 & 2 & 3 \\ 4 & 5 & 6 \\ 7 & 8 & 9 \end{pmatrix} x \begin{array}{c} \leq \\ \leq \\ \geq \end{array} \begin{pmatrix} 3 \\ 9 \\ 2 \end{pmatrix} \qquad \text{(P)}$$
$$x_1, x_2, x_3 \geq 0$$

and

$$\bar{x} = (1, 1, 0)^\top \qquad \bar{y} = (-1, -1, 0)^\top.$$

2 Recall that (3.9) is the IP formulation of the shortest path problem. Let (P) denote the LP relaxation of (3.9). The dual of (P) is given by (3.15).
(a) Write the complementary slackness conditions for (P) and (3.15).
(b) Using the CS theorem (Theorem 4.6), prove Proposition 3.6.

3 Recall that (3.24) is the IP formulation of the minimum cost matching problem. Let (P) denote the LP relaxation of (3.24). The dual of (P) is given by (3.30).
(a) Write the complementary slackness conditions for (P) and (3.30).
(b) Using the CS theorem (Theorem 4.6), prove Proposition 3.8.

4 (a) Suppose that the LP problem $\max\{c^\top x : Ax = b\}$ is feasible. Prove that it has an optimal solution if and only if $c^\top$ is in the row space of A.
     HINT: Consider the CS conditions.
(b) Consider the LP problem, $\max\{c^\top x : \mathbb{1}^\top x = 250\}$. Under what conditions on c does this problem have an optimal solution? Justify your answer.

5 Consider the following linear program where A is an $m \times n$ matrix:

$$\max \quad c^{\mathsf{T}} x$$
$$\text{subject to} \quad \quad \quad \quad \quad \quad (4.12)$$
$$Ax \le b.$$

(a) Write the dual (D) of (P).

Let $\bar{x}$ and $\bar{y}$ be feasible solutions to (P) and (D) respectively. We say that $\bar{x}$ and $\bar{y}$ *almost satisfy the CS conditions* if for every $i = 1, \dots, m$ at least one of the following holds:

(CS1) $\bar{y}_i = 0$,

(CS2) $row_i(A)\bar{x} = b_i$,

(CS3) $\bar{y}_i \le 1$ and $b_i - row_i(A)\bar{x} \le \frac{1}{1000}$.

Suppose that A has 100 rows and that $\bar{x}$ and $\bar{y}$ almost satisfy the CS conditions.

(b) Show that the value of an optimal solution of (P) cannot exceed the value of $\bar{x}$ by more than $\frac{1}{10}$.

(c) Show that the value of an optimal solution of (D) cannot be inferior to the value of $\bar{y}$ by more than $\frac{1}{10}$.

4.3.2 Geometry

Before we can characterize optimal solutions to linear programs geometrically, we will need a number of preliminary definitions.

Let $a^{(1)}, \dots, a^{(k)}$ be a set of vectors in $\mathbb{R}^n$. We define the *cone generated by* $a^{(1)}, \dots, a^{(k)}$ to be the set

$$C = \left\{ \sum_{i=1}^{k} \lambda_i a^{(i)} : \lambda_i \ge 0 \text{ for all } i = 1, \dots, k \right\},$$

i.e. C is the set of all points that can be obtained by multiplying each of $a^{(1)}, \dots, a^{(k)}$ by a nonnegative number and adding all of the resulting vectors together. We denote the set C by cone$\{a^{(1)}, \dots, a^{(k)}\}$. Note, we define the cone generated by an empty set of vectors to be the set consisting of the zero vector only, i.e. the set $\{0\}$.

Example 19 Consider vectors

$$a^{(1)} = \begin{pmatrix} 2 \\ -1 \end{pmatrix}, \quad a^{(2)} = \begin{pmatrix} 3 \\ 1 \end{pmatrix}, \quad \text{and} \quad a^{(3)} = \begin{pmatrix} 2 \\ 1 \end{pmatrix}.$$

In Figure 4.2, we represent each of the vectors $a^{(1)}, a^{(2)}, a^{(3)}$ by an arrow from the origin, and the infinite region containing $a^{(2)}$ bounded by the half lines from the origin determined by $a^{(1)}$ and $a^{(3)}$ respectively is cone $\{a^{(1)}, a^{(2)}, a^{(3)}\}$.

Let $P := \{x : Ax \le b\}$ be a polyhedron and let $\bar{x} \in P$. Let $J(\bar{x})$ be the row indices of A corresponding to the tight constraints of $Ax \le b$ for $\bar{x}$, i.e. $i \in J(\bar{x})$ when $row_i(A)\bar{x} = b_i$.

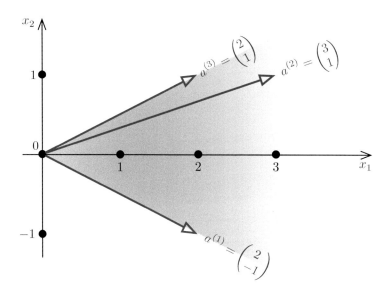

Figure 4.2 Illustration of Example 19.

We define the *cone of tight constraints* for $\bar{x}$ to be the cone C generated by the rows of A corresponding to the tight constraints, i.e. $C = \text{cone}\{row_i(A)^\top : i \in J(\bar{x})\}$.

Example 20 Consider for instance

$$\max \quad (\tfrac{3}{2}, \tfrac{1}{2})x$$

$$\text{s.t.}$$

$$\begin{pmatrix} 1 & 0 \\ 1 & 1 \\ 0 & 1 \end{pmatrix} x \leq \begin{pmatrix} 2 \\ 3 \\ 2 \end{pmatrix}. \qquad \begin{matrix} (1) \\ (2) \\ (3) \end{matrix} \qquad (4.13)$$

Note, $\bar{x} = (2, 1)^\top$ is a feasible solution. The tight constraints for $\bar{x}$ are constraints (1) and (2). Hence, the cone of tight constraints for $\bar{x}$ is

$$C := \text{cone}\left\{\begin{pmatrix} 1 \\ 0 \end{pmatrix}, \begin{pmatrix} 1 \\ 1 \end{pmatrix}\right\}.$$

We are now ready to present our geometric characterization of optimality.

THEOREM 4.7   *Let $\bar{x}$ be a feasible solution to $\max\{c^\top x : Ax \leq b\}$. Then $\bar{x}$ is optimal if and only if c is in the cone of tight constraints for $\bar{x}$.*

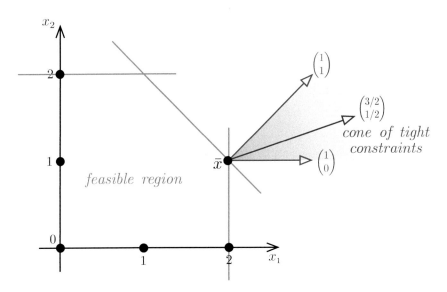

Figure 4.3 Geometric interpretation of optimality.

Example 20, continued. The objective function for (4.13) is $z = (\frac{3}{2}, \frac{1}{2})x$ and $(\frac{3}{2}, \frac{1}{2})^{\top} \in C$ as

$$\begin{pmatrix} 3/2 \\ 1/2 \end{pmatrix} = \bar{y}_1 \begin{pmatrix} 1 \\ 0 \end{pmatrix} + \bar{y}_2 \begin{pmatrix} 1 \\ 1 \end{pmatrix} \qquad \text{for } \bar{y}_1 = 1 \geq 0 \text{ and } \bar{y}_2 = \tfrac{1}{2} \geq 0. \tag{4.14}$$

It follows from Theorem 4.7 that $\bar{x} = (2, 1)^{\top}$ is an optimal solution of (4.13). We illustrate this result in Figure 4.3. Note, in that figure the cone of tight constraints is drawn with its origin shifted to the point $\bar{x}$.

We will use the linear program (4.13) with $\bar{x} = (2, 1)^{\top}$ to illustrate one direction of the proof of Theorem 4.7, namely that if the objective function vector c is in the cone of tight constraints for $\bar{x}$, then $\bar{x}$ is an optimal solution. Observe that because of Theorem 4.5, it will suffice to:

Step 1. find a feasible solution $\bar{y}$ for the dual of (4.13), and
Step 2. verify that $\bar{x}, \bar{y}$ satisfy the CS conditions.

The dual of (4.13) is given by (see Table 4.1)

$$\begin{aligned}
\min \quad & (2, \ 3 \ , 2)y \\
\text{subject to} \quad & \\
& \begin{pmatrix} 1 & 1 & 0 \\ 0 & 1 & 1 \end{pmatrix} y = \begin{pmatrix} 3/2 \\ 1/2 \end{pmatrix} \\
& y \geq 0.
\end{aligned} \tag{4.15}$$

Step 1. Let $\bar{y}_1 := 1, \bar{y}_2 = \frac{1}{2}$, and $\bar{y}_3 = 0$. Then (4.14) implies that

$$\begin{pmatrix} 3/2 \\ 1/2 \end{pmatrix} = \bar{y}_1 \begin{pmatrix} 1 \\ 0 \end{pmatrix} + \bar{y}_2 \begin{pmatrix} 1 \\ 1 \end{pmatrix} + \bar{y}_3 \begin{pmatrix} 0 \\ 1 \end{pmatrix} = \begin{pmatrix} 1 & 1 & 0 \\ 0 & 1 & 1 \end{pmatrix} \begin{pmatrix} \bar{y}_1 \\ \bar{y}_2 \\ \bar{y}_3 \end{pmatrix}.$$

Hence, $\bar{y} = (\bar{y}_1, \bar{y}_2, \bar{y}_3)^\top \geq \mathbb{0}$ is a feasible solution to (4.15). In this argument, we used the fact that $c = (3, 1)^\top$ is in the cone of tight constraints to obtain values for the dual variables of the tight constraints, and we assigned the value zero to all dual variables corresponding to nontight constraints.

Step 2. The CS conditions for $\bar{x}, \bar{y}$ are:

- $y_1 = 0$ OR $\boxed{x_1 = 2}$

- $y_2 = 0$ OR $\boxed{x_1 + x_2 = 3}$

- $\boxed{y_3 = 0}$ OR $x_2 = 2$.

We indicate by a $\square$ the part of the OR condition that is satisfied. In our particular case, the CS conditions are satisfied. Note, because of the way the dual variables were defined in Step 1, if the constraints are not tight, then the corresponding dual variable is zero. Hence, the fact that the CS conditions hold is no surprise.

We completed Step 1 and Step 2, hence $\bar{x} = (2, 1)^\top$ is optimal.

Proof of Theorem 4.7   We consider the following primal–dual pair:

$$\max\{c^\top x : Ax \leq b\}, \tag{P}$$
$$\min\{b^\top y : A^\top y = c, y \geq \mathbb{0}\}, \tag{D}$$

where A is an $m \times n$ matrix. The CS conditions for (P) and (D) are

$$y_i = 0 \quad OR \quad \text{row}_i(A)^\top x = b_i \qquad \text{(for every row index } i\text{).} \tag{$\star$}$$

Suppose c is in the cone of tight constraints for $\bar{x}$. We need to show that $\bar{x}$ is an optimal solution to (P). Because of Theorem 4.5, it will suffice to:

Step 1. find a feasible solution $\bar{y}$ for (D) and
Step 2. verify that $\bar{x}, \bar{y}$ satisfy the CS conditions ($\star$).

Step 1. Recall, $J(\bar{x})$ denotes the row indices of the tight constraints of $Ax \leq b$ for $\bar{x}$. Since c is in the cone of tight constraints for $\bar{x}$, i.e. $c \in \text{cone}\{\text{row}_i(A)^\top : i \in J(\bar{x})\}$, there exists $\bar{y}_i \geq 0$ for all $i \in J(\bar{x})$ such that

$$c = \sum \left(\bar{y}_i \ \text{row}_i(A)^\top : i \in J(\bar{x}) \right).$$

Let us assign the value zero to all dual variables corresponding to nontight constraints, i.e. we set $\bar{y}_i = 0$ for all row indices $i \notin J(\bar{x})$. Then

$$c = \sum \left(\bar{y}_i \, \text{row}_i(A)^{\top} : i = 1, \ldots, m \right) = A^{\top} (\bar{y}_1, \ldots, \bar{y}_m)^{\top}.$$

Hence, $\bar{y} := (\bar{y}_1, \ldots, \bar{y}_m)^{\top} \geq \mathbb{0}$ is a feasible solution to (D).

Step 2. Because of the way the dual variables where defined in Step 1, either $i \in J(\bar{x})$ and the constraints $\text{row}_i(\bar{x}) \leq b_i$ are tight for $\bar{x}$, or the corresponding dual variable $\bar{y}_i = 0$. Hence, the CS conditions in ($\star$) hold.
   We completed Step 1 and Step 2, hence $\bar{x}$ is optimal for (P).

   Suppose $\bar{x}$ is an optimal solution to (P). We need to show that c is in the cone of tight constraints for $\bar{x}$. Theorem 4.5 implies that there exists a feasible solution $\bar{y}$ of (D) such that the CS conditions ($\star$) hold for $\bar{x}, \bar{y}$. Define, $J := \{i : \bar{y}_i > 0\}$. Then

$$c = A^{\top} \bar{y} = \sum \left(\bar{y}_i \, \text{row}_i(A)^{\top} : i = 1, \ldots, m \right) = \sum \left(\bar{y}_i \, \text{row}_i(A)^{\top} : i \in J \right).$$

Hence, $c \in \text{cone}\{\text{row}_i(A)^{\top} : i \in J\}$. Finally, observe that if $i \in J$, then $\bar{y}_i > 0$, which implies by the CS conditions ($\star$) that constraint i of $Ax \leq b$ is tight for $\bar{x}$. Hence, $J \subseteq J(\bar{x})$, and c is in the cone of tight constraints for $\bar{x}$.   □

Exercises

1   Consider the following linear program (P):

$$\begin{array}{rrcrcll}
\max & 7x_1 & + & 3x_2 & & & \\
\text{s.t.} & & & & & & \\
& x_1 & - & x_2 & \leq & 1 & \quad (1) \\
& x_1 & + & x_2 & \leq & 5 & \quad (2) \\
& 2x_1 & - & 3x_2 & \leq & 1 & \quad (3).
\end{array}$$

Consider the feasible solution $\bar{x} = (3, 2)^{\top}$.
(a) Let J be the index set of the tight constraints for $\bar{x}$ and let

$$C := \text{cone}\{\text{row}_i(A)^{\top} : i \in J\}.$$

   Prove that $(7, 3)^{\top} \in C$ and deduce that $\bar{x}$ is an optimal solution to (P).
(b) Find the dual (D) of (P).
(c) Write the complementary slackness conditions for (P) and (D).
(d) Using the fact that $(7, 3)^{\top} \in C$, construct a feasible solution $\bar{y}$ to (D) such that $\bar{x}$ and $\bar{y}$ satisfy the complementary slackness conditions. Deduce that $\bar{x}$ is an optimal solution.

4.4　Farkas' lemma*

The following result gives a sufficient and necessary condition for a linear program in SEF to have a feasible solution.

THEOREM 4.8 (Farkas' Lemma).　*Let A be an m × n matrix and let b be a vector with m entries. Then exactly one of the following statements is true:*

(1) *the system $Ax = b, x \geq \mathbb{0}$ has a solution;*
(2) *there exists a vector y such that $A^\top y \geq \mathbb{0}$ and $b^\top y < 0$.*

Proof　Statements (1) and (2) cannot both be true.

Let us proceed by contradiction and suppose that there exists a solution $\bar{x}$ to $Ax = b, x \geq \mathbb{0}$ and that we can find $\bar{y}$ such that $\bar{y}^\top A \geq \mathbb{0}^\top$ and $\bar{y}^\top b < 0$. Since $A\bar{x} = b$ is satisfied, we must also satisfy $\bar{y}^\top A\bar{x} = \bar{y}^\top b$. Since $\bar{y}^\top A \geq \mathbb{0}^\top$ and $\bar{x} \geq \mathbb{0}$, it follows that $\bar{y}^\top A\bar{x} \geq 0$. Then $0 \leq \bar{y}^\top A\bar{x} = \bar{y}^\top b < 0$, a contradiction.

If (1) is not true, then (2) must be true.

Consider the following primal–dual pair (see Table 4.1):

$$\max\{\mathbb{0}^\top x : Ax = b, x \geq \mathbb{0}\}, \tag{P}$$

$$\min\{b^\top y : A^\top y \geq \mathbb{0}\}. \tag{D}$$

By hypothesis, (P) is infeasible. Observe that (D) is feasible since $A^\top \mathbb{0} \geq \mathbb{0}$. It follows from the possible outcomes for primal–dual pairs (see Table 4.2) that (D) is unbounded. In particular, there exists a feasible solution $\bar{y}$ of (D) with negative value, i.e. $A^\top \bar{y} \geq \mathbb{0}$ and $b^\top \bar{y} < 0$ as required.　□

Note, the fact that both (1) and (2) cannot hold at the same time was already proved in Proposition 2.1. In the remainder of the section, we give an alternate proof for the fact that "if (1) is not true, then (2) must be true" based on the two phase simplex algorithm. This proof will provide an algorithm that takes as input A, b and either finds: $\bar{x}$ as in (1) or find $\bar{y}$ as in (2).

　　Suppose that $Ax = b, x \geq \mathbb{0}$ has no solution and let us assume that $b \geq \mathbb{0}$ (we leave it as an exercise to modify the argument for an arbitrary vector b). Assume that A is an $m \times n$ matrix. The auxiliary problem is given by (see Section 2.6)

$$\max \quad d^\top (x_1, \ldots, x_{n+m})$$

$$\text{subject to}$$

$$\left(A \mid I \right) \begin{pmatrix} x_1 \\ \vdots \\ x_{n+m} \end{pmatrix} = b \tag{Q}$$

$$(x_1, \ldots, x_{n+m})^\top \geq \mathbb{0},$$

where d is the vector with $n + m$ entries such that $d_j = 0$ for $j = 1, \ldots, n$ and $d_j = -1$ for $j = n + 1, \ldots, n + m$. Let B be the optimal basis for (Q) obtained by the simplex algorithm. The objective function of the linear program obtained by rewriting (Q) in canonical form for the basis B is

$$\max \ \underbrace{\left[d^\top - \bar{y}^\top \left(A \mid I \right) \right]}_{=:\bar{d}^\top} (x_1, \ldots, x_{n+m})^\top + \bar{y}^\top b$$

for some vector $\bar{y}$ (see Proposition 2.4). The value of the objective function in (Q′) for the basic solution of B is equal to $b^\top \bar{y}$. Since $Ax = b, x \geq \mathbb{0}$ has no feasible solution, that value is negative[1] (see Remark 2.10). Hence

$$b^\top \bar{y} < 0. \tag{4.16}$$

Since the simplex algorithm terminated, we have that $\bar{d} \leq \mathbb{0}$, i.e.

$$d^\top - \bar{y}^\top \left(A \mid I \right) \leq \mathbb{0}^\top.$$

In particular, all entries of $\bar{d}$ corresponding to the columns of A are nonpositive. Thus

$$\mathbb{0}^\top - \bar{y}^\top A \leq \mathbb{0}^\top \qquad \text{or equivalently} \qquad A^\top y \geq \mathbb{0}. \tag{4.17}$$

Equation (4.16) and (4.17) now imply that $\bar{y}$ is the required certificate of infeasibility.

Exercises

1   (a) Prove that exactly one of the following statements holds:

     • there exists x such that $Ax \leq b, x \geq \mathbb{0}$;
     • there exists y such that $A^\top y \geq \mathbb{0}, y \geq \mathbb{0}$ and $b^\top y < 0$.

   (b) Prove that exactly one of the following statements holds:

     • there exists $x \neq \mathbb{0}$ such that $Ax = \mathbb{0}, x \geq \mathbb{0}$;
     • there exists y such that all entries of $A^\top y$ are positive.

     HINT: Consider the LP $\max\{\mathbb{1}^\top x : Ax = \mathbb{0}, x \geq \mathbb{0}\}$.

   (c) Prove that exactly one of the following statements holds:

     • there exists x such that $Ax \leq b$ and $A'x = b'$;
     • there exists y and z such that $y \geq \mathbb{0}, y^\top A + z^\top A' \geq \mathbb{0}$, and $y^\top b + z^\top b' < 0$.

   (d) Prove that exactly one of the following statements hold:

     • there exists x and x' such that $Ax + A'x' = b$ and $x \geq \mathbb{0}$;
     • there exists y such that $y^\top A \geq \mathbb{0}, y^\top A' = \mathbb{0}$, and $y^\top b < 0$.

---

[1] We consider a maximization problem here rather than a minimization problem as in Section 2.6.

2 The goal of this exercise is to prove the following:

THEOREM *For vectors $a^{(1)}, \ldots, a^{(n)}$ and b (all with same number of entries) exactly one of the following statements is true:*

(S1) $b \in cone\{a^{(1)}, \ldots, a^{(n)}\}$,

(S2) *there exists a halfspace $H = \{x : d^\top x \le 0\}$ that contains all of $a^{(1)}, \ldots, a^{(n)}$ but does not contain b.*

(a) Explain what the theorem says when all vectors $a^{(1)}, \ldots, a^{(n)}, b$ have two entries. Denote by A the matrix with columns $a^{(1)}, \ldots, a^{(n)}$.

(b) Show that (S1) holds if and only if there exists x such that $Ax = b$ and $x \ge 0$.

(c) Show that (S2) holds if and only if there exists y such that $A^\top y \ge 0$ and $b^\top y < 0$.

(d) Prove the theorem above using parts (b) and (c) and Farkas' lemma.

3 Consider the LP

$$\max\{c^\top x : Ax = b, x \ge 0\},$$

such that $A = -A^\top$ (skew-symmetric), $b = -c$. Prove that the dual of such an LP problem is equivalent to the original LP problem (such problems are called *self-dual*). Apply the duality theory we developed to such LP problems to prove as many interesting properties as you can.

4.5 Further reading and notes

Just as Dantzig was developing his ideas on linear programming, independently of Dantzig, von Neumann [1903–1957] was working on game theory. There is a very nice connection between LP duality theory and game theory for two-person zero-sum games. We will comment on this some more following the chapter on nonlinear optimization.

Duality theory for linear programming has close ties to work on solving systems of linear inequalities. Such a connection goes back at least to Joseph Fourier (recall Fourier series and Fourier transform) [1768–1830]. Fourier's paper from 1826 addresses systems of linear inequalities. The result known as Farkas' lemma came during the early 1900s. Hermann Minkowski [1864–1909] also had similar results in his pioneering work on convex geometry. There are many related results involving different forms of inequalities, some using strict inequalities, for example, theorems by Gordan and Stiemke. See Schrijver [58].

5 Applications of duality*

In Chapter 3, we have seen two examples for how the framework of duality theory leads to algorithmic insights. Both in the case of the shortest path problem and in that of the minimum cost perfect matching problem we presented efficient algorithms that are *primal–dual*. Such algorithms compute feasible solutions for the respective primal problems but also, at the same time, construct feasible solutions for their duals. The dual solution is in the end used as a certificate of optimality, but crucially provides algorithmic *guidance* in finding the primal solution. In this chapter, we discuss a few more select examples that demonstrate the diverse use of duality.

5.1 Approximation algorithm for set-cover

The figure on the right shows the floor plan of an art gallery. The management team of the gallery wants to place guards in their gallery such that (a) every point in the gallery is visible by at least one guard, and (b) the number of guards is as small as possible. The figure on the right shows a solution with six guards: guards are depicted as dots, and the portion of museum floor visible by a guard is shaded.

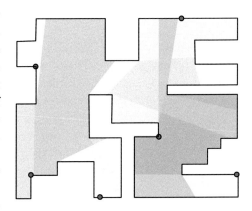

The art gallery problem is an instance of the more general *set-cover* problem. In an instance of this problem, we are given a universe $\mathscr{U} = \{e_1, \ldots, e_m\}$ of *elements*, and a collection of sets $\mathscr{S} = \{S_1, \ldots, S_n\}$ where each S_j is a subset of $\mathscr{U}$. Each set S_j has a nonnegative cost c_{S_j} associated with it, and the goal is to find a collection $\mathscr{C} \subseteq \mathscr{S}$ of sets of the smallest cost such that each element e_i is contained in at least one of the sets in $\mathscr{C}$. In the art gallery example $\mathscr{U}$ is the set of locations in the gallery[1], and S_i is the set of locations guard i can see from her position.

The problem has a natural IP formulation. We introduce a binary variable x_S for every set $S \in \mathscr{S}$ that takes on value 1 if set S is chosen in our solution $\mathscr{C}$ and 0 otherwise.

[1] We will need to assume that we are only concerned with a finite number of locations in the gallery.

The IP has a constraint for every element $e \in \mathscr{U}$ that forces a feasible solution to pick at least one set containing the element.

$$
\begin{aligned}
\min \quad & \sum \left(c_S x_S : S \in \mathscr{S} \right) \\
\text{subject to} \quad & \\
& \sum \left(x_S : S \in \mathscr{S} \text{ where } e \in S \right) \geq 1 \quad \left(e \in \mathscr{U} \right) \\
& x \geq 0, x \text{ integer.}
\end{aligned}
\tag{5.1}
$$

Following the proven strategy from Chapter 3, we could now simply look at the LP relaxation of the above IP, and derive its dual. We could then design an algorithm that finds a set-cover solution as well as a feasible dual solution whose value equals that of the cover. However, it is easy to construct an example where the IP and the dual of the LP relaxation have distinct optimum values (see Exercise 1 in Section 5.3). Moreover, it is unlikely that this scheme will work for any practical IP formulation of the set-cover problem. This is because it is widely believed that no efficient algorithm exists to solve all instances of the set cover problem (see Appendix A).

Now that we know that we should not expect an efficient exact algorithm for the set-cover problem, what *can* be done? We settle for the next best thing: we *approximate* the problem. Suppose we are given a class $\mathscr{I}$ of instances of some optimization (say minimization) problem. For a given instance $I \in \mathscr{I}$, let opt_I be the instance's unknown optimal value. We are looking for an efficient algorithm A that, given any instance $I \in \mathscr{I}$, computes a solution $A(I)$ whose value is at most $\alpha(I)\mathrm{opt}_I$, for some function α of the instance data. Such an algorithm is called an *α-approximation algorithm*, and α is sometimes called its *approximation-* or *performance guarantee*. For any instance I, $\alpha(I) \geq 1$, and if $\alpha(I) = 1$, then our algorithm returns an exact solution for instance I.

In the case of the set-cover problem, we will present two algorithms. Both algorithms employ linear programming duality in order to compare the computed solution to the value of the LP relaxation of (5.1). The first algorithm follows a primal–dual strategy, similar to what we have seen in Chapter 3. The second algorithm is a so-called *greedy* algorithm, and it produces a solution whose cost can once again be bounded within a certain factor of the objective value of a feasible dual solution.

5.1.1 A primal–dual algorithm

Let us start by giving the dual of the LP relaxation of (5.1). This LP has a variable y_e for every element $e \in \mathscr{U}$, and a constraint for every set $S \in \mathscr{S}$. The constraint for set S limits the total y-value of elements in S to be at most the cost c_S of the set.

$$\max \quad \sum (y_e \; : \; e \in \mathscr{U})$$

$$\text{subject to} \quad \sum (y_e \; : \; e \in S) \le c_S \quad (S \in \mathscr{S})$$

$$y \ge 0.$$

(5.2)

The next remark indicates how the existence of a solution to the previous LP allows us to prove that the value of a particular solution to the set-cover problem is within a certain factor of the optimum value.

Remark 5.1 *Consider an instance I of the set-cover problem, opt_I denotes its optimal value. Let $\mathscr{C}$ be a feasible cover with value $\bar{z} := \sum(c_S \; : \; S \in \mathscr{C})$ and let $\bar{y}$ be a solution to (5.2). If for some $\alpha \ge 1$, $\alpha \mathbb{1}^\top \bar{y} \ge \bar{z}$, then $\bar{z} \le \alpha \; opt_I$.*

Proof Let $\mathscr{C}'$ be an optimal solution to the set-cover problem and let $x'_S = 1$ if $S \in \mathscr{C}'$ and let $x'_S = 0$ otherwise. Then

$$\alpha \mathbb{1}^\top \bar{y} \ge \bar{z} \ge opt_I = c^\top x' \ge \mathbb{1}^\top \bar{y}, \tag{5.3}$$

where the first inequality is given in the hypothesis, the second inequality follows from the fact that opt_I is the optimum value, the equality by the definition of x' and the third inequality by weak duality (Theorem 4.1). Now (5.3) implies that both $\bar{z}$ and opt_I are within the interval bounded by $\mathbb{1}^\top \bar{y}$ and $\alpha \mathbb{1}^\top \bar{y}$ and the result follows. □

Consider a primal–dual pair (P) and (D) of LPs. Suppose that $\bar{x}$ is a feasible solution to (P) and that $\bar{y}$ is a feasible solution to (D). Complementary slackness (Theorem 4.6) says that if $\bar{x}$ and $\bar{y}$ satisfy the complementary slackness conditions, then $\bar{x}$ is optimal for (P) and $\bar{y}$ is optimal for (D). Recall the minimum cost perfect matching Algorithm 3.3 from Chapter 3.2. At the end of the algorithm, we have a perfect matching M and a feasible solution $\bar{y}$ to (3.30). Moreover, $\bar{x}$ (the vector where $\bar{x}_e = 1$ if $e \in M$ and $\bar{x}_e = 0$ otherwise) is a feasible solution to (3.24). In addition, all the edges in M are equality edges with respect to $\bar{y}$ or equivalently (see Exercise 3, Section 4.3.1) $\bar{x}$ and $\bar{y}$ satisfy complementary slackness, proving that $\bar{x}$ is optimal for (3.24), i.e. that M is a minimum cost matching. Thus, the strategy for the algorithm is to construct a pair of primal–dual solutions that satisfy complementary slackness conditions.

We will follow a similar strategy for set-cover problems, except that we will only impose that *some* of the complementary slackness conditions be satisfied. The complementary slackness conditions for set-cover are as follows:

(CS1) $x_S > 0$ implies $\sum (y_e \; : \; e \in S) = c_S$, and
(CS2) $y_e > 0$ implies $\sum (x_S \; : \; S \in \mathscr{S}$ where $e \in S) = 1$.

Since we cannot expect to find an integral solution x to (5.1), and a feasible solution y to (5.2) such that (CS1) and (CS2) hold, we will enforce condition (CS1) and *relax* condition (CS2).

Our algorithm works as follows: initially, let $y = \mathbb{0}$ be the trivial feasible solution to (5.2), and let $\mathscr{C} = \emptyset$ be an infeasible cover. The algorithm adds sets to $\mathscr{C}$ as long as there is an element $e \in \mathcal{U}$ that is not covered by any of the sets in $\mathcal{C}$. Whenever such an element e exists, we increase the corresponding dual variable y_e. In fact, we increase its value by the largest possible amount ϵ that maintains y's feasibility for (5.2); ϵ is the minimum slack of the dual constraints corresponding to sets containing e,

$$\epsilon = \min\left\{ c_S - \sum\left(y_{e'} : e' \in S\right) : S \in \mathcal{S} \text{ where } e \in S \right\}.$$

Let S be a set that attains the minimum on the right-hand side of the above expression. S is now added to $\mathcal{C}$, leading to e being covered. Furthermore, note that condition (CS1) is satisfied by set S: in the primal solution corresponding to $\mathcal{C}$, the variable for S has value 1, and its dual constraint is tight (i.e. $\sum(y_e : e \in S) = c_S$) for the updated vector y by our choice of ϵ (see Algorithm 5.6 for a formal description of the algorithm).

Algorithm 5.6 Primal–dual algorithm for set-cover

Input: Elements $\mathcal{U}$, sets $\mathcal{S}$, costs c_S for all $S \in \mathcal{S}$.
Output: A collection $\mathcal{C} \subseteq \mathcal{S}$ such that $\mathcal{U} \subseteq \bigcup_{S \in \mathcal{C}} S$, and a feasible dual y for (5.2).
1: $y = 0, \mathcal{C} = \emptyset$
2: **while** $\exists e \in \mathcal{U}$ that is not covered by any set in $\mathcal{C}$ **do**
3:    Increase y_e as much as possible, maintaining feasibility of y for (5.2)
4:    Let S be a tight set cover e
5:    Add S to $\mathcal{C}$
6: **end while**
7: **return** $\mathcal{C}$ and feasible dual y

We define the *frequency* f_e of element $e \in \mathcal{U}$ to be the number of sets in $\mathcal{S}$ that cover the element. The frequency f of the given set-cover instance is the maximum over all element frequencies; i.e. $f = \max_{e \in \mathcal{U}} f_e$. The next result shows that Algorithm 5.6 returns a feasible set-cover $\mathcal{C}$ that has value within a factor f of the optimum value.

THEOREM 5.2   *Algorithm 5.6 is an f-approximation algorithm.*

Proof   The algorithm clearly returns a feasible set-cover $\mathcal{C}$ if there is one. Notice that the cost of the final cover is

$$\sum_{S \in \mathcal{C}} c_S = \sum_{S \in \mathcal{C}} \sum_{e \in S} y_e, \tag{5.4}$$

where the equality follows directly from condition (CS1). Observe that a variable y_e may appear multiple times in the double-sum on the right-hand side; it appears for every set in $\mathcal{C}$ that covers e. The right-hand side of (5.4) can therefore be rewritten as

$$\sum_{e \in \mathcal{U}} y_e \,|\{S \in \mathcal{C} : e \in S\}| \leq f \mathbb{1}^\top y,$$

where the inequality follows from the fact that every element appears in at most f sets. The result now follows from Remark 5.1. ☐

A particularly nice special case of set-cover is the
vertex-cover problem. In an instance of this problem, we
are given a graph $G = (V, E)$, and a cost c_v for each vertex $v \in V$. The goal is to find a set $\mathscr{C} \subseteq V$ of smallest
total cost such that every edge in E has at least one of
its endpoints in $\mathscr{C}$. The black vertices in the graph on the
right form a vertex-cover of size 6. Note that the vertex-
cover problem is a special case of the set-cover problem:
the elements of this instance are the edges of G, and each
vertex $v \in V$ forms a set that covers all edges that are incident to v. Every edge is
incident to two vertices, and hence appears in two sets of the corresponding set-cover
instance. In other words, the frequency of the instance is $f = 2$. Hence, we obtain the
following immediate consequence:

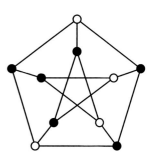

COROLLARY 5.3   *Algorithm 5.6 is a 2-approximation algorithm for the vertex-cover*
problem.

In other words, the algorithm returns a vertex-cover of cost at most twice the optimal
vertex-cover for any given instance of the problem.

5.1.2   Greed is good ... at least sometimes

In the last section, we developed an algorithm that computes a cover whose cost is at
most f times the cost of an optimal cover, where f is the maximum frequency of any
element in $\mathscr{U}$. Can we do better? Can we find $f' < f$, and an algorithm that finds a cover
of cost no more than f' opt_I for any instance I? This appears to be unlikely (more on this
later), but we can improve the guarantee obtained in the previous section sometimes.

Suppose you are faced with a set-cover instance with universe $\mathscr{U}$, and sets $\mathscr{S}$, all
of which have unit cost. Your goal is to find the smallest *number* of sets that jointly
cover all elements. What would your natural strategy be? Well, you would make locally
optimal decisions: you would first pick a set that covers most elements, in the next step
you would pick a set that covers most uncovered elements, and so on. It is easy to
extend this strategy to the setting where sets have non-unit costs. Then we would want
to balance between two objectives: cover as many previously uncovered elements, and
do this at the smallest possible cost. One way of balancing the two competing objectives
would be to always pick a set $S \in \mathscr{S}$ with smallest cost divided by the number of newly
covered elements. The resulting Algorithm 5.7 is known as the *greedy algorithm* for
set-cover.

Let us demonstrate Algorithm 5.7 on the set-cover instance given in Figure 5.1. The
instance has m elements and $m + 1$ sets. The shaded set S_i in the figure has a cost of
$1/i$ for all i, and set S has cost $1 + \epsilon$ for some tiny $\epsilon > 0$. In the first step, each of the
shaded sets S_i covers one element and has cost $1/i$, while the large set S has cost $1 + \epsilon$.
Thus, our algorithm first chooses set S_m, as its cost of newly covered elements ratio is
smaller than that of S. Similar in the second step, the algorithm chooses set S_{m-1}, and so
on. In the end, our algorithm will have picked up all the shaded sets, while the optimal

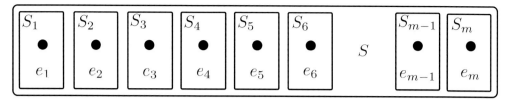

Figure 5.1 An instance of the set-cover problem. Black dots correspond to elements, and boxes to sets. Shaded set S_i has cost $1/i$, and the large set S has cost $1 + \epsilon$ for a fixed small $\epsilon > 0$.

Algorithm 5.7 Greedy algorithm for set-cover

Input: Elements $\mathcal{U}$, sets $\mathcal{S}$, costs c_S for all $S \in \mathcal{S}$.
Output: A collection $\mathcal{C} \subseteq \mathcal{S}$ such that $\mathcal{U} \subseteq \bigcup_{S \in \mathcal{C}} S$.
1: $\mathcal{C} = \emptyset$
2: **while** $\exists e \in \mathcal{U}$ that is not covered by any set in $\mathcal{C}$ **do**
3:    Choose set S that minimizes $c_S / |S \setminus \bigcup_{S' \in \mathcal{C}} S'|$
4:    Add S to $\mathcal{C}$
5: **end while**
6: **return** $\mathcal{C}$

---

solution clearly is to pick just the set S. The solution computed by the greedy algorithm has cost

$$H_m := \sum_{i=1}^{m} \frac{1}{i},$$

compared to an optimal cost of just $1+\epsilon$. Thus, we have shown that the greedy algorithm can not be better than H_m-approximate. We will now show that Algorithm 5.7 always returns a cover of cost at most H_m times that of an optimal solution.

THEOREM 5.4   *Algorithm 5.7 is an H_m-approximation algorithm for the set-cover problem.*

Proof We will prove this theorem by, once again, using linear programming duality. Suppose Algorithm 5.7 picks the sets

$$S_1, \ldots, S_p,$$

in this order, and let $\mathcal{C}$ be the cover consisting of these p sets. For all $1 \leq i \leq p$, let us also define

$$U_i := S_i \setminus \bigcup_{j=1}^{i-1} S_j$$

as the set of newly covered elements resulting from adding set S_i. For $1 \leq i \leq p$, and for all $e \in U_i$, define

$$y_e := \frac{c_{S_i}}{|U_i|}. \tag{5.5}$$

That is, we distribute the cost of set S_i evenly over the elements in U_i, and thus

$$\sum_{e \in \mathcal{U}} y_e = \sum_{i=1}^{p} c_{S_i}. \tag{5.6}$$

We will now show that y/H_m is feasible for (5.2). This, together with (5.6) and Remark 5.1, shows that the algorithm returns a solution of cost no more than H_m times the optimal value.

The dual (5.2) has a constraint for every set $S \in \mathcal{S}$ that limits the y-value of elements in S to the cost c_S of the set. We will show that each such constraint is satisfied by y/H_m. Let $e_1, \ldots, e_l$ be the elements of some set $S \in \mathcal{S}$, in the order in which they were covered by the greedy algorithm. Hence, e_i was covered at the same time or before element $e_{i'}$ for any $1 \le i \le i' \le l$. We claim that the value y_{e_i} picked according to (5.5) satisfies

$$y_{e_i} \le \frac{c_S}{l - i + 1},$$

for all $1 \le i \le l$. Suppose e_i was covered in iteration r, where Algorithm 5.7 chose a set S_r in step 3. Note that set S contained $l - i + 1$ uncovered elements when S_r was picked. The fact that S_r was picked regardless implies

$$\frac{c_{S_r}}{|S_r \setminus (S_1 \cup S_2 \cup \cdots \cup S_{r-1})|} \le \frac{c_S}{l - i + 1}.$$

The left-hand side of the inequality is exactly the value of y_{e_i}, and this proves the claim. Using this, the left-hand side of the dual constraint for set S can be bounded as follows:

$$\sum \left(y_{e_i} : 1 \le i \le l \right) \le \sum \left(\frac{c_S}{l - i + 1} : 1 \le i \le l \right) \le H_m c_S,$$

where the final inequality follows from the fact that S contains at most $m = |\mathcal{U}|$ elements. Thus, y/H_m is feasible for (5.2) as required. $\square$

5.1.3   Discussion

The two algorithms presented in this chapter are incomparable in terms of their approximation guarantee. Algorithm 5.6 is obviously superior when each element lies in a few sets, and Algorithm 5.7 shines when the maximum frequency is high. Examples for both situations are easy to come by, and we have in fact seen a low-frequency one in the vertex-cover problem. At present, no algorithm with a better approximation guarantee for the vertex cover problem is known, and it is known that no α-approximation with $\alpha < 10\sqrt{5} - 21 \approx 1.36$ can exist unless $NP=P$ [23]. If one assumes the less tried and tested *unique games conjecture* (e.g., see Williamson and Shmoys [69] for some more information), the 2-approximation algorithm given here is the best possible unless $NP=P$ [39].

The set-cover problem in general is harder to approximate than the vertex cover problem. In this section, we have presented two algorithms that compute covers $\mathcal{C}$ whose cost is within a certain factor of the optimal value of the LP relaxation of (5.1). How good an approximation algorithm can we obtain with such a strategy? Well, it appears that there are limits.

Let (IP) be an IP formulation for some given minimization problem, and let (P) be its LP relaxation. Let $\mathscr{I}$ be the set of all instances for this problem, and for some $I \in \mathscr{I}$ let opt_I^{IP} and opt_I^P be optimal values for (IP) and (P), respectively, for the given instance I. We then define the *integrality gap* of (P) as

$$\alpha(P) := \max_{I \in \mathscr{I}} \frac{\text{opt}_I^{IP}}{\text{opt}_I^P}.$$

Hence, α_P is the worst case multiplicative gap between the fractional and integral optimal values for (P). Notice that we know from weak duality (Theorem 4.1) that the value of any feasible dual solution for instance $I \in \mathscr{I}$ cannot be higher than opt_I^P. Furthermore, the value of an integral solution must be at least opt_I^{IP}. Thus, using only LP (P) and the above proof technique, we cannot hope to obtain an α-approximation algorithm with $\alpha < \alpha(P)$.

For set-cover it is easy to find example instances where $(1 - \epsilon) \ln(m) \, \text{opt}_I^P \leq \text{opt}_I^{IP}$ [2] for any fixed $\epsilon > 0$, (see [68] for an example), and this shows that the approximation guarantee obtained by Algorithm 5.7 is the best possible (up to constant factors). In fact, we can even show that there is a constant $c \in (0, 1)$ such that no $O((1 - c) \ln(m))$-approximation algorithm exists unless $NP=P$(see [25, 56, 2], and [69]).

5.2   Economic interpretation

After seeing the beautiful applications of duality theory to the shortest paths, minimum cost perfect matching and set-cover problems, it should be clear to the reader that we only scratched the surface in terms of similar applications and interpretations of dual variables.

Duality theory has very nice connections to mathematical economics as well. A very commonly used setting to explain some of this relationship is the class of production planning problems. Consider a factory that makes n products from m resources. The jth product earns a profit of c_j dollars per unit and consumes a_{ij} units of resource i per unit of product made. We have b_i units of resource i available. The production problem is to decide how many units to produce of each product such that total profit is maximized while not exceeding the available resources. If we let A denote the $m \times n$ matrix where entry (i,j) is equal to a_{ij}, we can write the production problem as the following LP:

$$\max\{c^\top x : Ax \leq b, x \geq 0\}, \tag{P}$$

where x_j is the number of units of product j made. The dual of (P) is given by:

$$\min\{b^\top y : A^\top y \geq c, y \geq 0\}. \tag{D}$$

The complementary slackness conditions for (P) and (D) are given by:

(CS1) For all $i = 1, \ldots, m$: $\sum_{j=1}^{n} a_{ij} x_j = b_i$ OR $y_i = 0$,
(CS2) For all $j = 1, \ldots, n$: $\sum_{i=1}^{m} a_{ij} y_i = c_j$ OR $x_j = 0$.

---

[2] Where m is the number of elements in the groundset $\mathscr{U}$.

In this setting, the optimal values $\bar{y}$ of (D) correspond to *shadow prices* of resources. These are "internal prices" which depend on the current optimal solution to (P), i.e. the optimal production plan. Assume that you are running the factory and that you own the resources. Then the shadow prices correspond to the minimum price rate at which you would start selling a portion of the resources without incurring a loss (and possibly gaining extra profit).

Let us try to interpret the constraints of (D) in that light. Constraints $y \geq 0$ indicate that the price of the resources are nonnegative. For every product j, we have a constraint $\sum_{i=1}^{m} a_{ij} y_i \geq c_j$. Note, that $a_{ij} y_i$ is the cost for resources of type i used to make one unit of product j. Thus, the constraint says that the total cost of the resources used to make one unit of product j should be at least as large as the value of one unit of product j. The objective function minimizes the total value of the resources. Let us now try to interpret the complementary slackness conditions. (CS1) says that if resource i is not completely used, then the price of the resources is zero (as you can sell a small amount $\epsilon > 0$ at a price of zero without decreasing the production value). (CS2) says that if the cost of the resources used to make one unit of product j is larger than the value of product j, then you should not produce any unit of product j (as you could sell the resources rather than producing product j).

Under the right conditions, the prices $\bar{y}$ allow us to compute the impact of increasing the available resources on the optimal value for (P), i.e. on the profit from the production. Let us consider the special case where we increase the amount of resource ℓ available in (P) by some small value $\epsilon > 0$ and we keep all other resources unchanged, i.e. we are trying to solve the problem

$$\max\{c^{\top}x : Ax \leq \tilde{b}, x \geq 0\}, \qquad (P_{\epsilon,\ell})$$

where

$$\tilde{b}_i = \begin{cases} b_i + \epsilon & \text{if } i = \ell \\ b_i & \text{otherwise.} \end{cases}$$

The dual problem is

$$\min\{\tilde{b}^{\top}y : A^{\top}y \geq c, y \geq 0\}. \qquad (D_{\epsilon,\ell})$$

Let $\bar{x}, \bar{y}$ be a pair of optimal solutions for (P) and (D) respectively. We know from complementary slackness (Theorem 4.6) that for each of (CS1) and (CS2) at least one of the OR conditions holds. An extreme point of $\{x : Ax \leq b, x \geq 0\}$ where A is a $m \times n$ matrix, is *nondegenerate* if exactly n of the $(m + n)$ inequalities are satisfied with equality at $\bar{x}$.

THEOREM 5.5   *Suppose $\bar{x}$ is a nondegenerate extreme point of the feasible region of (P) that is optimal. Then (D) has a unique optimal solution $\bar{y}$. Moreover, for every $\epsilon > 0$ small enough, the optimal objective value for $(P_{\epsilon,\ell})$ is obtained from the optimal objective value for (P) by adding $\epsilon \bar{y}_\ell$. In other words, the profit from the optimal production plan increases by the shadow price of resource ℓ.*

Proof Complementary slackness conditions applied to $\bar{x}$ yield a linear system of equations with a unique solution $\bar{y}$. Therefore, $\bar{y}$ is the unique optimal solution of (D). Note that it suffices to prove that $\bar{y}$ remains an optimal solution for $(D_{\epsilon,\ell})$ as the value for (D) is given by $b^\top \bar{y}$ and the value for $(D_{\epsilon,\ell})$, or equivalently $(P_{\epsilon,\ell})$, is then given by $\tilde{b}^\top \bar{y} = b^\top \bar{y} + \epsilon \bar{y}_\ell$. We leave it as an exercise to show that for $\epsilon > 0$ small enough, the constraints determining $\bar{x}$ still determine a feasible solution for $(P_{\epsilon,\ell})$ such that the index sets of inequalities satisfied by equality and those satisfied strictly are the same for $\bar{x}$ and the new feasible solution of $(P_{\epsilon,\ell})$. It follows by complementary slackness (Theorem 4.6) that $\bar{y}$ is an optimal solution for $(D_{\epsilon,\ell})$ as required. $\square$

We may also want to consider larger perturbations. What if we wanted to analyze what would happen to our profits as the availability of one or many of our resources were being continuously increased? We could again use the techniques of linear optimization and duality theory to prove that the rate of increase in our profits will be monotone nonincreasing per unit increase in the availability of resources. For example, if our profit goes up by \$10 when b_ℓ is increased from 500 to 501, our profit will go up by no more than \$10 when b_ℓ is increased by one provided $b_\ell \geq 500$. This allows us to prove a version of the economic principle of *law of diminishing marginal returns* in our setting.

In addition to such strong connections to mathematical economics, linear programming also has close historical ties to economics. During the early 1900s, in the area of mathematical economics, Leontief (1905–1999) and others were working on various problems that had connections to linear optimization. One of the most notable models is the *input–output systems*. Suppose there are n major sectors of a national economy (e.g., construction, labor, energy, timber and wood, paper, banking, iron and steel, food, real estate, etc.). Let a_{ij} denote the amount of inputs from sector i required to produce one unit of the product of sector j (everything is in same units, dollars). Let $A \in \mathbb{R}^{n \times n}$ denote the matrix of these coefficients. (A is called the input–output matrix.) Now given $b \in \mathbb{R}^n$, where b_i represents the outside demands for the output of sector i, we solve the linear system of equations $Ax + b = x$. If $(A - I)$ is nonsingular, we get a unique solution x determining the output of each sector. Indeed, to have viable system, we need every x_j to be nonnegative. Otherwise, the economy requires some imports and/or some kind of outside intervention to function properly. Leontief proposed this model in the 1930s. Then in 1973 it won him the Nobel Prize in Economics.

Kantorovich [1912–1986] won the Nobel Prize in Economics in 1975. Koopmans [1910–1985] who had worked on the transportation problem (a generalization of the maximum weight bipartite matching problem) in the 1940s, as well as on input–output analysis (similar to Leontief's interests) for production systems, shared the Nobel Prize in Economics in 1975 with Kantorovich. According to Dantzig, the transportation problem was formulated (and a solution method proposed) by Hitchcock in 1941. Later, the problem was also referred to as the *Hitchcock–Koopmans transportation problem.*

The transportation problem together with earlier work on electrical networks (by mathematicians, engineers, and physicists) laid some of the ground work for what was to become network flows and network design. Next we will see another beautiful application of duality theory in the area of network flows and network/graph connectivity.

5.3   The maximum-flow–minimum-cut theorem

Imagine the following situation: you and your friend are in far away parts of the world, and would like to conduct online video chats. Both of you own a computer, and the two hosts are connected by a common network. Video chats, in particular if they are conducted in high definition (HD) resolution, require a substantial amount of end-to-end network bandwidth (some manufacturers suggest as much as 1500 Mb/s). In this section, we will look at the following question: given two computers and a connecting network, how many bits per second can we simultaneously transmit from one to the other?

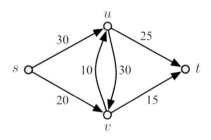

Let us model the network connecting the two computers as a *directed graph* $\vec{G} = (V, \vec{E})$ that has vertices V and arcs $\vec{E}$ which are ordered pairs of vertices. Recall that for an arc uv, u is the *tail* of the arc and v is the *head* of the arc. An arc is represented by an arrow directed from its tail to its head. The two computers correspond to specific vertices s and t in this graph, and every other vertex corresponds to a router or switch in the network. An arc qr represents a physical link between the two network entities corresponding to q and r, and it comes with a capacity $c_{qr} \geq 0$ that specifies the maximum number of bits per second that the link may carry. The figure above shows a very simple example.

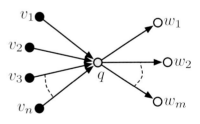

Our goal is to write an LP that computes the maximum transmission rate between s and t in the given network. In our model, we assume that data transmission is *lossless*; i.e. a data bit that is injected at vertex s vanishes only at vertex t. This yields the following *flow balance* condition in our network: consider a vertex $q \notin \{s, t\}$, and suppose that there are n arcs $v_1 q, \ldots, v_n q$ (coming into q) and m arcs $q w_1, \ldots, q w_m$ (going out of q). Any bit that travels on one of the n arcs with head q must subsequently traverse one of the m arcs with tail q.

We can express this as

$$f_x(q) := \sum \left(x_a : a \in \delta^+(q)\right) - \sum \left(x_a : a \in \delta^-(q)\right) = 0, \qquad (5.7)$$

where x_a denotes the number of bits on arc $a \in \vec{E}$, and $\delta^+(q)$ and $\delta^-(q)$ denote the set of arcs with tail q and head q, respectively. A vector $\bar{x}$ that satisfies (5.7) for all $q \in V \setminus \{s, t\}$ as well as $\mathbb{0} \leq x \leq c$ is called an s, t *flow*, and we let $f_x(s)$ be its *value*.

Our goal is to find an *st*-flow of maximum value, in other words we wish to solve the LP

$$
\begin{aligned}
\max \quad & f_x(s) \\
\text{subject to} \quad & \\
& f_x(q) = 0 \qquad (q \in V \setminus \{s, t\}) \\
& \mathbb{0} \leq x \leq c.
\end{aligned}
$$
(5.8)

A flow $\bar{x}$ is said to be *integer* if $\bar{x}$ is integer.

Example 21 The first number next to each arc a is the flow $\bar{x}_a$, the second is the capacity c_a. Note, $f_{\bar{x}}(u) = \bar{x}_{uw} + \bar{x}_{uv} - \bar{x}_{su} - \bar{x}_{zu} = 1 + 1 - (2 + 0) = 0$. Similarly, $f_{\bar{x}}(v) = f_{\bar{x}}(w) = f_{\bar{x}}(z) = 0$. As, $\mathbb{0} \leq \bar{x} \leq c$, $\bar{x}$ is a flow. Moreover, since $f_{\bar{x}}(s) = \bar{x}_{su} + \bar{x}_{sw} = 2 + 3$, the value of $\bar{x}$ is 5.

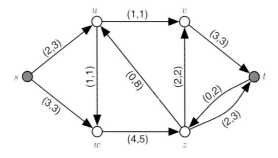

We have the following surprising result:

THEOREM 5.6   *Consider an st-flow problem with integer capacities $c \geq \mathbb{0}$. Then the value of a maximum st-flow is equal to the value of a maximum integer st-flow. Equivalently, among all optimal solutions to (5.8) there exists an optimal solution that is integer.*

We postpone the proof of this result. Consider an *st*-flow problem in a digraph $\overrightarrow{G} = (V, \overrightarrow{E})$ with capacities c. Let $U \subseteq V$ where $s \in U$ and $t \notin U$. We define $\delta^+(U)$ to be an *st*-cut of the directed graph $\overrightarrow{G}$.[3] $\delta^+(U)$ is the set of arcs with tail in U and head not in U. One of the computers in the motivating example corresponds to vertex $s \in U$ and the other to vertex $t \notin U$. Thus, all messages going from s to t have to use one of the arcs in $\delta^+(U)$. In particular, the *total capacity* $c\left(\delta^+(U)\right) := \sum \left(c_e : e \in \delta^+(U)\right)$ is an upper bound on the value of the *st*-flow. Surprisingly, this upper bound is tight, namely:

THEOREM 5.7 (Maximum-flow–minimum-cut).   *Consider an st-flow problem on a digraph $\overrightarrow{G} = (V, \overrightarrow{E})$ with capacities $c \geq \mathbb{0}$. Then the maximum value of an st-flow is equal to the minimum capacity of an st-cut.*

---

[3] This differs from the definition of *st*-cuts for graphs as we are not considering arcs in $\delta^-(U)$.

Example 21, continued. Consider $U = \{s, u, w\}$. Then $\delta^+(U) = \{uv, wz\}$ is an *st*-cut of capacity $c(\delta^+(U)) = c_{uv} + c_{wz} = 1 + 5$. Thus, 6 is an upper bound on the possible value of an *st*-flow. However, we can get a better upper bound by picking the *st*-cut $\delta^+(\{s, u\})$ as $c(\delta^+(\{s, u\})) = c_{sw} + c_{uv} + c_{uw} = 3 + 1 + 1 = 5$. As 5 is the flow value of the *st*-flow $\bar{x}$, it is a maximum *st*-flow and $\delta^+(\{s, u\})$ is a minimum capacity *st*-cut.

To prove Theorems 5.6 and 5.7, we will develop general tools that will allow us to prove that some LPs admit optimal solutions that are integer.

5.3.1   Totally unimodular matrices

In the following, we say that a matrix B is a *submatrix* of A if it is obtained by deleting some of A's rows and columns. A matrix A is *totally unimodular* (TU) if all of its square submatrices B have determinant in $\{0, +1, -1\}$. Note, that in particular it implies that all entries of a TU matrix must be 0, 1, or -1.

Let us first describe classes of TU matrices. Given a digraph $\overrightarrow{G}$, the *vertex-arc incidence matrix* of $\overrightarrow{G}$ is the matrix A where the columns correspond to arcs, the rows correspond to vertices, and

$$A_{v,e} = \begin{cases} +1 & \text{if } v \text{ is the tail of } e \\ -1 & \text{if } v \text{ is the head of } e \\ 0 & \text{otherwise.} \end{cases}$$

PROPOSITION 5.8   *Let A be the vertex-arc incidence matrix of a digraph. Then A is TU.*

Proof It suffices to show that $\det(B) = \pm 1$ for all $r \times r$ nonsingular submatrices of A. We will do this by induction on r. Clearly, the claim is true for $r = 1$ as every entry of A is either 0, or ± 1. Therefore, consider some nonsingular $r \times r$ submatrix B of A with $r \geq 2$, and assume that all nonsingular $r' \times r'$ submatrices B of A with $r' < r$ have determinant ± 1. Note that B cannot have a 0 column as otherwise $\det(B) = 0$.

Let us first assume that B has a column with a single nonzero entry. We may assume that column j has a single nonzero entry in row i. Then we can use Laplace's formula and expand the determinant of B along column j

$$\det(B) = (-1)^{i+j} B_{ij} \det(B^{ij}),$$

where B^{ij} is the submatrix of B obtained by removing row i and column j from B. B^{ij} is a nonsingular matrix of rank $r - 1$ and, using induction, has determinant ± 1. It now easily follows that the determinant of B is ± 1 as well. Thus, every column of B has exactly two nonzero entries. However, then B is singular as the sum of all rows of B is the zero vector.   □

We leave the proof of the next remark as an exercise.

Remark 5.9 *Let A be a TU matrix. Then the transpose $A^\top$ of A is TU. The matrix $[A|I]$ obtained from A by augmenting with the identity matrix is TU. The matrix $[A|0]$*

obtained from A by augmenting with the $\mathbb{0}$ matrix is TU. Moreover, any submatrix of A is TU.

THEOREM 5.10 *Suppose M is a TU matrix. For each of the following LPs, if the LP has an optimal solution, then it has an optimal solution that is integer.*

(1) $\max\{c^\top x : Mx = b, x \geq \mathbb{0}\}$, *assuming b is integer,*
(2) $\max\{d^\top x : Mx = \mathbb{0}, \mathbb{0} \leq x \leq c\}$, *assuming c is integer,*
(3) $\min\{c^\top \ell : M^\top y + \ell \geq d, \ell \geq \mathbb{0}\}$, *assuming d is integer.*

In fact, it will follow from the proof of (1) that every basic optimal solution is integer. Thus, if the Simplex algorithm terminates with an optimal solution, it will terminate with an integer optimal solution. Results (2) and (3) will be needed for the proofs of Theorems 5.6 and 5.7.

Proof of Theorem 5.10 Consider (1) first. We may assume that the rows of M are linearly independent (otherwise remove dependent rows). The problem is in SEF (standard equality form). Since there exists an optimal solution, it follows from Theorem 2.11 that there exists an optimal basic solution $\bar{x}$ for some basis B of M. Thus, $\bar{x}_N = \mathbb{0}$ and $\bar{x}_B$ is the unique solution to $M_B x_B = b$. It follows from Cramer's rule from linear algebra that for all $j \in B$

$$\bar{x}_j = \frac{\det(M_B^j)}{\det(M_B)},$$

where M_B^j is the matrix obtained from M_B by replacing the jth column by the vector b. Since b and M are integer matrices, so is M_B^j. Hence, $\det(M_B^j)$ is integer. Furthermore, since M is TU, $\det(M_B) \in \{-1, 1\}$. It follows that $\bar{x}_j$ is integer as required.

To prove (2) and (3), express the problem in SEF. Use Remark 5.9 to show that the resulting left-hand-side is TU. Then apply the result in (1). We leave the details as an exercise. □

5.3.2 Applications to *st*-flows

Consider an instance of the *st*-flow problem with digraph $\overrightarrow{G} = (V, \overrightarrow{E})$ and capacities $c \geq \mathbb{0}$. Let A be the vertex-arc incidence matrix of $\overrightarrow{G}$ and let M be the matrix obtained from A by removing rows s and t. Let d be the vector where $d_e = 1$ if $e \in \delta^+(s)$, $d_e = -1$ if $e \in \delta^-(s)$ and $d_e = 0$ otherwise. Then the LP (5.8) can be rewritten as follows:

$$\max \quad d^\top x$$

$$\text{subject to}$$

$$Mx = \mathbb{0}$$

$$\mathbb{0} \leq x \leq c.$$

(5.9)

By Proposition 5.8 and Remark 5.9, M is a TU matrix. Thus, Theorem 5.10(2) implies immediately Theorem 5.6. In the remainder of this chapter, we will prove the maximum-flow–minimum-cut theorem (Theorem 5.7).

The dual of (5.9) has a dual variable y_v for every row v of M, i.e. $v \in V \setminus \{s, t\}$ and a dual variable ℓ_e for every row e of I, i.e. $e \in \overrightarrow{E}$. Following Table 4.1, the dual is given by

$$\begin{aligned}
\min \quad & c^\top \ell \\
\text{subject to} \quad & \\
& M^\top y + \ell \geq d \\
& \ell \geq 0.
\end{aligned}$$

(5.10)

Note that (5.10) has one constraint for every arc e. We wish to rewrite that LP but before we proceed, let us first simplify our life, and assume that vertex s has no in-arcs, and that t has no out-arcs. This assumption is easily seen to be benign: add new vertices s' and t', and infinite capacity arcs $s's$ and tt' to $\overrightarrow{G}$. Note that the old graph has an st-flow of value γ if and only if the new graph has an $s't'$-flow of the same value. Thus, we need to distinguish between four types of arcs only, namely: E_1, the set of arcs with tail and head distinct from s, t; E_2, the set of arcs with tail s but head distinct from t; E_3, the set of arcs with tail distinct from s but head t; E_4, consists of arc st if it exists or the empty set otherwise.

Hence, (5.10) can be written as

$$\begin{aligned}
\min \quad & \sum (c_e \ell_e : e \in \overrightarrow{E}) \\
\text{subject to} \quad & \\
& y_u - y_v + \ell_{uv} \geq 0 && (uv \in E_1) \\
& \qquad -y_v + \ell_{sv} \geq 1 && (sv \in E_2) \\
& y_u \qquad + \ell_{ut} \geq 0 && (ut \in E_3) \\
& \qquad\quad \ell_{st} \geq 1 && (st \in E_4) \\
& \qquad\quad \ell \geq 0.
\end{aligned}$$

(5.11)

We can unify constraints corresponding to E_1, E_2, E_3, E_4 by introducing constants $y_s = -1, y_t = 0$, and by adding $y_s = -1$ to both sides of constraints E_2, adding $-y_t = 0$ to both sides of constraints E_3, and adding $y_s - y_t = -1$ to both sides of constraint E_4. We then obtain

$$\begin{aligned}
\min \quad & \sum (c_e \ell_e : e \in \overrightarrow{E}) \\
\text{subject to} \quad & \\
& y_s = -1, y_t = 0 \\
& y_u - y_v + \ell_{uv} \geq 0 && \left(uv \in \overrightarrow{E}\right) \\
& \qquad\quad \ell \geq 0.
\end{aligned}$$

(5.12)

We are now ready for the proof of the maximum-flow–minimum-cut theorem.

Proof of Theorem 5.7   Note that $x = 0$ is a feasible solution to (5.8), and that $\ell = 1$ and $y_u = 0$ for all $u \neq s$ is a feasible solution to (5.12). If an LP and its dual both have feasible solutions, then they both have optimal solutions (see Corollary 4.2). It follows that (5.12) has an optimal solution (note that in addition to the duality theory we used the above arguments which show equivalence of various LP formulations). By Proposition 5.8 and Remark 5.9, M is a TU matrix. Theorem 5.10(3) implies there exists an optimal solution $\bar{\ell}, \bar{y}$ to (5.10) (respectively (5.12)) that is integer.

Let $W = \{u \in V : \bar{y}_u \leq -1\}$. Then $s \in W$ but $t \notin W$. It follows that $\delta^+(W)$ is an st-cut of the digraph $\vec{G}$. Since $c, \bar{\ell} \geq 0$

$$\sum\left(c_e\bar{\ell}_e : e \notin \delta^+(W)\right) \geq 0. \tag{5.13}$$

Since $\bar{y}, \bar{\ell}$ are feasible for (5.12), for every arc uv where $u \in W$ and $v \notin W$ we have $\bar{\ell}_{uv} \geq \bar{y}_v - \bar{y}_u$. By definition of W and since $\bar{y}$ is integer, $\bar{y}_v \geq 0$ and $\bar{y}_u \leq -1$. Thus, $\bar{\ell}_{uv} \geq 1$. It follows that

$$\sum\left(c_e\bar{\ell}_e : e \in \delta^+(W)\right) \geq c\left(\delta^+(W)\right). \tag{5.14}$$

Combining (5.13) and (5.14) we obtain that the value of $\bar{y}, \bar{\ell}$ is at least $c(\delta^+(W))$.

Consider y', ℓ' where $y'_u = -1$ if $u \in W$ and $y'_u = 0$ otherwise and where $\ell'_e = 1$ if $e \in \delta^+(W)$ and $\ell'_e = 0$ otherwise. Then it can be readily checked that y', ℓ' is a feasible solution to (5.12) of value $c(\delta^+(W))$. Since $\bar{y}, \bar{\ell}$ is an optimal solution of value at least $c(\delta^+(W))$, it follows that the optimal value to (5.12) is exactly $c(\delta^+(W))$. It follows from strong duality (Theorem 4.3) that the optimal value of (5.8) (i.e. the maximum st-flow value) is equal to $c(\delta^+(W))$. $\qquad\square$

Exercises

1 Consider the IP (5.1) and the dual of its LP relaxation (5.2). Construct a simple instance of the set-cover problem for which the optimal values to the aforementioned IP and LP are distinct.

2 Consider the LP problem $\max\{c^\top x : Ax \leq b, x \geq 0\}$. Suppose it has an optimal solution x^*.
(a) Define $b(\theta) := b + \theta e_1$ and let (P_θ) be the LP problem obtained from the above by replacing b by $b(\theta)$. Prove that for all $\theta \geq 0$, (P_θ) has an extreme point that is optimal.
(b) Let $z^*(\theta)$ denote the optimal value of (P_θ). Prove that $z^* : \mathbb{R}_+ \to \mathbb{R}$ is a piecewise linear function of θ.
(c) Prove that the following set is convex:

$$\left\{ \begin{pmatrix} \theta \\ z \end{pmatrix} \in \mathbb{R}^2 : \theta \geq 0, z \leq z^*(\theta) \right\}.$$

3 We discussed the shortest paths problem as well as the maximum-flow problem. In this exercise, we will try to formulate the shortest paths problem using a network flow idea. Given s and t, consider formulating the problem of finding a shortest path from s to t as the problem of finding one unit flow from s to t of minimum cost, where the cost of sending one unit of flow along an arc is the length of that arc. Note that for every vertex except s and t in the graph, we must have *flow conservation* (if there is flow coming into the vertex, it must leave the vertex). For s, one unit of flow must leave it, and one unit of flow must arrive at t. Using the variables

$$x_{ij} := \begin{cases} 1, & \text{if there is a unit flow on the arc } (i,j); \\ 0, & \text{otherwise} \end{cases}$$

and the above reasoning, formulate the shortest path problem as an IP. Then consider the LP relaxation of your IP, write down the dual and interpret the strong duality theorem for this pair of primal–dual LP problems.

4 (Advanced) Consider a digraph $\overrightarrow{G} = (V, \overrightarrow{E})$ with nonnegative integer arc capacities c and a pair of distinct vertices s and t. We propose an alternate formulation for the st-flow problem. Assign one variable x_P for every directed st-path, and consider

max $\sum (x_P : P$ is a directed st-path$)$

subject to

$$\sum (x_P : e \in P \text{ and } P \text{ is a directed } st\text{-path}) \le c_e \qquad \left(e \in \overrightarrow{E}\right) \qquad \text{(P)}$$

$$x \ge 0.$$

(a) Show that the optimum value to (P) is the value of the maximum st-flow.
(b) Write the dual (D) of (P) with variables y.
(c) Explain what the maximum-flow–minimum-cut Theorem 5.7 says for the pair (P) and (D).
(d) Write the CS (complementary slackness) conditions for the x variables.
(e) Write the CS conditions for the y variables.
(f) Find an intuitive explanation as to why the CS condition for the y variables must hold. HINT: What can you say about the capacity usage of arcs in a minimum cut?
(g) Find an intuitive explanation as to why the CS condition for the x variables must hold. HINT: What can you say about how the flow along each directed st-path is routed.

5 Sophie, a student at a private university decides that she needs to save money on food to pay her tuition. Yet she does not want to compromise her health by not obtaining the recommended daily allowance of various nutrients. Let $1, \ldots, n$ correspond to different food options available at the cafeteria, and let $1, \ldots, m$ be essential nutrients. Sophie collects the following information: for $j = 1, \ldots, n$, c_j is the cost of a single serving of food j; for $i = 1, \ldots, m$, b_i is the daily recommended allowance of nutrient i, and for $j = 1, \ldots, n$ and $i = 1, \ldots, m$, a_{ij} is the number of units of nutrient i a single serving of food j contains.

(a) Help Sophie formulate a linear program (P) that will minimize her daily food expenses while guaranteeing that she gets the recommended doses of essential nutrients.
(b) Write the dual (D) of (P).
(c) Write the complementary slackness conditions.
(d) Give an economic interpretation of the dual variables and of the dual.
(e) Give an economic interpretation of the complementary slackness conditions.

6   Solving integer programs

In Chapter 2, we have seen how to solve LPs using the simplex algorithm, an algorithm that is still widely used in practice. In Chapter 3, we discussed efficient algorithms to solve the special class of IPs describing the shortest path problem and the minimum cost perfect matching problem in bipartite graphs. In both these examples, it is sufficient to solve the LP relaxation of the problem.

Integer programming is widely believed to be a difficult problem (see Appendix A). Nonetheless, we will present algorithms that are guaranteed to solve IPs in finite time. The drawback of these algorithms is that the running time may be exponential in the worst case. However, they can be quite fast for many instances, and are capable of solving many large-scale, real-life problems.

These algorithms follow two general strategies. The first attempts to reduce IPs to LPs – this is known as the *cutting plane* approach and will be described in Section 6.2. The other strategy is a divide and conquer approach and is known as *branch and bound* and will be discussed in Section 6.3. In practice, both strategies are combined under the heading of *branch and cut*. This remains the preferred approach for all general purpose commercial codes.

In this chapter, in the interest of simplicity we will restrict our attention to *pure IPs* where *all* the variables are required to be integer. The theory developed here extends to *mixed IPs* where only *some* of the variables are required to be integer, but the material is beyond the scope of this book.

6.1   Integer programs versus linear programs

In this section, we show that in principle the problem of solving an IP can be reduced to the problem of solving an LP. This reduction is in general very time consuming and thus does not directly lead to an efficient algorithm to solve IPs. It will however provide a framework for understanding the idea of cutting planes.

Consider a (possibly infinite) set S of points in $\mathbb{R}^n$. The *convex hull* of S, denoted $\mathrm{conv}(S)$, is defined as the smallest convex set that contains S. In particular, $S = \mathrm{conv}(S)$ if and only if S is a convex set. We illustrate this concept in a couple of examples.

Example 22 Consider the case where S contains the following three points in $\mathbb{R}^2$:

$$x^{(1)} = \begin{pmatrix} 0 \\ 2 \end{pmatrix} \qquad x^{(2)} = \begin{pmatrix} 1 \\ 0 \end{pmatrix} \qquad \text{and} \qquad x^{(3)} = \begin{pmatrix} 3 \\ 1 \end{pmatrix}.$$

Then the convex hull of S is given by the triangle with vertices $x^{(1)}, x^{(2)}, x^{(3)}$ as indicated in Figure 6.1. Note, that this triangle is convex and contains S. Moreover, any convex set that contains S must contain all the points in this triangle.

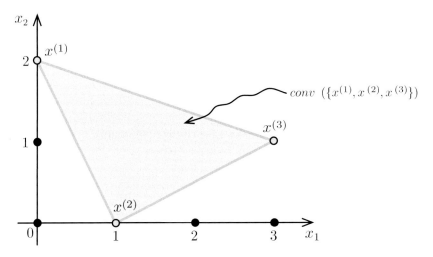

Figure 6.1 Illustration of Example 22.

Example 23 Consider the following polyhedron:

$$P = \left\{ \begin{pmatrix} x_1 \\ x_2 \end{pmatrix} : \begin{pmatrix} -1 & 0 \\ 1 & -1 \\ -1 & -3 \end{pmatrix} \begin{pmatrix} x_1 \\ x_2 \end{pmatrix} \leq \begin{pmatrix} -3/2 \\ 5/2 \\ -3 \end{pmatrix} \begin{matrix} (1) \\ (2) \\ (3) \end{matrix} \right\}.$$

The polyhedron P is represented in Figure 6.2. Each of the constraints (1), (2), (3) corresponds to a halfspace as indicated in the figure. Let us define S to be the set of all integer points in P. Note, that in this case S is an infinite set of points. Then the convex hull of S is described by another polyhedron

$$Q = \left\{ \begin{pmatrix} x_1 \\ x_2 \end{pmatrix} : \begin{pmatrix} -1 & 0 \\ 1 & -1 \\ 0 & -1 \end{pmatrix} \begin{pmatrix} x_1 \\ x_2 \end{pmatrix} \leq \begin{pmatrix} -2 \\ 2 \\ -1 \end{pmatrix} \begin{matrix} (a) \\ (b) \\ (c) \end{matrix} \right\}.$$

The polyhedron Q is represented in Figure 6.2. Each of the constraints (a), (b), (c) corresponds to a halfspace as indicated in the figure.

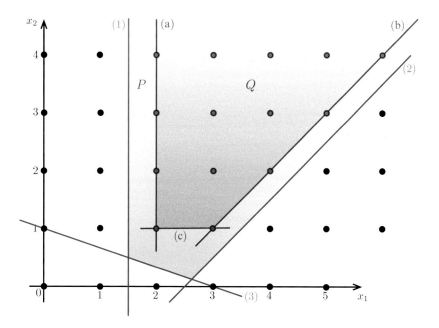

Figure 6.2 Illustration of Example 23.

In Example 23, we saw that the convex hull of the set of all integer points in a polyhedron is a polyhedron. This is no accident; indeed, we have the following *fundamental theorem of integer programming* due to Meyer [48] that relates feasible solutions of IPs to feasible solutions of LPs:

THEOREM 6.1   *Consider the following polyhedron $P = \{x \in \mathbb{R}^n : Ax \leq b\}$ where all entries of A and b are rational numbers. Let S be the set of all integer points in P. Then the convex hull of S is a polyhedron Q described by a matrix and a vector where all entries are rational.*

We omit the proof of this result in this book (e.g., see [14]). The condition that all entries of A and b be rational numbers cannot be excluded from the hypothesis.

Consider now an IP of the following form:

$$\max \quad c^\top x$$
$$\text{subject to}$$
$$Ax \leq b \tag{6.1}$$
$$x \text{ integer.}$$

Let us assume that all entries in the matrix A and the vector b are rational. This is a natural assumption since numbers stored on a computer can only be recorded with finite precision and hence are rational. Let S denote the set of all integer points satisfying the constraints $Ax \leq b$; i.e. S is the set of feasible solutions to the IP (6.1). It follows from Theorem 6.1 that the convex hull of S is a polyhedron; i.e. that $\mathrm{conv}(S) = \{x \: : \: A'x \leq b'\}$ for some matrix A' and vector b' where all entries of A' and b' are rational.

Let us define the following LP:

$$\max \quad c^\top x$$
$$\text{subject to} \qquad\qquad\qquad\qquad (6.2)$$
$$A'x \leq b'.$$

THEOREM 6.2   *The following hold for the IP (6.1) and the LP (6.2):*

1. *(6.1) is infeasible if and only if (6.2) is infeasible,*
2. *(6.1) is unbounded if and only if (6.2) is unbounded,*
3. *every optimal solution to (6.1) is an optimal solution to (6.2),*
4. *every optimal solution to (6.2) that is an extreme point is an optimal solution to (6.1).*

We omit the proof of these results in this book.

Example 24 We illustrate the previous theorem. Suppose that (6.1) is of the form

$$\max \quad (1, 1)x$$
$$\text{subject to}$$

$$\begin{pmatrix} 2 & 1 \\ 1 & 2 \\ -1 & -4 \\ -1 & 0 \end{pmatrix} x \leq \begin{pmatrix} 7 \\ 7 \\ -4 \\ -\frac{1}{2} \end{pmatrix} \qquad \begin{matrix} (1) \\ (2) \\ (3) \\ (4) \end{matrix} \qquad \text{(IP)}$$

$$x \text{ integer.}$$

We describe the feasible region of (IP) in Figure 6.3. Each of the constraints (1), (2), (3), (4) corresponds to a halfspace as indicated in the figure. Thus, the set of all feasible solutions to (IP) is given by

$$S = \left\{ \begin{pmatrix} 1 \\ 1 \end{pmatrix}, \begin{pmatrix} 1 \\ 2 \end{pmatrix}, \begin{pmatrix} 1 \\ 3 \end{pmatrix}, \begin{pmatrix} 2 \\ 1 \end{pmatrix}, \begin{pmatrix} 2 \\ 2 \end{pmatrix}, \begin{pmatrix} 3 \\ 1 \end{pmatrix} \right\}.$$

It follows that (6.2) is of the form

$$\max \quad (1, 1)x$$
$$\text{subject to}$$

$$\begin{pmatrix} -1 & 0 \\ 0 & -1 \\ 1 & 1 \end{pmatrix} x \leq \begin{pmatrix} -1 \\ -1 \\ 4 \end{pmatrix} \qquad \begin{matrix} (a) \\ (b) \\ (c). \end{matrix} \qquad \text{(P)}$$

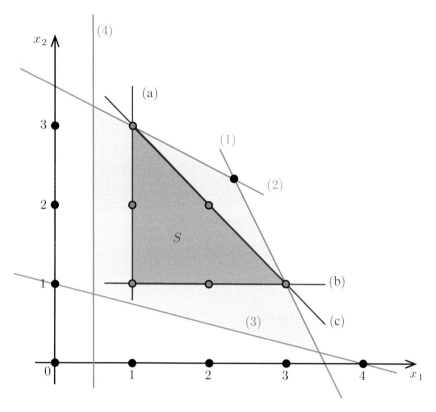

Figure 6.3 Illustration of Example 24.

Each of the constraints (a), (b), (c) corresponds to a halfspace as indicated in the previous figure. Since the objective function of (IP) is $z = x_1 + x_2$, the set of all optimal solutions to (IP) is given by $(3, 1)^\top$, $(2, 2)^\top$, and $(1, 3)^\top$. It can be readily checked that each of these points is also an optimal solution to (P) as was predicted by Theorem 6.2(c). Note, however, that every point in the line segment between the points $(1, 3)^\top$ and $(3, 1)^\top$ is an optimal solution to (P). In particular, it is not true that every optimal solution to (P) is an optimal solution to (IP). As predicted by Theorem 6.2(d), the extreme points $(1, 3)^\top$, $(3, 1)^\top$ are optimal solutions to (IP).

Theorem 6.2 tells us that in a theoretical sense integer programming reduces to linear programming: it suffices to compute an optimal solution that is an extreme point of (6.2). The catch here is that the system $A'x \leq b'$ in the LP (6.2) is in general much larger (exponentially larger) than the system $Ax \leq b$ in the IP (6.1), hence it cannot be described completely in practice. What we do instead is to try to approximate the description of the polyhedron $\{x : A'x \leq b'\}$ using a limited number of constraints. One way of proceeding is described in the next section.

Exercises

1 Consider the IP

$$\max \qquad x_1 + 10x_2$$
$$\text{subject to}$$
$$\begin{pmatrix} 1 & -20 \\ 1 & 20 \end{pmatrix} x \le \begin{pmatrix} 0 \\ 20 \end{pmatrix}$$
$$x \ge 0, \ x \text{ integer.}$$

(a) What is the minimum Euclidean distance, in $\mathbb{R}^2$, between the optimal solution of the IP and of its LP relaxation? Explain.

(b) Give an example of an IP for which the minimum Euclidean distance between its optimal solution and that of its LP relaxation is at least 100.

(c) Give an example of an IP which has no feasible integer solutions, but its LP relaxation has a feasible set in $\mathbb{R}^2$ of area at least 100.

2 Consider the IP problem

$$\max \{x_1 + 6x_2 : x_1 + 9x_2 \le 10, 2x_1 \le 19, x \ge 0 \text{ and integer}\}.$$

(a) Graph the feasible region of the LP relaxation of IP; indicate the feasible solutions of the IP on the graph. Then find the optimal solutions of the LP relaxation and the IP.

(b) Focus on the optimal solution of the LP relaxation. What are the closest (in the sense of Euclidean distance) feasible solutions of the IP? Are any of these closest feasible IP solutions optimal for the IP? Now replace the objective function with $\max x_1 + 8.5x_2$. Answer the same questions again. What related conclusions can you draw from these examples?

(c) Create a family of IP problems with two nonnegative integer variables parameterized by an integer $k \ge 10$ (in the above example $k = 10$) generalizing the above example and its highlighted features. Then answer the parts (a) and (b) for all values of k.

3 Let $S \subseteq \mathbb{R}^n$, show there exists a unique set R that is inclusion-wise minimal and satisfies the property that $S \subseteq R$ and that R is convex. In other words, the term "convex hull" is well-defined.

6.2   Cutting planes

Let us motivate our approach with an example.

Example 25 Suppose we wish to solve the following IP:

$$\text{max} \quad (2,5)x$$

subject to

$$\begin{pmatrix} 1 & 4 \\ 1 & 1 \end{pmatrix} x \leq \begin{pmatrix} 8 \\ 4 \end{pmatrix} \qquad \begin{matrix} (1) \\ (2) \end{matrix} \qquad \text{(IP)}$$

$$x \geq \mathbb{0} \text{ integer.}$$

As we do not know how to deal with the integrality conditions, we shall simply ignore them. Thus, we solve the LP relaxation (LP1) obtained by removing integrality conditions from (IP). We obtain the optimal solution $x^{(1)} = (8/3, 4/3)^\top$. Unfortunately, $x^{(1)}$ is not integral, and thus it is not a feasible solution to (IP).

We wish to find a valid inequality $(\star)$ which satisfies the following properties:

(I) $(\star)$ is valid for (IP); i.e. every feasible solution of (IP) satisfies $(\star)$,
(II) $x^{(1)}$ does not satisfy $(\star)$.

An inequality that satisfies both (I) and (II) is called a *cutting plane* for $x^{(1)}$. We will discuss in the next section how to find such a cutting plane but for the time being let us ignore that problem and suppose that we are simply given such an inequality

$$x_1 + 3x_2 \leq 6. \qquad (\star)$$

We add $(\star)$ to the system (LP1) to obtain a new LP (LP2). Because $(\star)$ satisfies (I), every feasible solution to (IP) is a feasible solution to (LP2). Moreover, (II) implies that $x^{(1)}$ is not feasible for (LP2), so in particular the optimal solution to (LP2) will be distinct from $x^{(1)}$. In our case, (LP2) has an optimal solution $x^{(2)} = (3, 1)^\top$. Note that $x^{(2)}$ is integral. Since it maximizes the objective function over all feasible solutions of (LP2), it also maximizes the objective function over all feasible solutions of (IP). Hence, $x^{(2)}$ is optimal for (IP).

We describe the feasible region of (LP1) in Figure 6.4 (i). Each of the constraints (1), (2) corresponds to a halfspace as indicated in the figure. We also indicate the optimal solution $x^{(1)} = (8/3, 4/3)^\top$. Since $x^{(1)}$ was not integral, we added a cutting plane $(\star)$: $x_1 + 3x_2 \leq 6$ and obtained (LP2). We describe the feasible region of (LP2) in Figure 6.4 (ii). Constraint $(\star)$ corresponds to a halfspace as indicated in the figure. We also indicate the optimal solution $x^{(2)} = (3, 1)^\top$ of (LP2).

Let us formalize the procedure. Consider the following pair of optimization problems:

$$\text{max}\{c^\top x \, : \, x \in S_1\}, \qquad \text{(P1)}$$

$$\text{max}\{c^\top x \, : \, x \in S_2\}. \qquad \text{(P2)}$$

If $S_2 \supseteq S_1$, then we say that (P2) is a *relaxation* of (P1). For instance, (P1) may be an IP and (P2) its LP relaxation.

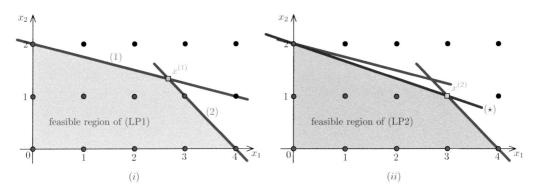

Figure 6.4 Illustration of Example 25.

Remark 6.3 *Suppose (P2) is a relaxation of (P1). Then:*

(a) *if (P2) is infeasible, (P1) is infeasible, and*
(b) *if $\bar{x}$ is optimal for (P2) and $\bar{x}$ is feasible for (P1), $\bar{x}$ is optimal for (P1).*
(c) *if $\bar{x}$ is an optimal solution for (P2), $c^{\top}\bar{x}$ is an upper bound for (P1).*

Proof (a) If (P2) is infeasible, i.e. $S_2 = \emptyset$, then $S_1 = \emptyset$, i.e. (P1) is infeasible. (b) Suppose $\bar{x}$ is an optimal solution to (P2) that is also a feasible solution to (P1). Then $\bar{x}$ maximizes $c^{\top}x$ among all $x \in S_2$, so in particular it maximizes $c^{\top}x$ among all $x \in S_1$. Hence, $\bar{x}$ is an optimal solution to (P1). (c) Since $S_2 \supseteq S_1$, the optimal value to (P2) is at least as large as the optimal value of (P1). □

Algorithm 6.8 shows the cutting plane method for solving the IP

$$
\begin{aligned}
\max \quad & c^{\top}x \\
\text{subject to} \quad & \\
& Ax \leq b \\
& x \geq \mathbb{0}, \quad x \text{ integer.}
\end{aligned}
\tag{IP}
$$

It follows from Remark 6.3 that the algorithm gives correct answers whenever it terminates. If cutting planes are chosen carefully, the algorithm can also be shown to terminate (e.g., see Chapter 9 of [5]). We did not discuss the possibility of (LP) being unbounded. In this case, it is possible for (IP) to be unbounded or it being infeasible, but we do not know which case occurs. In practice, we are often in situations where we know that the (IP) is not unbounded.

6.2.1   Cutting planes and the simplex algorithm

We have yet to show how to find cutting planes. Let us revisit Example 25 from the previous section. After introducing slack variable x_3 for constraint (1) and x_4 for constraint (2), we can rewrite (IP) as follows:

Algorithm 6.8 Cutting plane algorithm

1: **loop**
2: Let (LP) denote $\max\{c^\top x : Ax \le b\}$.
3: **if** (LP) is infeasible **then**
4:   **stop** (IP) is infeasible.
5: **end if**
6: Let $\bar{x}$ be the optimal solution to (LP).
7: **if** $\bar{x}$ is integral **then**
8:   **stop** $\bar{x}$ is an optimal solution to (IP).
9: **end if**
10: Find a cutting plane $a^\top x \le \beta$ for $\bar{x}$
11: Add constraint $a^\top x \le \beta$ to the system $Ax \le b$
12: **end loop**

$$\max \qquad (2,5,0,0)x$$

subject to

$$\begin{pmatrix} 1 & 4 & 1 & 0 \\ 1 & 1 & 0 & 1 \end{pmatrix} x = \begin{pmatrix} 8 \\ 4 \end{pmatrix} \qquad \text{(IP')}$$

$$x \ge 0, \text{ integer.}$$

Observe, that since $x_3 = 8 - x_1 - 4x_2$ and $x_4 = 4 - x_1 - x_2$ and since x_1, x_2 are integer in (IP) we must have x_3, x_4 integer as well. Thus, (IP') is equivalent to (IP) in Example 25. Let (LP1') denote the LP relaxation of (IP'). Using the simplex algorithm, we find the optimal basis $B = \{1, 2\}$ and rewrite (LP1') in canonical form for the optimal basis $B = \{1, 2\}$ to obtain

$$\max \qquad 12 + (0, 0, -1, -1)x$$

subject to

$$\begin{pmatrix} 1 & 0 & -1/3 & 4/3 \\ 0 & 1 & 1/3 & -1/3 \end{pmatrix} x = \begin{pmatrix} 8/3 \\ 4/3 \end{pmatrix}$$

$$x \ge 0.$$

The corresponding basic solution is $(8/3, 4/3, 0, 0)^\top$. It implies that $(8/3, 4/3)^\top$ is the optimal solution for (LP1) in Example 25. We will use the previous LP to find a cutting plane. Consider any constraint of the previous LP where the right-hand side is fractional. In this case, we have a choice and select the first constraint, namely

$$x_1 - \frac{1}{3}x_3 + \frac{4}{3}x_4 = \frac{8}{3}.$$

Every feasible solution to (LP1') satisfies the above constraint, and hence clearly also

$$x_1 - \frac{1}{3}x_3 + \frac{4}{3}x_4 \le \frac{8}{3}.$$

Observe that every variable is nonnegative. Thus, if we replace any of the coefficient for the variables in the previous equation by a smaller value, the resulting inequality will still be valid for (LP1'). In particular

$$x_1 + \left\lfloor -\frac{1}{3} \right\rfloor x_3 + \left\lfloor \frac{4}{3} \right\rfloor x_4 = x_1 - x_3 + x_4 \leq \frac{8}{3}$$

is valid for (LP1'), where $\lfloor \alpha \rfloor$ denotes the largest integer smaller or equal to α. Let $\bar{x}$ be any feasible solution to (IP'). Then $\bar{x}$ is a feasible solution to (LP1') and $\bar{x}_1 - \bar{x}_3 + \bar{x}_4 \leq \frac{8}{3}$. However, as $\bar{x}$ is integer, $\bar{x}_1 - \bar{x}_3 + \bar{x}_4$ is integer, it follows that $\bar{x}$ satisfies

$$x_1 - x_3 + x_4 \leq \left\lfloor \frac{8}{3} \right\rfloor = 2. \tag{6.3}$$

Moreover, $(8/3, 4/3, 0, 0)^\top$ does not satisfy (6.3). Hence, (6.3) is a cutting plane. Recall, that (IP') implies that $x_3 = 8 - x_1 - 4x_2$ and $x_4 = 4 - x_1 - x_2$. Substituting this in (6.3) yields the constraint $x_1 + 3x_2 \leq 6$ which was the cutting plane ($\star$) given in Example 25.

To proceed further, we add a slack variable x_5 to the constraint (6.3) and modify the LP relaxation (LP1') of (IP') by adding the resulting constraint. Note that by definition, $x_5 = 2 - x_1 + x_3 - x_4$ and hence x_5 will be integer, provided $x_1, x_3,$ and x_4 are. We thus obtain

$$\begin{aligned}
\max \quad & (2, 5, 0, 0, 0)x \\
\text{subject to} \quad & \\
& \begin{pmatrix} 1 & 4 & 1 & 0 & 0 \\ 1 & 1 & 0 & 1 & 0 \\ 1 & 0 & -1 & 1 & 1 \end{pmatrix} x = \begin{pmatrix} 8 \\ 4 \\ 2 \end{pmatrix} \qquad \text{(LP2')} \\
& x \geq 0.
\end{aligned}$$

We solve (LP2') using the two-phase simplex algorithm (we need to use phase I, since we do not have a feasible basis), and rewrite (LP2') in canonical form for the optimal basis $B = \{1, 2, 3\}$ to obtain

$$\begin{aligned}
\max \quad & 11 + \left(0, 0, 0, -\frac{1}{2}, -\frac{3}{2} \right) x \\
\text{subject to} \quad & \\
& \begin{pmatrix} 1 & 0 & 0 & 3/2 & -1/2 \\ 0 & 1 & 0 & -1/2 & 1/2 \\ 0 & 0 & 1 & 1/2 & -3/2 \end{pmatrix} x = \begin{pmatrix} 3 \\ 1 \\ 1 \end{pmatrix} \\
& x \geq 0.
\end{aligned}$$

The basic optimal solution is $(3, 1, 1, 0, 0)^\top$. Since it is integer, it follows (see Remark 6.3) that it is an optimal solution to (IP'). Note that it corresponds to the optimal solution $(3, 1)^\top$ of (IP) in Example 25.

Let us generalize this argument. Consider an IP

$$
\begin{aligned}
\max \quad & c^\top x \\
\text{subject to} \quad & \\
& Ax = b \\
& x \geq 0, \text{ integer.}
\end{aligned}
\tag{IP}
$$

We solve the LP relaxation (LP) of (IP) using the simplex procedure. If (LP) has no solution, then neither does (IP). Suppose we obtain an optimal basis B of (LP). We rewrite (LP) in canonical form for B to obtain an LP of the form

$$
\begin{aligned}
\max \quad & \bar{z} + \bar{c}_N^\top x_N \\
\text{subject to} \quad & \\
& x_B + \bar{A}_N x_N = \bar{b} \\
& x \geq 0.
\end{aligned}
$$

The corresponding basic solution $\bar{x}$ is given by $\bar{x}_B = \bar{b}$ and $\bar{x}_N = 0$. If $\bar{b}$ is integer, then so is $\bar{x}$, and $\bar{x}$ is an optimal solution to (IP) (see Remark 6.3). Thus, we may assume that $\bar{b}_i$ is fractional for some index i. Let ℓ denote the ith basic variable. Constraint i is of the form

$$
x_\ell + \sum_{j \in N} \bar{A}_{ij} x_j = \bar{b}_i.
$$

As this constraint is valid for (LP) with equality, it is also valid with $\leq$. Observe that every variable is nonnegative. Hence, the following inequality will be valid for (LP)

$$
x_\ell + \sum_{j \in N} \lfloor \bar{A}_{ij} \rfloor x_j \leq \bar{b}_i.
$$

For $\bar{x}$ feasible for (IP), the LHS of the previous equation is an integer. Hence, the following constraint is valid for (IP):

$$
x_\ell + \sum_{j \in N} \lfloor \bar{A}_{ij} \rfloor x_j \leq \lfloor \bar{b}_i \rfloor.
\tag{$\star$}
$$

Remark 6.4 *Constraint ($\star$) is a cutting plane for the basic solution $\bar{x}$.*

Proof It suffices to show that $\bar{x}$ does not satisfy ($\star$). Since $\bar{x}_j = 0$ for all $j \in N$, the left-hand side of ($\star$) is $\bar{x}_\ell = \bar{b}_i$. As $\bar{b}_i$ is fractional, $\bar{b}_i > \lfloor \bar{b}_i \rfloor$. Then the left-hand side is larger than the right-hand side for $\bar{x}$. $\qquad\square$

Exercises

1 Suppose that we solve an LP relaxation of a pure IP and that for the optimal basis $B = \{1, 2, 3\}$ of (P), we have the following canonical form:

$$\max \quad \left(0 \quad 0 \quad 0 \quad \tfrac{17}{12} \quad \tfrac{1}{12} \quad \tfrac{5}{12} \right) x + 140$$

subject to

$$\begin{pmatrix} 1 & 0 & 0 & \tfrac{111}{60} & -\tfrac{13}{12} & -\tfrac{1}{60} \\ 0 & 1 & 0 & \tfrac{1}{10} & \tfrac{1}{2} & -\tfrac{1}{10} \\ 0 & 0 & 1 & 2 & \tfrac{11}{2} & -\tfrac{1}{12} \end{pmatrix} x = \begin{pmatrix} \tfrac{9}{2} \\ 1 \\ \tfrac{11}{10} \end{pmatrix}$$

$$x \geq 0.$$

Indicate which constraints lead to cutting planes and for each such constraints generate the corresponding cutting plane.

2 Consider the following LP:

$$\max\{c^\top x : Ax = b, x \geq 0\}. \tag{P}$$

Let (D) be the dual of (P). Suppose you found an optimal solution $\bar{x}$ of (P) and an optimal solution $\bar{y}$ of (D). You are being told however that you forgot to include one important constraint

$$\sum_{j=1}^{n} a_j x_j \leq \beta. \tag{$\star$}$$

(a) Construct a new LP (P$'$) by adding a slack variable to ($\star$) and by adding the resulting constraint to (P).
(b) Show that if $\bar{x}$ satisfies ($\star$) then $\bar{x}$ is an optimal solution to (P$'$).
(c) Compute the dual (D) of (P) and the dual (D$'$) of (P$'$).
(d) Show how to compute a feasible solution y' of (D$'$) from your solution $\bar{y}$ of (D). Note, this works whether or not $\bar{x}$ satisfies ($\star$).

3 Consider the following IP:

$$\max \quad (1, 0, 0)x$$

subject to

$$\left(1 \quad 1 \quad \tfrac{1}{2} \right) x = \tfrac{3}{2} \tag{IP}$$

$$x \geq 0, x \text{ integer.}$$

Denote by (P1) the LP relaxation of (IP):
(a) Solve (P1) with the simplex (this is trivial). Denote by $x^{(1)}$ the optimal basic solution.
(b) Find a cutting plane ($\star$) for $x^{(1)}$.

Denote by (P2) the LP in standard equality form obtained from (P1) by adding constraint ($\star$) and by adding a slack variable x_4:

(c) Solve (P2) with the simplex, starting with the basis $B = \{3, 4\}$. Denote by $x^{(2)}$ the optimal basic solution.

(d) Using (c), deduce an optimal solution to (IP).

4 Consider the following LP:

$$\max\{c^\top x : Ax = b, x \geq 0\}. \tag{P}$$

Suppose that (P) is the LP relaxation of a pure IP, and suppose that (P) is written in canonical form for the optimal basis B. Let $\bar{x}$ be the basic solution corresponding to B. Let $k \in N$, then constraint k of $Ax = b$ is of the form

$$x_\ell + \sum_{j \in N} A_{kj} x_j = b_k,$$

where ℓ is the kth basic variable. Define

$$N^* := \{j \in N : A_{kj} \text{ is not an integer}\}$$

and suppose b_k is not integer.

(a) Prove that

$$\sum_{j \in N^*} x_j \geq 1$$

is a cutting plane for $\bar{x}$.

(b) Prove that

$$\sum_{j \in N} x_j \geq 1$$

is a cutting plane for $\bar{x}$.

(c) Compare the cutting planes obtained in (a) and (b). Which one do you think is a better cut? Explain why and prove your claims.

5 Let $G = (V, E)$ be a graph with edge weights w_e for all $e \in E$. Recall that (3.24) is the IP formulation of the minimum weight matching problem. Let $S \subseteq V$ be a subset of the vertices of G where $|S|$ is odd.

(a) Show (by combining constraints and using rounding) that the following inequality is valid for the (IP), i.e. that all feasible solutions to the (IP) satisfy this constraint

$$\sum (x_{ij} : ij \in E \text{ and } i \in S, j \in S) \leq \left\lfloor \frac{|S|}{2} \right\rfloor.$$

In other words, the sum of all variables corresponding to edges with both ends in S is at most the number of vertices in S divided by two and rounded down.

(b) Find an intuitive explanation as to why this inequality is valid.

6.3   Branch and bound

In this section, we introduce a solution method for IPs, called *branch and bound*. It is a "divide and conquer" approach for solving IP problems. We illustrate some of the main ideas using a production example.

Every year the University of Waterloo readies itself for the annual homecoming festivities. The university expects hundreds of alumni and their families to visit its campus for the occasion, and campus stores in particular expect to be busier than usual. Two items promise to be in especially high demand during these times: the university's infamous *pink tie*, as well as the (almost) equally popular *pink bow tie*. With only a few weeks to go until homecoming, it is high time to replenish stocks for these two items. The university manufactures its ties from its own special cloth of which it currently has 45ft^2 in stock. For ten pink bow ties, it needs 9ft^2 of cloth, and producing ten pink ties requires 11ft^2 of the raw material. In addition to cloth, the manufacturing process also requires labor which is particularly short during the next few weeks: only a total of six work days are available. Manufacturing of the ties is done in batches of ten. Producing ten pink bow ties takes 1.5 days, and producing ten pink ties takes a day. The university expects a profit of $6 for a pink bow tie, and a profit of $5 for a pink tie. How much of each product should it produce?

As in Chapter 1, we can formulate this problem as an IP. We introduce variables x and y where $10x$ and $10y$ is the number of bow ties and ties produced. In the following IP, the first constraint imposes labor restriction and the second imposes restriction due to limited material. The objective function computes the profit.

$$\max \quad 60x + 50y$$
$$\text{subject to}$$
$$\begin{pmatrix} 1.5 & 1 \\ 9 & 11 \end{pmatrix} \begin{pmatrix} x \\ y \end{pmatrix} \leq \begin{pmatrix} 6 \\ 45 \end{pmatrix} \tag{6.4}$$
$$x, y \geq 0, \ x, y \text{ integer.}$$

How can we solve this IP? Let us use what we know: linear programming! We drop the integrality constraints and consider the LP relaxation of (6.4). Remark 6.3(b) states that if we solve the LP relaxation of an IP, and obtain a solution, all of whose variables are integers, then it is also an optimal solution of the original IP. Unfortunately, solving the LP relaxation of (6.4) yields the fractional solution $x = 2.8, y = 1.8$ with value 258. Remark 6.3(c) implies 258 is upper bound on the value of any feasible solution to (6.4).

We will now use the fractional solution $(2.8, 1.8)^\top$ to *partition* the feasible region of (6.4). In the following, we let *Subproblem 1* denote the LP relaxation of (6.4). We observe that every integer feasible solution for (6.4) must have either $x \leq 2$ or $x \geq 3$, and that the current fractional solution does not satisfy either one of these constraints. Thus, we now *branch* on variable x and create two additional subproblems:

Subproblem 2 Subproblem 1 + Constraint $x \geq 3$.
Subproblem 3 Subproblem 1 + Constraint $x \leq 2$.

Clearly, in order to find an optimal solution to (6.4) it suffices to find the best integer solution in both Subproblem 2 and Subproblem 3. None of the two subproblems contains $(2.8, 1.8)^\top$ and this solution can therefore not re-occur when solving the LP relaxation of either of these subproblems. We now choose any one of the above two subproblems to process next. Arbitrarily, we pick Subproblem 2. Solving the problem yields the optimal solution $x = 3$ and $y = 1.5$ with value 255. This solution is still not integral (as y is fractional), but because of Remark 6.3(c), 255 gives us a new, tighter upper bound on the maximum value of any integral feasible solution for this subproblem.

The subproblem structure explored by the branch and bound algorithm is depicted in Figure 6.5. Each of the subproblems generated so far is shown as a box. These boxes are referred to as *branch and bound nodes*. The two nodes for Subproblems 2 and 3 are connected to their parent node, and the corresponding edges are labeled with the corresponding constraints that were added to Subproblem 1. The entire figure is commonly known as the *branch and bound tree* generated by the algorithm.

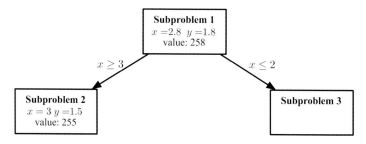

Figure 6.5 The branch and bound tree after two iterations.

The optimal solution for Subproblem 2 is still not integral as the value of y is fractional. We branch on y, and generate two more subproblems:

Subproblem 4 Subproblem 2 + Constraint $y \leq 1$.
Subproblem 5 Subproblem 2 + Constraint $y \geq 2$.

Running the simplex algorithm on Subproblem 5, we quickly find that the problem is infeasible. Hence, this subproblem has no fractional feasible solution, and thus no integral one either. Solving Subproblem 4, we find the solution $x = \frac{10}{3}$ and $y = 1$ of value 250. We generate two more subproblems by branching on x:

Subproblem 6 Subproblem 4 + Constraint $x \leq 3$.
Subproblem 7 Subproblem 4 + Constraint $x \geq 4$.

Solving Subproblem 6 yields integral solution $x = 3$, $y = 1$ of value 230. Solving Subproblem 7 gives integral solution $x = 4$, $y = 0$ of value 240, which is the current best. Figure 6.6 shows the current status.

So far, we have found a feasible integral solution of value 240 for Subproblem 7. Thus, 240 is a lower bound for (6.4). We continue exploring the tree at Subproblem 3. Solving the subproblem yields solution $x = 2$ and $y = 2.45$ and value 242.73. Branching on y gives two more subproblems:

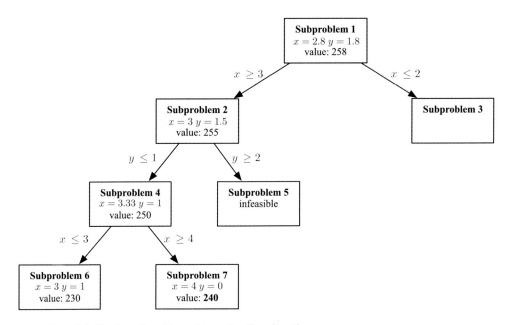

Figure 6.6 The branch and bound tree after three iterations.

Subproblem 8 Subproblem 3 + Constraint $y \leq 2$.
Subproblem 9 Subproblem 3 + Constraint $y \geq 3$.

Solving Subproblem 8 gives integral solution $x = 2, y = 2$ of value 220; this is inferior to our current lower bound of 240 and can thus be discarded. Solving Subproblem 9 gives the fractional solution $x = 4/3$, and $y = 3$ of value 230. Since 230 is an upper bound for any integer solution to Subproblem 9, but our current lower bound is 240, we cannot find a better solution within this subproblem. We therefore might as well stop branching here.

Formally, whenever the optimal value of the current subproblem is at most the value of the best integral solution that we have already found, then we may stop branching at the current node. We say: we *prune* the branch of the branch and bound tree at the current node.

We are done as no unexplored branches of the branch and bound tree remain. The final tree is shown in Figure 6.7. The optimal solution to the original IP (6.4) is therefore $x = 4$, $y = 0$ and achieves a value of 240. Therefore, in order to optimize profit, the university is best advised to invest all resources into the production of pink bow ties!

We conclude this section with a brief discussion. The branch and bound algorithm presented here can be viewed as a *smart* enumeration algorithm: it uses linear programming to partition the space of feasible solutions. For example, in the pink tie example discussed above, it is instructive to see that any feasible integral solution occurs in one of the subproblems corresponding to the leaves of the tree in Figure 6.7. The algorithm is smart as it uses the optimal LP value of a subproblem to possibly prune it. Some-

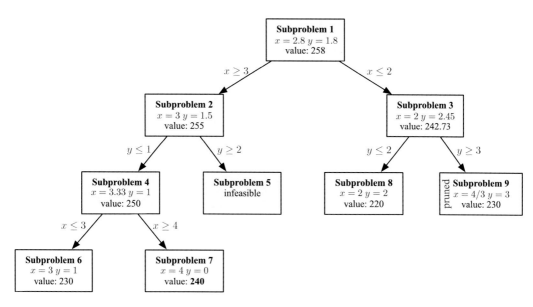

Figure 6.7 The final branch and bound tree.

times, such pruning can save exploring a large number of potential solutions hidden in a subproblem.

The algorithm as described is quite flexible in many ways, and in our description we made many arbitrary choices. For example, if we solve a subproblem, and the solution has more than one fractional variable, how do we decide which variable to branch on? And once we have branched on a variable, which of the generated problems do we solve next? The *depth-first search* style exploration chosen in our example, where newly generated subproblems are explored first, is popular as it leads to integral solutions quickly. However, many other strategies have been analyzed.

Why do we not simply present the strategy that works best? Well, such a strategy likely does not exist, as is it widely believed that no efficient algorithm exists to solve all IPs (see Appendix A). In practice, however, branch and bound is nearly always superior to simple enumeration of feasible solutions, and is used in some form or the other in most commercial codes.

We conclude with one last comment regarding implementation. In this section, we reduced the problem of finding an optimal production plan for a tie production problem to that of solving a series of nine linear programming problems. The reader may notice that these problems are extremely similar; i.e. branching on a variable merely adds a single constraint to an LP for which we know an optimal basic solution. Can we use such an optimal basic solution for a parent problem to compute solutions for the generated subproblems faster? The answer is yes! The so-called *dual* simplex algorithm applies in situations where a single constraint is added to an LP, and where the old optimal solution is rendered infeasible by this new constraint (see Exercise 2 in Section 6.2). In many cases, the algorithm *reoptimizes* quickly from the given primal-infeasible starting point.

Exercises

1 Suppose we use a branch and bound scheme to solve an IP that is a maximization problem. The following figure describes the partial enumeration tree. For each subproblem, we indicate the value of the objective function as well as whether the solution is an integer solution or not. Indicate in the figure which subproblems we still need to solve (if any). In particular, can we already deduce the value of the optimal solution to the IP?

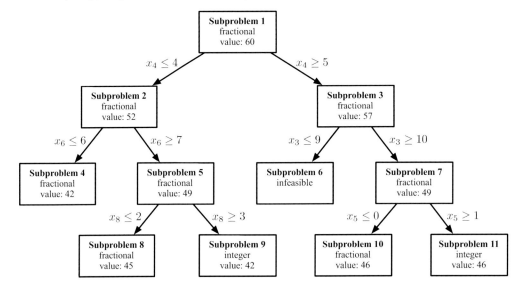

2 Use branch and bound to solve the following LP:

$$max \qquad 18x_1 + 10x_2 + 6x_3 + 4x_4$$

subject to

$$12x_1 + 10x_2 + 8x_3 + 6x_4 \leq 18$$
$$0 \leq x_i \leq 1, x_i \text{ integer} \qquad\qquad (i = 1, 2, 3, 4).$$

At each iteration select the active subproblem with the largest upper bound. Branch on fractional variables, i.e. if $x_i = 1/3$ create subproblem $x_i \leq 0$ and $x_i \geq 1$.

You do not need to use the simplex algorithm to find a solution to each of the relaxations. We illustrate a simple technique on the relaxation of the original problem. Consider the ratios

$$\frac{18}{12} > \frac{10}{10} > \frac{6}{8} > \frac{4}{6}.$$

To maximize the objective function, we set x_1 as large as possible; then x_2 as large as possible, etc. In this case, it yields, $x_1 = 1$, $x_2 = 6/10$ and $x_3 = x_4 = 0$. Proceed in a similar way to solve each of the other relaxations.

3 Consider the IP

$$\max \left\{ -x_{n+1} \; : \; 2x_1 + 2x_2 + \cdots + 2x_n + x_{n+1} = n, \; x \in \{0,1\}^{n+1} \right\},$$

where n is an odd positive integer. Prove that the branch and bound algorithm (without using cuts) will have to examine at least $2^{\lfloor \frac{n}{2} \rfloor}$ subproblems before it can solve the main problem, in the worst case.

6.4   Traveling salesman problem and a separation algorithm*

Consider a traveling salesman who needs to visit cities $1, 2, 3, \ldots, n$ in some order and end up at the city where he started. Cost of traveling from city i to city j is given by c_{ij}. The goal is to find a *tour* of these n cities, visiting every city exactly once, such that the total travel costs are minimized. This is a very well-studied IP problem called the *traveling salesman problem* (TSP).

Let us start formulating the problem as an IP. Let the variable x_{ij} take the value 1, if the chosen tour of the cities includes going directly from city i to city j. Otherwise, x_{ij} will take the value zero. It is clear that every tour must enter every city exactly once and every tour must leave every city exactly once. Note that we only have the variables x_{ij} for distinct pairs i, j. Then we can write the following objective function and the constraints:

$$\min \quad \sum_{i=1}^{n} \sum_{j=1, j \neq i}^{n} c_{ij} x_{ij}$$

subject to

$$\sum_{j=1, j \neq i}^{n} x_{ij} \;=\; 1 \qquad (i \in \{1, \ldots, n\}) \tag{6.5}$$

$$\sum_{i=1, i \neq j}^{n} x_{ij} \;=\; 1 \qquad (j \in \{1, \ldots, n\})$$

$$x_{ij} \;\in\; \{0,1\} \qquad (i \in \{1, \ldots, n\}, \, j \in \{1, \ldots, n\}).$$

This formulation may seem very familiar – this is because it is! We have seen it as a formulation of the 'assignment problem' in Section 1.3.1. While this is a correct formulation of the assignment problem, it is not a correct formulation of TSP. Suppose $n = 8$ and consider the $\bar{x}$ vector which is zero in every entry except $\bar{x}_{12} = \bar{x}_{23} = \bar{x}_{34} = \bar{x}_{45} = \bar{x}_{51} = \bar{x}_{67} = \bar{x}_{78} = \bar{x}_{86} = 1$. We can verify that $\bar{x}$ is a feasible solution of the above IP. However, $\bar{x}$ is not a feasible solution of the TSP, since it does not represent a tour of the cities $1, 2, \ldots, 8$ ($\bar{x}$ represents union of two disjoint subtours, one for the cities $1, 2, 3, 4, 5$ and another for $6, 7, 8$ – see Figure 6.8).

This leads us to the observation that *every tour must leave every nontrivial subset of the cities* $1, 2, \ldots, n$. Notice that we already have constraints for subsets of size 1 and subsets of size $(n-1)$. Therefore, we add to the formulation the following *subtour elimination constraints*:

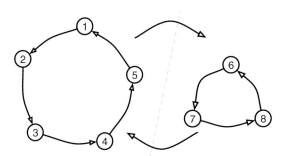

Figure 6.8 TSP subtours.

$$\sum \left(x_{ij} : i \in U, j \notin U \right) \geq 1,$$

where U is a subset of the cities $\{1, \ldots, n\}$ with at least 2 and at most $(n-2)$ cities in it.

Now we can verify that with the addition of these subtour elimination constraints, we have a correct IP formulation of TSP.

A more careful look at this new formulation reveals that the number of constraints in this formulation is perhaps too large. The number of subtour elimination constraints is the total number of subsets of $\{1, \ldots, n\}$ of size at least two and at most $(n-2)$

$$\binom{n}{2} + \binom{n}{3} + \cdots + \binom{n}{n-2} = 2^n - 2(n+1).$$

Since this number is an exponential function of the number of cities n, even for moderate values of n it can become intractable to even write down every constraint in the LP relaxation of this IP.

Wouldn't it be great if we could solve this LP without writing every constraint down? It turns out that in this case we can solve the LP by utilizing a cutting plane scheme and our knowledge of maximum-flow–minimum-cut problems and by avoiding having to consider every subtour elimination constraint explicitly.

Suppose we are given $\bar{x}$, which is a feasible solution of the assignment problem (6.5). $\bar{x}$ may or may not satisfy some of the subtour elimination constraints. Recall that a *directed graph* $\overrightarrow{G} = (V, \overrightarrow{E})$ has vertices V and arcs $\overrightarrow{E}$ that are ordered pairs of vertices. We set up a directed graph on the vertex set $\{1, 2, \ldots, n\}$ based on $\bar{x}$. For every pair i, j such that $\bar{x}_{ij} > 0$, we include an arc $\overrightarrow{ij}$ and an arc $\overrightarrow{ji}$ in our graph, each with capacity $\bar{x}_{ij}$. Let us call this directed graph $\overrightarrow{G}$. Suppose our $\bar{x}$ violates the subtour elimination constraint for set U. Then pick a vertex $u \in U$ and $v \notin U$. Consider the maximum-flow problem on $\overrightarrow{G}$, where the capacities are given by $\bar{x}_{ij}$, the source is $s := u$, and the sink is $t := v$. Since $\bar{x}$ violates the subtour elimination constraint for U, the capacity of the cut determined by U is strictly less than 1. On the other hand, if $\bar{x}$ satisfies all subtour elimination constraints, then the value of the maximum flow from u to v would be at least 1 for every pair of vertices u and v in $\overrightarrow{G}$.

This discussion suggests an algorithm. The only obvious problem is that we do not know ahead of time whether $\bar{x}$ satisfies all the subtour elimination constraints, or, if it violates a subtour elimination constraint, we do not know for which U. However, we can set up the above directed graph $\overrightarrow{G}$ and try every distinct pair u, v as our s and t. For each such pair, we solve the underlying maximum-flow–minimum-cut problem. There are $\binom{n}{2}$ pairs, so we solve $O(n^2)$ problems (in fact, if we are a bit more careful, we see that solving $2(n-1)$ problems suffices). If for every one of these problems the objective value is at least 1, then we have proven that the given $\bar{x}$ satisfies all subtour elimination constraints (in a cutting plane scheme, $\bar{x}$ would be an optimal solution of the current LP relaxation since it satisfies all the subtour elimination constraints; in this case, it is an optimal solution of the LP relaxation given by all the subtour elimination constraints). The only remaining option is that for one of the maximum-flow–minimum-cut problems we solve we find that the objective value is strictly less than 1. In this case, we find a corresponding minimum cut of capacity strictly less than 1, and this cut defines our set U for which $\bar{x}$ violates the corresponding subtour elimination constraint

$$\sum \left(x_{ij} : i \in U, j \notin U\right) \geq 1.$$

We add this constraint to our current LP relaxation and find an optimal solution yielding the next $\bar{x}$ to consider.

Given $\bar{x}$, the above algorithm either verifies that $\bar{x}$ satisfies all the subtour elimination constraints or delivers a subtour elimination constraint which is violated by $\bar{x}$. Such algorithms are called *separation algorithms*. (The constraint delivered by the algorithm *separates* $\bar{x}$ from the polyhedron defined by the subtour elimination constraints.)

There are more efficient ways of finding a minimum cut in a graph (rather than solving very many maximum-flow problems). Also, there are techniques for making this idea into a polynomial-time algorithm to solve the LP relaxation of the above IP formulation of TSP. These are beyond the scope of this introductory book.

Exercises

1 Consider the IP formulation (6.5). We claimed that it was an instance of the bipartite assignment problem from the Introduction. However, since we excluded variables $x_{i,i}$ in the formulation (6.5), this formulation does not seem to be exactly the same as an instance of the assignment problem. Suppose you have an algorithm to solve the assignment problem and you are given an instance of the IP (6.5). How would you define $c_{i,i}$ so that using the algorithm for the assignment problem you can solve the IP (6.5)?

2 Consider the traveling salesman problem with cities $\{1, 2, \ldots, n\}$ and costs of travel c_{ij}. Write an IP formulation of TSP using only the variables

$$x_{ijk} := \begin{cases} 1, & \text{if on the } k\text{th leg of the trip, salesman goes from city } i \text{ to city } j; \\ 0, & \text{otherwise.} \end{cases}$$

Your IP formulation should have $O(n^3)$ constraints.

3 Consider the IP formulation of the TSP using the subtour elimination constraints. Instead of adding subtour elimination constraints, add new variables $u_2, u_3, \ldots, u_n$ and the constraints

$$nx_{ij} + u_i - u_j \leq n - 1, \ \forall \text{ distinct pairs } i \neq 1, j \neq 1.$$

Further add the constraints that $u_2, u_3, \ldots, u_n$ are all nonnegative. Prove that this is also a correct formulation of TSP.

4 Consider the "global" maximum-flow–minimum-cut problem that we had to solve to determine whether a given vector $\bar{x}$ satisfies all subtour elimination constraints. Prove that this global maximum-flow–minimum-cut problem can be solved by solving $2(n-1)$ maximum-flow problems on the same graph G.

5 Given a graph $G = (V, E)$, a *Hamiltonian cycle* is a tour of the vertices of G which only uses the edges in E. Suppose we want to solve the problem of determining whether a given graph G has a Hamiltonian cycle and if it does to find one. Suppose that you have an algorithm to solve the TSP. How would you use the algorithm for TSP to solve the Hamiltonian cycle problem efficiently? (You are allowed to run the TSP algorithm only once, and only with $n \le |V|$.) Prove all your claims.

6 Suppose we are given the data for a TSP with nonnegative travel costs $c_{ij} = c_{ji}$ also satisfying the *triangle inequalities*

$$c_{ij} + c_{jk} \ge c_{ik}, \text{ for every distinct triple } i, j, k.$$

That is, the costs are nonnegative and the cost of going from city i to city k is cheapest when we travel directly. Moreover, the cost of going from i to j is the same as the cost of going from j to i. Such instances of TSP are called *metric-TSP* (since c_{ij} are consistent with a metric). Given a metric-TSP instance, construct an efficient algorithm to find a minimum cost tree T on the vertices $\{1, 2, \ldots, n\}$ such that T contains all of $\{1, 2, \ldots, n\}$ (such a tree is called a *spanning tree*, since it contains all the vertices in the underlying graph). Try to use T with some additional ideas to make up a tour whose total cost is guaranteed to be at most twice the cost of the optimal TSP tour. Prove all your claims.

6.5   Further reading and notes

When we study pure IPs or mixed IPs, we usually assume that the data are rational numbers (no irrational number is allowed in the data). This is a reasonable assumption. Besides, allowing irrational numbers in data can cause some difficulties, e.g., (IP) may have an optimal solution and the LP relaxation may be unbounded (consider the problem $\max\{x_1 : x_1 - \sqrt{2}x_2 = 0, x_1, x_2 \text{ integer}\}$). Or, we may need infinitely many cutting planes to reach a solution of the IP (to construct an example for this, suitably modify the previous example).

In practice, for hard IP problems, we use much more sophisticated rules for branching, pruning, choosing the next subproblem to solve, etc. Moreover, as we mentioned before, we use a hybrid approach, called branch and cut, which uses cutting planes for the original IP as well as some of the subproblems. In addition, in many hard cases, we adjust our pruning strategy so that instead of guaranteeing an optimal solution to the IP, we lower our standards and strive for generating feasible solutions to the IP that are provably within say 10% of the optimum value (or 2% of the optimum value, etc.). For further background in integer programming, see, for instance Wolsey [71].

The traveling salesman problem is one of the most well-known problems in optimization. It has very wide-ranging applications, from scheduling to hardware design to obviously transportation/distribution systems, just to name a few. This problem has also been a prime example of a combinatorial optimization problem for developing new theoretical, algorithmic, and computational techniques over many decades. There are many books dedicated to the subject (see [45, 15]). There are more efficient ways of finding a "global" minimum cut in a graph (rather than solving very many maximum-flow problems). See, for instance, Karger [34] and the references therein. Also, there are techniques for making this idea into a polynomial-time algorithm to solve the LP relaxation of the above IP formulation of TSP (see [30]). These are beyond the scope of this introductory book.

7 Nonlinear optimization

In this chapter, we will show that solving general NLP is likely to be difficult, even when the problem has small size. One reason is that the feasible region (the set of all feasible solutions) of an NLP is not always convex. We therefore, turn our attention to the special case where the feasible region is convex. We discuss optimality conditions and give a brief overview of a primal–dual polynomial algorithm for linear programming based on ideas from nonlinear optimization.

Key concepts covered in this chapter include: convex functions, level sets, epigraphs, subgradients, Lagrangians, the Karush–Kuhn–Tucker theorem, and primal–dual interior-point algorithms.

7.1 Some examples

Recall the definition of NLP, given in Section 1.6. Let us look at some examples of NLP.

Example 26 Suppose $n = 2$ and $m = 4$ in (1.35) and that for $x = (x_1, x_2)^\top \in \mathbb{R}^2$ we have

$$f(x) := x_2, \quad \text{and}$$

$$g_1(x) := -x_1^2 - x_2 + 2, \quad g_2(x) := x_2 - \frac{3}{2}, \quad g_3(x) := x_1 - \frac{3}{2}, \quad g_4(x) := -x_1 - 2.$$

The feasible region is a subset of $\mathbb{R}^2$. It corresponds to the union of the two shaded regions in Figure 7.1. For instance, $g_1(x) = -x_1^2 - x_2 + 2 \leq 0$ or equivalently $x_2 \geq 2 - x_1^2$. Thus, the solution to the first constraint of the NLP is the set of all points above the curve indicated by g_1 in the figure. As we are trying to find among all points $x = (x_1, x_2)^\top$ in the feasible region the one that minimizes $f(x) = x_2$, the unique optimal solution will be the point $a = (-2, -2)^\top$. Observe that the feasible region is not convex, indeed it is not

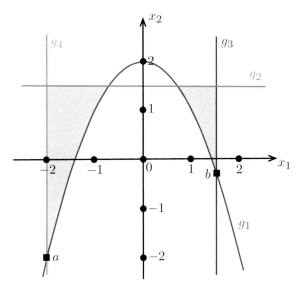

Figure 7.1 Feasible region for Example 26.

even "connected" (i.e. there does not exist a continuous curve contained in the feasible region joining points a and b of the feasible region).

Example 27 Suppose $n = 2$ and $m = 3$ in (1.35) and that for $x = (x_1, x_2)^\top \in \mathbb{R}^2$ we have

$$f(x) := -x_1 - x_2, \quad \text{and}$$

$$g_1(x) := -x_1 + x_2^2, \quad g_2(x) := -x_2 + x_1^2, \quad g_3(x) := -x_1 + \frac{1}{2}.$$

The feasible region is a subset of $\mathbb{R}^2$. It corresponds to the shaded region in Figure 7.2. For instance, $g_2(x) = -x_2 + x_1^2 \le 0$ or equivalently $x_2 \ge x_1^2$. Thus, the solution to the second constraint of the NLP is the set of all points above the curve indicated by g_2 in the figure. For g_1, we interchange the role of x_1 and x_2 in g_2. We will prove that the feasible solution $a = (1, 1)^\top$ is an optimal solution to the NLP. Observe that the feasible region is convex in this case.

Observe that if in (1.35) every function f and g_i is *affine* (i.e. a function of the form $a^\top x + \beta$ for a given vector a and a given constant β), then we have $f(x) = c^\top x + c_0$, and $g_i(x) = a_i^\top x - b_i$, for every $i \in \{1, 2, \ldots, m\}$ and we see that our nonlinear optimization problem (1.35) becomes a linear optimization problem. Thus, NLPs generalize linear

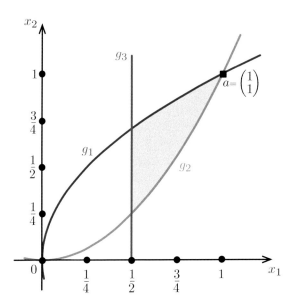

Figure 7.2 Feasible region for Example 27.

programs, but, as we will show in the next section, they are much harder to solve than linear programs.

7.2   Some nonlinear programs are very hard

7.2.1   NLP versus 0,1 integer programming

Nonlinear optimization programs can be very hard in general. Starting with a linear programming problem, even if we only allow very simple, and just mildly nonlinear functions in our formulations, we can run into very difficult optimization problems. For instance, suppose that for every $j \in \{1, \ldots, n\}$ we have the constraints

$$x_j^2 - x_j \leq 0,$$
$$-x_j^2 + x_j \leq 0.$$

Then the constraints define the same feasible region as the quadratic equations: $x_j^2 = x_j$, for every $j \in \{1, 2, \ldots, n\}$. Therefore, the feasible region of these constraints is exactly the 0,1 vectors in $\mathbb{R}^n$. Now if we also add the constraints $Ax \leq b$, we deduce that every 0,1 integer programming problem can be formulated as an NLP. In other words, 0,1 integer programming is reducible to nonlinear programming. Moreover, this reduction is clearly polynomial (for the notion of polynomial reduction, see Appendix A.4.1). Therefore, solving an arbitrary instance of an NLP is at least as hard as solving an arbitrary instance of a 0,1 integer programming problem. In particular, as 0,1 feasibility is *NP-hard*, so is nonlinear programming feasibility (see Appendix A.4.2).

7.2.2 Hard small-dimensional instances

We might be tempted to think that if the number of variables in the NLP is very small, perhaps then solving NLP would be easy. However, this is not the case as we will illustrate.

Pierre de Fermat in 1637, conjectured the following result:

THEOREM 7.1   *There do not exist integers $x, y, z \geq 1$ and integer $n \geq 3$ such that $x^n + y^n = z^n$.*

Fermat wrote his conjecture in the margin of a journal, and claimed to have a proof of this result but that it was too large to fit in the margin. This conjecture became known as *Fermat's last theorem*. The first accepted proof of this result was published in 1995, some 358 years after the original problem was proposed. We will show that a solution to a very simple looking NLP with only four variables has the following key property: the optimal objective value of zero is attained, if and only if Fermat's last theorem is false. Hence, solving this particular NLP is at least as hard as proving Fermat's last theorem! In this NLP, see (1.35), we have four variables ($n = 4$) and four constraints ($m = 4$). For $x = (x_1, x_2, x_3, x_4)^\top \in \mathbb{R}^4$

$$f(x) := \left(x_1^{x_4} + x_2^{x_4} - x_3^{x_4}\right)^2 + (\sin \pi x_1)^2 + (\sin \pi x_2)^2 + (\sin \pi x_3)^2 + (\sin \pi x_4)^2,$$

and

$$g_1(x) := 1 - x_1, \quad g_2(x) := 1 - x_2, \quad g_3(x) := 1 - x_3, \quad g_4(x) := 3 - x_4.$$

Observe that the feasible region of this NLP is given by

$$S := \{x \in \mathbb{R}^4 : x_1 \geq 1, x_2 \geq 1, x_3 \geq 1, x_4 \geq 3\}.$$

Note that $f(x)$ is a sum of squares. Therefore, $f(x) \geq 0$ for every $x \in \mathbb{R}^4$ and it is equal to zero if and only if every term in the sum is zero, i.e.

$$x_1^{x_4} + x_2^{x_4} = x_3^{x_4} \text{ and } \sin \pi x_1 = \sin \pi x_2 = \sin \pi x_3 = \sin \pi x_4 = 0.$$

The latter string of equations is equivalent to x_j being integer for every $j \in \{1, 2, 3, 4\}$. Moreover, the feasibility conditions require $x_1 \geq 1, x_2 \geq 1, x_3 \geq 1, x_4 \geq 3$. Therefore, $f(\bar{x}) = 0$ for some $\bar{x} \in S$ if and only if $\bar{x}_j$ is a positive integer for every j with $\bar{x}_4 \geq 3$, and $\bar{x}_1^{\bar{x}_4} + \bar{x}_2^{\bar{x}_4} = \bar{x}_3^{\bar{x}_4}$. That is, if and only if Fermat's last theorem is false. Surprisingly, it is not difficult to prove (and it is well-known) that the infimum of f over S is zero. Thus, the difficulty here lies entirely in knowing whether the infimum can be attained.

We just argued that some nonlinear optimization problems can be very hard even if the number of variables is very small (e.g., at most 10) or even if the nonlinearity is bounded (e.g., at most quadratic functions). However, carefully isolating the special nice structures in some classes of nonlinear programming problems and exploiting these structures allow us to solve many large-scale nonlinear programs in practice.

7.3   Convexity

Consider an NLP of the form given in (1.35) and denote by S its feasible region. We say that $\bar{x} \in S$ is *locally optimal* if for some positive $d \in \mathbb{R}$ we have that $f(\bar{x}) \leq f(x)$ for every $x \in S$ where $\|x - \bar{x}\| \leq d$, i.e. no feasible solution of the NLP that is within distance d of $\bar{x}$ has better value than $\bar{x}$. We sometimes call an optimal solution to the NLP, a *globally optimal* solution. It is easy to verify that if S is convex and f is a linear function (or more generally a *convex function* defined in the next section), then locally optimal solutions are globally optimal (see Exercise 2). However, when S is not convex (or when f is not a convex function), we can have locally optimal solutions that are not globally optimal. This is illustrated in Example 26. There, b is locally optimal, yet $a \neq b$ is the only globally optimal solution.

A natural scheme for solving an optimization problem is as follows: find a feasible solution, and then repeatedly either (i) show that the current feasible solution is globally optimal, using some optimality criteria, or (ii) try to find a better feasible solution (here better might mean one with better value for instance–though this may not always be possible). The simplex algorithm for linear programming follows this scheme. Both steps (i) and (ii) may become difficult when the feasible region is not convex. We will therefore turn our attention to the case where the feasible region is convex. In this section, we establish sufficient conditions for the feasible region of an NLP to be convex (see Remark 7.4).

7.3.1   Convex functions and epigraphs

We say that the *function* $f: \mathbb{R}^n \to \mathbb{R}$ is *convex* if for every pair of points $x^{(1)}, x^{(2)} \in \mathbb{R}^n$ and for every $\lambda \in [0, 1]$

$$f\left(\lambda x^{(1)} + (1 - \lambda)x^{(2)}\right) \leq \lambda f\left(x^{(1)}\right) + (1 - \lambda)f\left(x^{(2)}\right).$$

In other words, f is convex if for any two points $x^{(1)}, x^{(2)}$ the unique linear function on the line segment between $x^{(1)}$ and $x^{(2)}$ that takes the same value for $x^{(1)}$ and $x^{(2)}$ as f dominates the function f. An example of a convex function is given in Figure 7.3. An example of a nonconvex function is given in Figure 7.4.

Example 28 Consider the function $f: \mathbb{R} \to \mathbb{R}$ where $f(x) = x^2$. Let $a, b \in \mathbb{R}$ be arbitrary, and consider an arbitrary $\lambda \in [0, 1]$. To prove that f is convex, we need to verify[1] that

$$[\lambda a + (1 - \lambda)b]^2 \overset{?}{\leq} \lambda a^2 + (1 - \lambda)b^2.$$

---

[1]   Applying the definition is not the only way to prove that a function is convex. In this particular case, for instance, we can compute its second derivative and observe that it is nonnegative.

Clearly, we may assume that $\lambda \notin \{0, 1\}$, i.e. that $0 < \lambda < 1$. After expanding and simplifying the terms, it suffices to verify that

$$\lambda(1 - \lambda)2ab \overset{?}{\leq} \lambda(1 - \lambda)(a^2 + b^2),$$

or equivalently as $\lambda, (1 - \lambda) > 0$ that $a^2 + b^2 - 2ab \geq 0$, which is clearly the case as $a^2 + b^2 - 2ab = (a - b)^2$, and the square of any number is nonnegative.

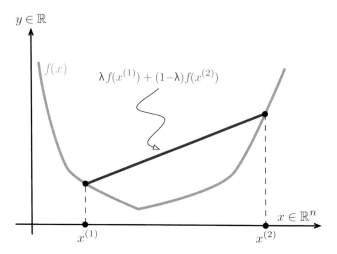

Figure 7.3 Convex function.

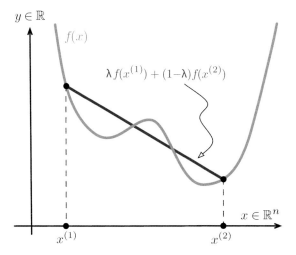

Figure 7.4 Nonconvex function.

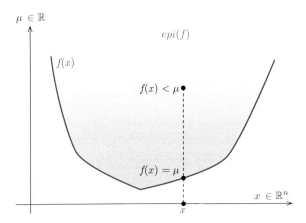

Figure 7.5 Convex epigraph.

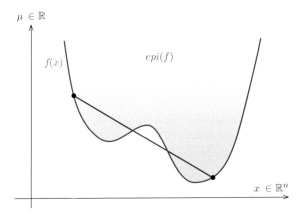

Figure 7.6 Nonconvex epigraph.

The concepts of *convex functions* and *convex sets* are closely related through the notion of *epigraph of a function*. Given $f : \mathbb{R}^n \to \mathbb{R}$, define the *epigraph of f* as

$$\mathrm{epi}(f) := \left\{ \begin{pmatrix} \mu \\ x \end{pmatrix} \in \mathbb{R} \times \mathbb{R}^n : f(x) \leq \mu \right\}.$$

In both Figures 7.5 and 7.6, we represent a function $f : \mathbb{R} \to \mathbb{R}$ and its corresponding epigraph (represented as the shaded region going to infinity in the up direction). The following result relates convex functions and convex sets:

PROPOSITION 7.2   *Let $f : \mathbb{R}^n \to \mathbb{R}$. Then f is convex if and only if $\mathrm{epi}(f)$ is a convex set.*

Observe in the previous proposition that the epigraph is living in an $(n+1)$-dimensional space. The function in Figure 7.5 is convex as its epigraph is convex. However, the function in Figure 7.6 is not convex as its epigraph is not convex.

Proof of Proposition 7.2   Suppose $f: \mathbb{R}^n \to \mathbb{R}$ is convex. Let $\begin{pmatrix} \mu_1 \\ u \end{pmatrix}, \begin{pmatrix} \mu_2 \\ v \end{pmatrix} \in \mathrm{epi}(f)$ and $\lambda \in [0,1]$. We have

$$f(\lambda u + (1-\lambda)v) \leq \lambda f(u) + (1-\lambda)f(v) \leq \lambda \mu_1 + (1-\lambda)\mu_2,$$

which implies $\begin{pmatrix} \lambda \mu_1 + (1-\lambda)\mu_2 \\ \lambda u + (1-\lambda)v \end{pmatrix} \in \mathrm{epi}(f)$. Note that in the above the first inequality uses the convexity of f and the second inequality uses that facts $\lambda \geq 0$, $(1-\lambda) \geq 0$ and $\begin{pmatrix} \mu_1 \\ u \end{pmatrix}, \begin{pmatrix} \mu_2 \\ v \end{pmatrix} \in \mathrm{epi}(f)$.

Now suppose that $\mathrm{epi}(f)$ is convex. Let $u, v \in \mathbb{R}^n$ and $\lambda \in [0,1]$. Then $\begin{pmatrix} f(u) \\ u \end{pmatrix}, \begin{pmatrix} f(v) \\ v \end{pmatrix} \in \mathrm{epi}(f)$. Hence

$$\lambda \begin{pmatrix} f(u) \\ u \end{pmatrix} + (1-\lambda) \begin{pmatrix} f(v) \\ v \end{pmatrix} \in \mathrm{epi}(f).$$

This implies (by the definition of $\mathrm{epi}(f)$), $f(\lambda u + (1-\lambda)v) \leq \lambda f(u) + (1-\lambda)f(v)$. Therefore, f is convex. □

7.3.2   Level sets and feasible region

Let $g : \mathbb{R}^n \to \mathbb{R}$ be a convex function and let $\beta \in \mathbb{R}$. We call the set

$$\{x \in \mathbb{R}^n : g(x) \leq \beta\}$$

a *level set* of the function g.

Remark 7.3 *The level set of a convex function is a convex set.*

In Figure 7.7, we represent a convex function with a convex level set. In Figure 7.8, we represent a nonconvex function with a nonconvex level set. We leave it as an exercise to show however, that it is possible to have a nonconvex function where every level set is convex.

Proof of Remark 7.3   Let $g : \mathbb{R}^n \to \mathbb{R}$ be a convex function and let $\beta \in \mathbb{R}$. We need to show that $S := \{x \in \mathbb{R}^n : g(x) \leq \beta\}$ is convex. Let $x^{(1)}, x^{(2)} \in S$ and let $\lambda \in [0,1]$. Then

$$g(\lambda x^{(1)} + (1-\lambda)x^{(2)}) \leq \underbrace{\lambda}_{\geq 0}\, \underbrace{g(x^{(1)})}_{\leq \beta} + \underbrace{(1-\lambda)}_{\geq 0}\, \underbrace{g(x^{(2)})}_{\leq \beta}) \leq \beta,$$

where the first inequality follows from the fact that g is a convex function. It follows that $\lambda x^{(1)} + (1-\lambda)x^{(2)} \in S$. Hence, S is convex as required. □

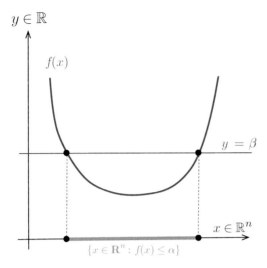

Figure 7.7 Convex level set.

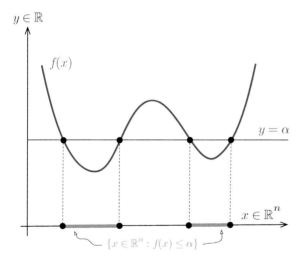

Figure 7.8 Nonconvex level set.

Consider the NLP defined in (1.35). We say that it is a *convex NLP* if $g_1, \ldots, g_m$ and f are all convex functions. It follows in that case from Remark 7.3 that for every $i \in \{1, \ldots, n\}$ the level set $\{x \in \mathbb{R}^n : g_i \leq 0\}$ is a convex set. Since the intersection of convex sets is a convex set (see Remark 2.17), we deduce that the feasible region

$$\{x \in \mathbb{R}^n : g_1(x) \leq 0, g_2(x) \leq 0, \ldots, g_m(x) \leq 0\} \tag{7.1}$$

is convex as well. Hence:

Remark 7.4 *The feasible region of a convex NLP is a convex set.*

When $g_1, \ldots, g_m$ are all affine functions, then the feasible region (7.1) is a polyhedron. Moreover, in that case, the functions are clearly convex. Hence, the previous result implies in particular that every polyhedron is a convex set, which was the statement of Proposition 2.18.

Exercises

1 (a) Let $g_1, g_2, g_3 : \mathbb{R} \to \mathbb{R}$ be defined by

$$g_1(x) := -x, \quad g_2(x) := 2, \quad g_3(x) := x.$$

Plot these functions on $\mathbb{R}^2$. Identify on your plot, the function $\hat{g} : \mathbb{R} \to \mathbb{R}$ defined by

$$\hat{g}(x) := \max\{g_1(x), g_2(x), g_3(x)\}, \quad \forall x \in \mathbb{R}.$$

Prove that $\hat{g}$ is a convex function.

(b) Suppose $g_1, g_2, \ldots, g_m : \mathbb{R}^n \to \mathbb{R}$ are given convex functions. Define

$$\hat{g}(x) := \max\{g_1(x), g_2(x), \ldots, g_m(x)\}, \quad \forall x \in \mathbb{R}^n.$$

Prove that $\hat{g}$ is a convex function.

2 Denote by (P) the following NLP

$$\text{minimize } f(x) \text{ subject to } g_i(x) \le 0 \text{ for } i = 1, 2, \ldots, m,$$

where each of the functions $f, g_1, \ldots, g_m$ is convex.
(a) Prove that the set of feasible solutions of (P) is a convex set.
(b) Let $\hat{x}$ be a locally optimal feasible solution of (P). Prove that $\hat{x}$ is globally optimal.

3 Denote by (P) the following NLP

$$\text{minimize } -f(x) \text{ subject to } g_i(x) \le 0 \text{ for } i = 1, 2, \ldots, m,$$

where each of the functions $f, g_1, \ldots, g_m$ is convex. Prove that "if (P) has an optimal solution then it has one that is an extreme point of the feasible region."

7.4   Relaxing convex NLPs

Consider an NLP of the form (1.35) and let $\bar{x}$ be a feasible solution. We say that constraint $g_j(x) \le 0$ is *tight* for $\bar{x}$ if $g_j(\bar{x}) = 0$ (see also Section 2.8.3). We will show that under the right circumstances, we can replace in a convex NLP a tight constraint by a linear constraint such that the resulting NLP is a relaxation of the original NLP (see Corollary 7.6). This will allow us in Section 7.5 to use our optimality conditions for linear programs to derive a sufficient condition for a feasible solution to be an optimal solution to a convex NLP. The key concepts that will be needed in this section are that of subgradients and supporting halfspaces.

7.4.1   Subgradients

Let $g : \mathbb{R}^n \to \mathbb{R}$ be a convex function and let $\bar{x} \in \mathbb{R}^n$. We say that $s \in \mathbb{R}^n$ is a *subgradient of f at $\bar{x}$* if for every $x \in \mathbb{R}^n$ the following inequality holds:

$$f(\bar{x}) + s^\top (x - \bar{x}) \le f(x).$$

Denote by $h(x)$ the function $f(\bar{x}) + s^\top (x - \bar{x})$. Observe that $h(x)$ is an affine function ($\bar{x}$ is a constant). Moreover, we have that $h(\bar{x}) = f(\bar{x})$ and $h(x) \le f(x)$ for every $x \in \mathbb{R}^n$. Hence, the function $h(x)$ is an affine function that provides a lower bound on $f(x)$ and approximates $f(x)$ around $\bar{x}$ (see Figure 7.9).

Example 29 Consider $g : \mathbb{R}^2 \to \mathbb{R}$ where for every $x = (x_1, x_2)^\top$ we have $g(x) = x_2^2 - x_1$. It can be readily checked (see Example 28) that g is convex. We claim that $s := (-1, 2)^\top$ is a subgradient of g at $\bar{x} = (1, 1)^\top$. We have $h(x) := g(\bar{x}) + s^\top (x - \bar{x}) = 0 + (-1, 2)(x - (1, 1)^\top) = -x_1 + 2x_2 - 1$. We need to verify, $h(x) \le f(x)$ for every $x \in \mathbb{R}^2$, i.e. that

$$-x_1 + 2x_2 - 1 \overset{?}{\le} x_2^2 - x_1,$$

or equivalently that $x_2^2 - 2x_2 + 1 = (x_2 - 1)^2 \ge 0$ which clearly holds as the square of any number is nonnegative.

7.4.2   Supporting halfspaces

Recall the definitions of hyperplanes and halfspaces from Section 2.8.1. Consider a convex set $C \subseteq \mathbb{R}^n$ and let $\bar{x} \in C$. We say that the halfspace $F := \{x \in \mathbb{R}^n : s^\top x \le \beta\}$

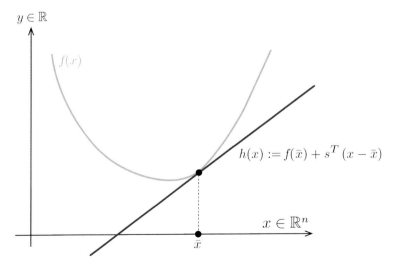

Figure 7.9 Subgradient.

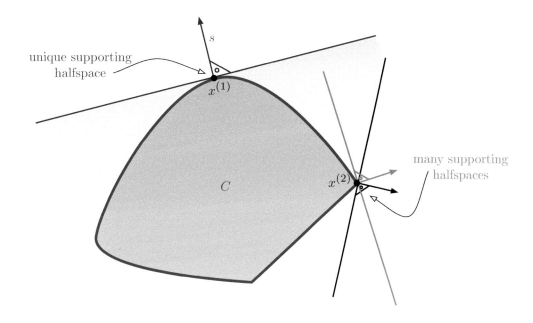

Figure 7.10 Supporting halfspace.

($s \in \mathbb{R}^n$ and $\beta \in \mathbb{R}$) is a *supporting halfspace of C at* $\bar{x}$ if the following conditions hold:

(1) $C \subseteq F$.
(2) $s^\top \bar{x} = \beta$, i.e. $\bar{x}$ is on the hyperplane that defines the boundary of F.

In Figure 7.10, we represent a convex set $C \subseteq \mathbb{R}^2$. For the point $x^{(1)} \in C$, there is a unique supporting halfspace. For the point $x^{(2)} \in C$, there are an infinite number of different supporting halfspaces – we represent two of these.

The following remark relates subgradients and supporting halfspaces:

Remark 7.5 *Let* $g : \mathbb{R}^n \to \mathbb{R}$ *be a convex function, let* $\bar{x} \in \mathbb{R}^n$ *such that* $g(\bar{x}) = 0$, *and let* $s \in \mathbb{R}^n$ *be a subgradient of* f *at* $\bar{x}$. *Denote by C the level set* $\{x \in \mathbb{R}^n : g(x) \leq 0\}$ *and by F the halfspace* $\{x \in \mathbb{R}^n : g(\bar{x}) + s^\top(x - \bar{x}) \leq 0\}$.[2] *Then F is a supporting halfspace of C at* $\bar{x}$.

Proof We need to verify conditions (1), (2) of supporting halfspaces. **(1)** Let $x' \in C$. Then $g(x') \leq 0$. Since s is a subgradient of g at $\bar{x}$, $g(\bar{x}) + s^\top(x' - \bar{x}) \leq g(x')$. It follows that $g(\bar{x}) + s^\top(x' - \bar{x}) \leq 0$, i.e. $x' \in F$. Thus, $C \subseteq F$. **(2)** $s^\top \bar{x} = s^\top \bar{x} - g(\bar{x})$ as $g(\bar{x}) = 0$. □

This last remark is illustrated in Figure 7.11. We consider the function $g : \mathbb{R}^2 \to \mathbb{R}$, where $g(x) = x_2^2 - x_1$ and the point $\bar{x} = (1, 1)^\top$. We saw in Example 29 that the vector

---

[2] F is clearly a halfspace since we can rewrite it as $\{x : s^\top x \leq s^\top \bar{x} - g(\bar{x})\}$ and $s^\top \bar{x} - g(\bar{x})$ is a constant.

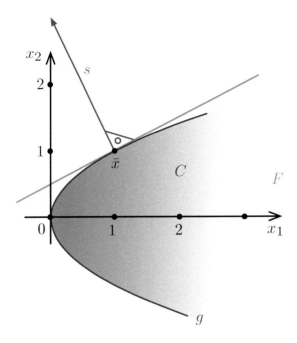

Figure 7.11 Subgradient and supporting halfspace.

$s = (-1, 2)^\top$ is a subgradient for g at $\bar{x}$. Then $F = \{x \in \mathbb{R}^2 : -x_1 + 2x_2 - 1 \leq 0\}$. We see in the figure that F is a supporting halfspace of C at $\bar{x}$ as predicted.

We deduce the following useful tool from the previous remark:

COROLLARY 7.6   *Consider an NLP of the form given in (1.35). Let $\bar{x}$ be a feasible solution and suppose that constraint $g_i(x) \leq 0$ is tight for some $i \in \{1, \ldots, m\}$. Suppose g_i is a convex function that has a subgradient s at $\bar{x}$. Then the NLP obtained by replacing constraint $g_i(x) \leq 0$ by the linear constraint $s^\top x \leq s^\top \bar{x} - g_i(\bar{x})$ is a relaxation of the original NLP.*

7.5   Optimality conditions for the differentiable case

In this section, we consider convex NLPs of the form (1.35) that satisfy the additional condition that all of the functions f and $g_1, \ldots, g_m$ are differentiable. In that setting, we can characterize (see Theorem 7.9) when a feasible solution $\bar{x}$ is an optimal solution (assuming the existence of a Slater point).

7.5.1   Sufficient conditions for optimality

We claim that it is sufficient to consider NLPs where the objective function is linear, i.e. of the form

$$\min \quad z = c^\top x$$

$$\text{s.t.}$$

$$
\begin{aligned}
g_1(x) &\le 0, \\
g_2(x) &\le 0, \\
&\vdots \\
g_m(x) &\le 0.
\end{aligned}
\tag{7.2}
$$

This is because problem (1.35) is reducible to problem (7.2). To prove this fact, we can proceed as follows: given a NLP of the form (1.35) introduce a new variable x_{n+1} and add the constraint $f(x) \le x_{n+1}$, to obtain the NLP

$$\min \quad z = x_{n+1}$$

$$\text{s.t.}$$

$$
\begin{aligned}
f(x) - x_{n+1} &\le 0, \\
g_1(x) &\le 0, \\
g_2(x) &\le 0, \\
&\vdots \\
g_m(x) &\le 0.
\end{aligned}
$$

This NLP is of the form (7.2), and minimizing x_{n+1} is equivalent to minimizing $f(x)$.

Let $\bar{x}$ be a feasible solution to the NLP (7.2) and assume that it is a convex NLP. Let us derive sufficient conditions for $\bar{x}$ to be an optimal solution to (7.2). Define, $J(\bar{x}) := \{i : g_i(\bar{x}) = 0\}$. That is, $J(\bar{x})$ is the index set of all constraints that are tight at $\bar{x}$. Suppose for every $i \in J(\bar{x})$ we have a subgradient s_i of the function g_i at the point $\bar{x}$. Then we construct a linear programming relaxation of the NLP as follows: first we omit every constraint that is not tight at $\bar{x}$, and for every $i \in J(\bar{x})$ we replace (see Corollary 7.6) the constraint $g_i(x) \le 0$ by the linear constraint $s_i^\top x \le s_i^\top \bar{x} - g(\bar{x})$. Since the objective function is given by "$\min c^\top x$", we can rewrite it as "$\max -c^\top x$." The resulting linear program is thus

$$\max \quad z = -c^\top x$$

$$\text{s.t.} \tag{7.3}$$

$$s_i^\top x \le s_i^\top \bar{x} - g(\bar{x}) \quad \text{for all } i \in J(\bar{x}).$$

Theorem 4.7 says that $\bar{x}$ is an optimal solution to (7.3), and hence of (7.2) as it is a relaxation of (7.3), when $-c$ is in the cone of the tight constraints, i.e. if $-c \in \text{cone}\{s_i : i \in J(\bar{x})\}$. Hence, we proved the following:

PROPOSITION 7.7   *Consider the NLP (7.2) and assume that $g_1, \ldots, g_m$ are convex functions. Let $\bar{x}$ be a feasible solution and suppose that for all $i \in J(\bar{x})$ we have a subgradient s_i at $\bar{x}$. If $-c \in \text{cone}\{s_i : i \in J(\bar{x})\}$, then $\bar{x}$ is an optimal solution.*

Thus, we have sufficient conditions for optimality. Theorem 7.9 which we give in the next section, essentially says that when the NLP satisfies an additional assumption (involving the existence of a *strictly feasible* solution), then these conditions are also necessary. We illustrate the previous proposition with an example.

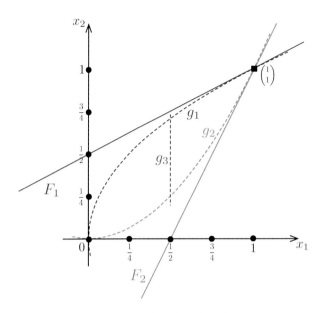

Figure 7.12 Linear programming relaxation of NLP.

Example 27 continued. Let us use the previous proposition to show that the point $\bar{x} = (1, 1)^{\top}$ is an optimal solution to the NLP given in Example 27. In this case, $J(\bar{x}) = \{1, 2\}$ and the feasible region for the linear programming relaxation (7.3) corresponds to the shaded region in Figure 7.12. In Example 29, we showed that the subgradient of g_1 at $\bar{x}$ is $(-1, 2)^{\top}$. Similarly, we can show that the subgradient of g_2 at $\bar{x}$ is $(2, -1)^{\top}$. In this example, we have $c = (-1, -1)^{\top}$, thus Proposition 7.7 asks us to verify that

$$-c = \begin{pmatrix} 1 \\ 1 \end{pmatrix} \overset{?}{\in} \text{cone} \left\{ \begin{pmatrix} -1 \\ 2 \end{pmatrix}, \begin{pmatrix} 2 \\ -1 \end{pmatrix} \right\},$$

which is the case since

$$\begin{pmatrix} 1 \\ 1 \end{pmatrix} = 1 \times \begin{pmatrix} -1 \\ 2 \end{pmatrix} + 1 \times \begin{pmatrix} 2 \\ -1 \end{pmatrix}.$$

7.5.2   Differentiability and gradients

Let $f \colon \mathbb{R}^n \to \mathbb{R}$ and $\bar{x} \in \mathbb{R}^n$ be given. If there exists $s \in \mathbb{R}^n$ such that

$$\lim_{h \to 0} \frac{f(\bar{x} + h) - f(\bar{x}) - s^{\top} h}{\|h\|_2} = 0,$$

we say that f is *differentiable* at $\bar{x}$ and call the vector s the *gradient* of f at $\bar{x}$. We denote s by $\nabla f(\bar{x})$. We will use the following without proof:

PROPOSITION 7.8   *Let $f \colon \mathbb{R}^n \to \mathbb{R}$ be a convex function and let $\bar{x} \in \mathbb{R}^n$. If the gradient $\nabla f(\bar{x})$ exists then it is a subgradient of f at $\bar{x}$.*

Note that in the above definition of the gradient, h varies over all vectors in $\mathbb{R}^n$. Under some slightly more favorable conditions, we can obtain the gradient $\nabla f(\bar{x})$ via partial derivatives of f at $\bar{x}$. For example, suppose that for every $j \in \{1, 2, \ldots, n\}$ the partial derivatives $\frac{\partial f}{\partial x_j}$ exist and are continuous at every $x \in \mathbb{R}^n$. Then

$$\nabla f(x) = \left[\frac{\partial f(x)}{\partial x_1}, \frac{\partial f(x)}{\partial x_2}, \ldots, \frac{\partial f(x)}{\partial x_n} \right]^\top,$$

and the gradient of f at $\bar{x}$ is given by $\nabla f(\bar{x})$.

Example 29 continued. Consider $g : \mathbb{R}^2 \to \mathbb{R}$ where for every $x = (x_1, x_2)^\top$ we have $g(x) = x_2^2 - x_1$. Earlier in Example 29, we gave a proof that $(-1, 2)^\top$ is a subgradient of g at $\bar{x} = (1, 1)^\top$. We now give an alternate proof based on the previous discussion. The partial derivatives of g exists and $\nabla g(x) = (2x_2, -1)^\top$. Evaluating at $\bar{x}$, we deduce that $(2, -1)^\top$ is the gradient of g at $\bar{x}$. Since g is convex, Proposition 7.8 implies that $(2, -1)^\top$ is also a subgradient of g at $\bar{x}$.

7.5.3 A Karush–Kuhn–Tucker theorem

To state the optimality theorem, we need a notion of the "strictly feasible point" for NLP. More rigorously, we say that the NLP has a *Slater point*, if there exists x' such that $g_i(x') < 0$ for every $i \in \{1, \ldots, m\}$, i.e. every inequality is satisfied strictly by x'. For instance, in Example 27 the point $\left(\frac{3}{4}, \frac{3}{4} \right)^\top$ is a Slater point.

We can now state our optimality theorem:

THEOREM 7.9 (Karush–Kuhn–Tucker theorem based on the gradients). *Consider a convex NLP of the form (1.35) that has a Slater point. Let $\bar{x} \in \mathbb{R}^n$ be a feasible solution and assume that $f, g_1, g_2, \ldots, g_m$ are differentiable at $\bar{x}$. Then $\bar{x}$ is an optimal solution of NLP if and only if*

$$-\nabla f(\bar{x}) \in \mathrm{cone}\{\nabla g_i(\bar{x}) : i \in J(\bar{x})\}. \tag{$\star$}$$

We illustrate the previous theorem in Figure 7.13. In that example, the tight constraints for $\bar{x}$ are $g_1(x) \le 0$ and $g_2(x) \le 0$. We indicate the cone formed by $\nabla g_1(\bar{x}), \nabla g_2(\bar{x})$ (translated to have $\bar{x}$ as its origin). In this example, $-\nabla f(\bar{x})$ is in that cone, hence the feasible solution $\bar{x}$ is in fact optimal.

Suppose that in Theorem 7.9 the function $f(x)$ is the linear function $c^\top x$, i.e. the NLP is of the form (7.2). Then $\nabla f(\bar{x}) = c$ and the sufficiency of Theorem 7.9 follows immediately from Proposition 7.7, i.e. we have shown that if the condition $(\star)$ holds, then $\bar{x}$ is indeed an optimal solution. The essence of the theorem is to prove the reverse direction, i.e. that $\bar{x}$ will only be optimal when $(\star)$ holds. This is when the fact that the problem has a Slater point comes into play. Observe, that when f and $g_1, \ldots, g_m$ are all affine functions, then Theorem 7.9 becomes the optimality theorem for linear programs (Theorem 4.7).

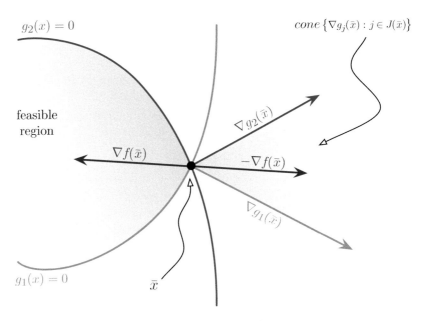

Figure 7.13 Karush–Kuhn–Tucker theorem based on gradients.

Theorem 4.7 was a restatement of the complementary slackness (Theorem 4.6). Similarly, we leave it as an exercise to check that condition $(\star)$ can be restated as follows, there exists $y \in \mathbb{R}^m_+$ such that the following conditions hold:

$$\nabla f(\bar{x}) + \sum_{i=1}^{m} \bar{y}_i \nabla g_i(\bar{x}) = \mathbb{0} \qquad \text{(dual feasibility, together with } y \in \mathbb{R}^m_+\text{)}$$

$$\bar{y}_i g_i(\bar{x}) = 0, \quad \forall i \in \{1, 2, \ldots, m\} \qquad \text{(complementary slackness).}$$

Exercises

1 Consider the following NLP:

$$(NLP) \quad \min \quad -7x_1 - 5x_2$$
$$\text{s.t.}$$
$$2x_1^2 + x_2^2 + x_1 x_2 - 4 \leq 0,$$
$$x_1^2 + x_2^2 - 2 \leq 0,$$
$$-x_1 + \tfrac{1}{2} \leq 0,$$

and the vector $\bar{x} := (1, 1)^\top$. Write down the optimality conditions for $\bar{x}$ for this NLP as described in the Karush–Kuhn–Tucker theorem. Using these conditions and the theorem, prove that $\bar{x}$ is optimal. Note, you may use the fact that the functions defining the objective function and the constraints are convex and differentiable without proving it.

2 Consider the following NLP:

$$(NLP) \quad \min \qquad\qquad x_3$$
$$\text{s.t.}$$
$$x_1 + x_2 \le 0,$$
$$x_1^2 - 4 \le 0,$$
$$x_1^2 - 2x_1 + x_2^2 - x_3 + 1 \le 0,$$

and the vector $\bar{x} := \begin{bmatrix} \frac{1}{2}, -\frac{1}{2}, \frac{1}{2} \end{bmatrix}^\top$. Write down the optimality conditions for $\bar{x}$ for this NLP as described in the Karush–Kuhn–Tucker theorem. Using these conditions and the theorem, prove that $\bar{x}$ is optimal.

3 Consider the following NLP:

$$(NLP) \quad \min \quad 2x_1 \arctan(x_1) - \ln\left(x_1^2 + 1\right) + x_2^4 + (x_3 - 1)^2$$
$$\text{s.t.}$$
$$x_1^2 + x_2^2 + x_3^2 - 4 \le 0,$$
$$-x_1 \le 0.$$

Prove that this NLP is convex. Using the Karush–Kuhn–Tucker theorem and some elementary observations, find an optimal solution. Prove all your claims.

4 (Advanced) Let $u, w \in \mathbb{R}^n$ be given such that u_j and w_j are positive for each j. Consider the following NLP:

$$(P) \quad \min \quad -\sum_{j=1}^{n} w_j \ln\left(x_j\right)$$
$$\text{s.t.}$$
$$u^\top x = n,$$
$$-x \le 0.$$

(a) Prove that this NLP is convex.
(b) Using the Karush–Kuhn–Tucker theorem (on possibly a slight modification of (P)), find an optimal solution (in terms of u and w).
(c) Prove that the solution you found is the unique optimal solution.

7.6 Optimality conditions based on Lagrangians

So far, we have made a lot of progress by utilizing the idea of relaxation when we face a very difficult problem. In the previous sections for nonlinear programming, the relaxation approach we replaced nonlinear functions with their linear approximations. In the case of integer programming, in the earlier chapters we simply removed the integrality constraint and dealt with it in other ways. Now let us try to apply this second approach to NLP. In a given NLP, many or all of the constraints $g_i(x) \le 0$ may be very difficult.

We may assign each of these constraints a coefficient (just like in our discussion of LP duality!) $y_i \geq 0$ and consider the function

$$f(x) + \sum_{i=1}^{m} y_i g_i(x).$$

Since $y \geq 0$, for every feasible solution x we have $\sum_{i=1}^{m} y_i g_i(x) \leq 0$. Therefore, the minimum value of $\left[f(x) + \sum_{i=1}^{m} y_i g_i(x) \right]$ over all $x \in \mathbb{R}^n$ gives a lower bound on the optimal objective value of the original NLP. Notice that if $\bar{x}$ is a feasible solution of NLP, then $f(\bar{x})$ is an upper bound on the optimal objective value of the NLP. On the other hand, if we pick some nonnegative vector $y \in \mathbb{R}^m$ and compute the minimum value of the relaxation

$$\min \left\{ f(x) + \sum_{i=1}^{m} y_i g_i(x) : x \in \mathbb{R}^n \right\},$$

we obtain a lower bound on the optimal objective value of the NLP.

Let us define the *Lagrangian* $L : \mathbb{R}^n \times \mathbb{R}^m \to \mathbb{R}$ *for NLP* as

$$L(x, y) := f(x) + \sum_{i=1}^{m} y_i g_i(x).$$

Note that the Lagrangian encodes all the information about the problem NLP:

- Setting all y variables to zero in $L(x, y)$, we obtain $f(x)$. That is

$$L(x, 0) = f(x), \forall x.$$

- Setting y to unit vectors and using the above, we can obtain all the constraint functions. That is, for every $i \in \{1, 2, \ldots, m\}$

$$L(x, e_i) - L(x, 0) = g_i(x).$$

Perturbing y along a unit vector, we have the same result for every y

$$L(x, y + e_i) - L(x, y) = g_i(x).$$

Let z^* denote the optimal objective objective value of *NLP*. Since the minimum value of the Lagrangian at every nonnegative y gives a lower bound on z^*, we also have

$$\max_{y \in \mathbb{R}^m_+} \min_{x \in \mathbb{R}^n} \{L(x, y)\} \leq z^*.$$

The next theorem summarizes the fundamental duality result based on the Lagrangian.

THEOREM 7.10 (Karush–Kuhn–Tucker theorem based on the Lagrangian). *Consider a convex NLP of the form (1.35) that has a Slater point. Then a feasible solution $\bar{x}$ is*

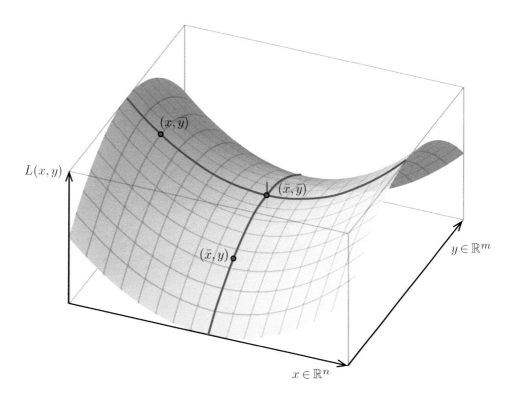

Figure 7.14 Saddle-point function.

an optimal solution if and only if there exists $\bar{y} \in \mathbb{R}^m_+$ such that the following conditions hold:

$$L(\bar{x}, y) \leq L(\bar{x}, \bar{y}) \leq L(x, \bar{y}), \quad \forall x \in \mathbb{R}^n, \ \forall y \in \mathbb{R}^m_+ \quad \text{(saddle-point)}$$
$$\bar{y}_i g_i(\bar{x}) = 0, \quad \forall i \in \{1, 2, \ldots, m\} \qquad \text{(complementary slackness)}.$$

We illustrate the saddle-point condition of the Lagrangian in Figure 7.14. Among all points $(x, \bar{y})$, the point $(\bar{x}, \bar{y})$ minimizes $L(x, y)$. Among all points $(\bar{x}, y)$, the point $(\bar{x}, \bar{y})$ maximizes $L(x, y)$.

Now let us see a proof that if $\bar{x} \in \mathbb{R}^n$ satisfies the conditions of the above theorem for some $\bar{y}$, then it is an optimal solution of the NLP in the convex case.

First, we prove that $\bar{x}$ is feasible. We saw that the Lagrangian encodes all the information on the constraints of NLP and we can extract this information using unit vectors. Let us fix $i \in \{1, 2, \ldots, m\}$ and apply the first inequality in the saddle-point condition using $y := \bar{y} + e_i$:

$$0 \geq L(\bar{x}, \bar{y} + e_i) - L(\bar{x}, \bar{y}) = g_i(\bar{x}).$$

Hence, $\bar{x}$ is feasible in NLP.

Next we use the second inequality in the saddle-point condition and complementary slackness. We have

$$L(\bar{x}, \bar{y}) \le L(x, \bar{y}), \forall x \in \mathbb{R}^n,$$

which is equivalent to

$$f(\bar{x}) + \sum_{i=1}^{m} \bar{y}_i g_i(\bar{x}) \le f(x) + \sum_{i=1}^{m} \bar{y}_i g_i(x), \forall x \in \mathbb{R}^n.$$

Consider all feasible x (i.e. $g_i(x) \le 0$ for every i). Since $\bar{y}_i \ge 0$ for every i, we have $\sum_{i=1}^{m} \bar{y}_i g_i(x) \le 0$. Hence, our conclusion becomes

$$f(\bar{x}) + \sum_{i=1}^{m} \bar{y}_i g_i(\bar{x}) \le f(x), \forall x \text{ feasible in } NLP.$$

Using the fact that complementary slackness holds at $(\bar{x}, \bar{y})$, we observe that $\sum_{i=1}^{m} \bar{y}_i g_i(\bar{x})$ is zero and we conclude that

$$f(\bar{x}) \le f(x), \forall x \text{ feasible in } NLP.$$

Indeed, this combined with feasibility of $\bar{x}$ means $\bar{x}$ is optimal in NLP.

Note that the above results motivate a notion of *Lagrangian dual* of the (NLP)

$$\max \left\{ \min \left\{ L(x, y) : x \in \mathbb{R}^n \right\} : y \ge 0 \right\}.$$

So the Lagrangian dual is maximization of a function $h(y)$ subject to the constraint that y is a nonnegative vector. Here

$$h(y) := \min \left\{ L(x, y) : x \in \mathbb{R}^n \right\}.$$

Let us work out the Lagrangian dual of the linear programming problem

$$\min \quad c^\top x$$
$$\text{subject to}$$
$$Ax - b \le 0.$$

For this example of the (NLP), the Lagrangian is

$$L(x, y) = c^\top x + y^\top (Ax - b).$$

To compute $h(y)$ more explicitly, we need to optimize over x. So let us rewrite the Lagrangian

$$L(x, y) = \left(c^\top + y^\top A \right) x - b^\top y.$$

Note that since x can be any vector in $\mathbb{R}^n$, unless $A^\top y = -c$ we can find a sequence $\{x^{(k)}\}$ such that $L\left(x^{(k)}, y\right) \to -\infty$. Since we would like to maximize $h(y)$, we would not choose a y violating $A^\top y = -c$. Therefore, the Lagrangian dual which is minimization of $h(y)$ subject to $y \geq \mathbb{0}$ becomes

$$\max \qquad -b^\top y$$

subject to

$$A^\top y = -c$$
$$y \geq \mathbb{0}.$$

We leave it to the reader to verify that this Lagrangian dual is equivalent to the linear programming dual (as defined in Chapter 4) of the original LP problem. Therefore, Lagrangian duality can be seen as a generalization of linear programming duality.

The last two theorems can be strengthened via replacing the assumption on the existence of a Slater point by weaker, but more technical, assumptions. Moreover, the last two theorems can be generalized to NLP problems that are not convex. In the general case, the first theorem only characterizes local optimality, and only provides a necessary condition for that.

7.7   Nonconvex optimization problems

For nonlinear optimization problems that are nonconvex, our work and ideas can be extended in a few ways. We will focus on two principal approaches:

(1) When a problem fails to be convex, some of our ideas/methods/theorems may still apply "locally." One way is to focus on "critical points" and then weed-out local maxima and saddle points, etc.
(2) We can consider an abstraction of our main approach to integer programming. Namely, we can consider convex relaxations of the feasible solution set. Since we may assume, without loss of generality that the objective function is linear, we will argue that by optimizing the linear objective function over the convex hull of the feasible region, we can recover the optimal objective value of the original NLP.

7.7.1   The Karush–Kuhn–Tucker theorem for nonconvex problems*

Recall the beautiful geometric characterization of optimality provided in the version of the Karush–Kuhn–Tucker theorem with the gradients (Theorem 7.9). It turns out that even if our NLP is nonconvex, we can still use the underlying geometric condition to help characterize local optimality.

A very important distinction is that the geometric condition $(\star)$ in the statement of Theorem 7.9 is no longer sufficient. To understand why this is so, consider a function $f : [0, 1] \to [0, 1]$. Assume that f is differentiable on the whole interval $[0, 1]$. Suppose that we would like to compute the minimum value of f. Note that we have no constraints except $x \in [0, 1]$. A good starting point would be to consider all critical points of f (i.e. points where the derivative of f is zero, $f'(x) = 0$). We would also consider the boundary points $x = 0$ and $x = 1$ (see Figure 7.15). For the example on the left, the local

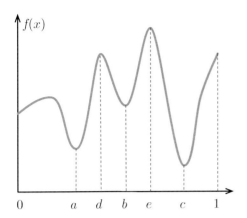
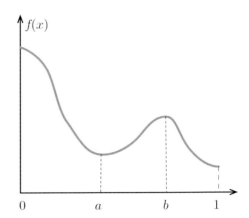

Figure 7.15 Local minima.

minimizers are $x = 0$, $x = a$, $x = b$, and $x = c$ (which is also the global minimizer). On the other hand, $x = d$, $x = e$ and $x = 1$ are local maximizers ($x = e$ is also the global maximizer). For the example on the right, $x = a$ is a local minimizer and $x = b$ is a local maximizer. However, the global minimum and global maximum values of the function are attained at the boundary points $x = 1$ and $x = 0$, respectively.

Computing the value of f in each of these points and picking the minimum would be one approach to solving this problem. Note, however that some of the critical points may correspond to local maxima! This is just for functions on the real line, as we consider functions $f : \mathbb{R}^2 \to \mathbb{R}$ we also start seeing *saddle points* as in Figure 7.14 manifesting themselves as critical points.

THEOREM 7.11 (Karush–Kuhn–Tucker theorem for nonconvex problems).   *Consider an NLP of the form (1.35). Let $\bar{x} \in \mathbb{R}^n$ be a feasible solution and assume that $f, g_1, g_2, \dots, g_m$ are continuously differentiable at $\bar{x}$ and that $\{\nabla g_i(\bar{x}) : i \in J(\bar{x})\}$ is linearly independent. If $\bar{x}$ is a local minimizer of f over the feasible region of NLP, then*

$$-\nabla f(\bar{x}) \in \mathrm{cone}\{\nabla g_i(\bar{x}) : i \in J(\bar{x})\}. \qquad (\star)$$

So, the local version of the Karush–Kuhn–Tucker theorem provides an important necessary condition for local optimality of a feasible solution $\bar{x}$. While this statement may seem too weak at a first glance, this necessary condition for optimality is widely used in theory of nonconvex NLPs, as well as in the design of algorithms for solving NLPs. One common theme along these lines is to write the necessary conditions for local minimizers of NLP as a system of nonlinear equations and try to solve the nonlinear system iteratively, by using simpler, local approximations of it.

Now we sketch a proof of this theorem. The proof, when specialized to the convex optimization case, also completes the proof of Theorem 7.9. Let

$$F := \{x \in \mathbb{R}^n : g_1(x) \leq 0, \dots, g_m(x) \leq 0\}$$

denote the feasible region. Consider two local approximations to the feasible region F at the point $\bar{x} \in F$.

The first one is the *local convex conic approximation to F at $\bar{x}$* defined as:

$$K_F(\bar{x}) := \left\{ d \in \mathbb{R}^n : \nabla g_i(\bar{x})^\top d \le 0, \forall i \in J(\bar{x}) \right\}.$$

Recall that convex functions are globally underestimated by linear approximations provided by their gradients. So, for convex optimization problems, $K_F(\bar{x})$ provides an outer approximation to the feasible region. We may expect this approximation to give a decent characterization of local geometry of the feasible region in a neighborhood of $\bar{x}$.

We are interested in whether $\bar{x}$ is a local minimizer. Therefore, we are interested in proving that no matter how small, there does not exist a local feasible move from $\bar{x}$ to improve the objective function. In the case of linear programming, in particular the simplex method, if $\bar{x}$, is an extreme point of the feasible region, it suffices to consider the directions emanating from $\bar{x}$ along the edges of the feasible region. (All other local feasible moves are nonnegative linear combinations of these directions, since the objective function is linear, it suffices to consider only the edges.) However, in NLP the local geometry of the feasible region is determined by nonlinear curves, surfaces, etc. Thus, we are led to consider all feasible local nonlinear moves from $\bar{x}$. We can represent such moves with directed curves rooted at $\bar{x}$ which we call *arcs*. The second local approximation is given by the *cone of tangents of feasible arcs of F emanating from $\bar{x}$*, denoted by $T_F(\bar{x})$. In the definition of this cone, we also include the limits of such tangents.

Now we arrive at two local approximations to the feasible region by convex cones. On the one hand, $K_F(\bar{x})$ clearly provides the kind of approximation that is amenable to use in establishing the geometric optimality condition in the Karush–Kuhn–Tucker theorem, via duality. On the other hand, $T_F(\bar{x})$ provides the kind of approximation that is easily used in verifying whether $\bar{x}$ is a local minimizer. Moreover, it is not too hard to observe that $K_F(\bar{x})$ contains $T_F(\bar{x})$. So, to finish the proof, it suffices to show $T_F(\bar{x})$ contains $K_F(\bar{x})$. This part of the proof is when we use a condition such as linear independence of gradients (or the existence of a Slater point in the convex case). These assumptions can be weakened (and replaced by a more technical condition) but it cannot be omitted altogether. Consider for example the NLP given by

$$\begin{array}{ll} \min & x_1 \\ \text{s.t.} & \\ & x_1^2 + x_2 \le 0, \\ & -x_2 \le 0. \end{array}$$

$\bar{x} := \mathbb{0}$ is the unique feasible solution and therefore is the unique local and global minimizer. However

$$-\nabla f(\bar{x}) \notin \text{cone}\left\{ \nabla g_i(\bar{x}) : i \in J(\bar{x}) \right\}$$

since

$$-\begin{bmatrix} 1 \\ 0 \end{bmatrix} \notin \text{cone}\left\{ \begin{bmatrix} 0 \\ 1 \end{bmatrix}, \begin{bmatrix} 0 \\ -1 \end{bmatrix} \right\}.$$

Indeed, the Slater condition fails for this problem. The cone $T_F(\bar{x})$ is empty (there are no feasible moves), and the cone $K_F(\bar{x})$ is the whole x_1-line. However, whenever $K_F(\bar{x})$

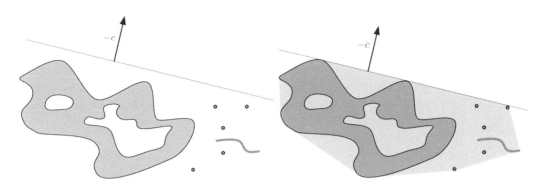

Figure 7.16 Left, nonconvex feasible region. Right, convex hull of the feasible region.

is full dimensional (as guaranteed by the Slater or a similar condition), $K_F(\bar{x}) = T_F(\bar{x})$ and the conclusion of the KKT Theorem follows.

7.7.2   Convex relaxation approach to nonconvex problems*

Another approach to nonconvex NLPs is to use a generalization of the ideas we used for integer programs. As we showed earlier in this chapter, without loss of generality, we may assume that our NLP has a linear objective function. Then under mild assumptions, optimizing the objective function over the feasible region of NLP and optimizing the objective function over the *convex hull* of the feasible region of the NLP give the same value! See, Figure 7.16, where the objective function is to minimize $c^\top x$.

Suppose we are given a nonlinear optimization problem that is not a convex optimization problem. As we observed at the beginning of this chapter, we may assume that the objective function is linear (we can achieve this by introducing a new variable and a new constraint). Thus, our problem is

$$
\begin{aligned}
\min \quad & z = c^\top x \\
\text{s.t.} \quad & \\
& g_1(x) \ \leq \ 0, \\
& g_2(x) \ \leq \ 0, \\
& \ \vdots \quad \vdots \quad \vdots \\
& g_m(x) \ \leq \ 0.
\end{aligned}
\tag{7.4}
$$

Recall, the feasible region is

$$
F = \left\{ x \in \mathbb{R}^n : g_1(x) \leq 0, \ldots, g_m(x) \leq 0 \right\}.
$$

In one of the nice cases, when F is a *compact set* (to be defined soon), we can compute unclear the optimal value of the nonconvex problem (7.4) by computing the optimal value of the convex optimization problem of maximizing $c^\top x$ over the convex hull of F. Moreover, if we can find an optimal solution of the latter which is an extreme point of the convex hull of F, then we have found an optimal solution of the nonconvex problem (7.4). However, the ideas in this paragraph are very hard to implement efficiently (if it were possible to implement them efficiently, we could solve almost any optimization problem efficiently). Nevertheless, in some special cases with the addition of other ideas which exploit the special structure of the hard nonconvex problem at hand, some progress towards at least approximate solutions may be achieved with these fundamental approaches. For an introduction to some of these topics, some notions from elementary analysis and set topology will be useful. This is what we develop next.

If a sequence of points $x^{(1)}, x^{(2)}, \ldots$ in $\mathbb{R}^n$ satisfies

$$\left\| x^{(k)} - \bar{x} \right\|_2 \to 0, \text{ as } k \to +\infty,$$

or equivalently

$$\lim_{k \to +\infty} x_j^{(k)} \to \bar{x}_j, \text{ for all } j = 1, \ldots, n,$$

then we say that $\{x^{(k)}\}$ *converges* to $\bar{x}$ and write $x^{(k)} \to \bar{x}$. Now we are ready to define the *closure* of a set S in $\mathbb{R}^n$

$$\mathrm{cl}(S) := \left\{ x \in \mathbb{R}^n : x = \lim_{k \to +\infty} x^{(k)}, \text{ for some } \{x^{(k)}\} \subseteq S \right\}.$$

A subset S of $\mathbb{R}^n$ is called *closed* if it is equal to its own closure, i.e. $S = \mathrm{cl}(S)$. Intersection of any collection of closed sets is a closed set. The closure of S is the smallest closed set containing S. A subset S of $\mathbb{R}^n$ is called *compact* if S is closed and bounded.

Now we are ready to state the Bolzano–Weierstrass theorem which characterizes compact sets in $\mathbb{R}^n$ in terms of the behavior of sequences in them.

THEOREM 7.12   *Let $S \subset \mathbb{R}^n$ be a compact set. Then for every sequence $\{x^{(k)}\} \subseteq S$ there exists a subsequence $\{x^{(\ell)}\}$ which converges and $\lim_{\ell \to +\infty} x^{(\ell)} \in S$.*

Next we note that continuity of a function and the closedness of a set are connected notions. Let $S \subseteq \mathbb{R}^n$. Recall, $f : S \to \mathbb{R}$ is *continuous* on S if for every sequence $\{x^{(k)}\}$ in S with a limit point $\bar{x} \in S, f(x^{(k)}) \to f(\bar{x})$. If S is closed and f is continuous on S, then the level set of f

$$\{x \in S : f(x) \leq \mu\}$$

for every $\mu \in \mathbb{R}$ is closed.

Since linear functions on $\mathbb{R}^n$ are continuous, we conclude that halfspaces are closed sets. Since the intersection of any collection of closed sets is closed, every polyhedron is closed. Recall that in our definition of LP problems, we did not allow strict inequalities. Indeed, if we were to allow them, then even in the cases when the optimal objective value is finite, we could not guarantee the existence of optimal solutions (e.g., consider maximizing x_1 subject to $0 \leq x_1 < 1$; even though the optimal objective value is 1, there is no feasible solution with objective value equal to 1).

An important general theoretical tool to establish the existence of optimal solutions of optimization problems is the following theorem of Weierstrass:

THEOREM 7.13   *Let $S \subset \mathbb{R}^n$ be a nonempty compact set. Also let $f : S \to \mathbb{R}$ be a continuous function on S. Then f attains its minimum as well as its maximum on S.*

Another fundamental result related to our discussion is a classical theorem of Minkowski:

THEOREM 7.14   *Every compact convex set in $\mathbb{R}^n$ is equal to the convex hull of its extreme points.*

Minkowski's theorem characterizes compact convex sets in $\mathbb{R}^n$ by convex combinations of their elements. There is also a dual version of this characterization: every compact convex set in $\mathbb{R}^n$ is the intersection of all closed halfspaces containing it. In fact, using only supporting halfspaces suffices. However, in many cases (outside of the set being a polyhedron) we may need infinitely many halfspaces to express a compact convex set.

Given a nonempty compact set $S \subset \mathbb{R}^n$ and $c \in \mathbb{R}^n$, we have

$$\max \left\{ c^\top x : x \in S \right\} = \max \left\{ c^\top x : x \in \text{conv}(S) \right\}.$$

In the light of these fundamental results, we see that at least in principle minimization of a continuous function over a compact set can be expressed as a minimization of a linear function over a compact convex set. Even though the latter is a convex optimization problem, in general we may only hope to solve some relaxations of the latter, which in turn provides some approximation to the original nonconvex problem.

Exercises

1   Call a feasible point $\bar{x}$ a *KKT point* if $\bar{x}$ satisfies the geometric condition

$$-\nabla f(\bar{x}) \in \text{cone} \left\{ \nabla g_i(\bar{x}) : i \in J(\bar{x}) \right\}$$

for the NLP. Compute all KKT points and determine an optimal solution of the following NLP. Why is this NLP nonconvex?

$$\min \quad x_1^2 - x_2^2$$
$$\text{subject to}$$
$$x_2 \le 1$$
$$-x_2 \le 0.$$

2   Consider a convex NLP. Suppose that Slater condition holds for the NLP. Let $\bar{x}$ be a feasible solution. Prove that there exists $d \in \mathbb{R}^n$ such that

$$\nabla g_i(\bar{x})^\top d < 0, \forall i \in J(\bar{x}).$$

Aside: The above condition is sufficient to establish the KKT Theorem (replacing the Slater point assumption). This condition is known as *Mangasarian–Fromowitz constraint qualification.*

3 (a) A convex function $f : \mathbb{R}^n \rightarrow \mathbb{R}$ is called *strictly convex* if for every pair of distinct points $x^{(1)}, x^{(2)} \in \mathbb{R}^n$ and for every $\lambda \in (0, 1)$

$$f\left(\lambda x^{(1)} + (1 - \lambda)x^{(2)}\right) < \lambda f(x^{(1)}) + (1 - \lambda)f(x^{(2)}).$$

Prove that the function $f : \mathbb{R}^n \rightarrow \mathbb{R}$ given by $f(x) := \|x\|^2 = x^\top x$ is strictly convex.

(b) Let $F \subseteq \mathbb{R}^n$ be a closed convex nonempty set. Prove that for every $u \in \mathbb{R}^n$, there exists a unique point in F which is closest to u. That is, a unique point $\bar{x}$ in F minimizing the function $\|u - \bar{x}\|$.

4 (a) Let $F \subset \mathbb{R}^n$ be a compact convex nonempty set. Prove that for every $u \in \mathbb{R}^n$, there exists a point in F which is farthest from u.

(b) Give three examples of F and u satisfying the above conditions such that in one example the farthest point from u is unique, in another example, there are exactly two farthest points and on the third example there are infinitely many farthest points.

(c) Do the farthest points have a specific geometric property with respect to F? Prove your claims.

5 Compute the convex hull of $S_1 \cup S_2$, where

$$S_1 := \{0\} \subset \mathbb{R}^3,$$

and

$$S_2 := \left\{x \in \mathbb{R}^3 : x_1^2 + x_2^2 \leq 1, x_3 = 1\right\}.$$

6 Consider the following NLP:

$$(P) \quad \min \quad x_3$$
$$\text{s.t.}$$
$$(x_1 - 1)^2 + x_2^2 + x_3^2 - 1 \geq 0,$$
$$(x_1 + 1)^2 + x_2^2 + x_3^2 - 1 \geq 0,$$
$$x_1^2 + x_2^2 + x_3^2 - 4 \leq 0.$$

Call the feasible region F.

(a) Prove that the feasible region F of (P) is not convex.

(b) Compute the convex hull of the feasible region F and express this convex hull as the intersection of a halfspace and another simple convex set. Prove that your characterization of the convex hull of F is correct.

(c) Find the set of optimal solutions minimizing x_3 over conv(F). Describe the subset of these solutions that are optimal solutions of (P).

7 (Advanced) We saw that if our nonconvex optimization problem can be expressed as a minimization of a continuous function over a nonempty compact set, then it has an optimal solution. In this exercise, we explore a related way of establishing the existence of an optimal solution. A function $f : \mathbb{R}^n \to \mathbb{R}$ is called *coercive*, if its level sets are bounded for every $\mu \in \mathbb{R}$.

(a) Let $f : \mathbb{R}^n \to \mathbb{R}$ be a continuous and coercive function. Prove that f attains its minimum value in $\mathbb{R}^n$.

(b) We would like to generalize the result from the previous part of this problem so that it applies to the NLP

$$\min f(x) \text{ subject to: } x \in F,$$

where

$$F := \left\{ x \in \mathbb{R}^n : g_1(x) \leq 0, \ldots, g_m(x) \leq 0 \right\}.$$

Find some sufficient conditions for g_i guaranteeing that F is closed and nonempty. Then define the notion of coercivity of f over a set F (rather than the whole $\mathbb{R}^n$) so that you can prove that under these conditions the NLP always has an optimal solution.

8 (Advanced) Consider the nonconvex NLP problem

$$\min \quad \sum_{j=1}^{n} x_j^2$$

subject to

$$\prod_{j=1}^{n} x_j = \beta$$

$$x \geq 0.$$

Apply a dynamic programming approach to solve this problem.

HINT: Focus on the difficult looking, nonconvex constraint. If you already know the value of x_1, what would this constraint become?

7.8 Interior-point method for linear programs*

In this section, we introduce a new breed of algorithms, namely *interior-point methods*. Recall that when we start from a basic feasible solution of an LP, the simplex method generates a sequence of basic feasible solutions eventually leading to either a basic feasible solution that is optimal or to a feasible direction along which the objective function value is unbounded. Since each basic feasible solution is an extreme point of the underlying feasible region and each (nondegenerate) simplex iteration corresponds to moving from one extreme point to another that is connected by an edge (line segment) on the boundary of the feasible region, the simplex method considers only the feasible solutions that lie on the boundary of the underlying feasible region. In contrast, the interior-point methods generate iterates which satisfy all inequality constraints strictly. Hence, geometrically speaking, all iterates lie in the relative interior of the feasible region.

Given $\bar{x} \in \mathbb{R}^n$ and $\delta > 0$, the *open Euclidean ball* centered at $\bar{x}$ with radius δ is defined as

$$B(\bar{x}; \delta) := \left\{ x \in \mathbb{R}^n : \|x - \bar{x}\|_2 < \delta \right\}.$$

For every subset S of $\mathbb{R}^n$, we define the *interior* of S by

$$\text{int}(S) := \left\{ x \in \mathbb{R}^n : B(x; \delta) \subseteq S, \text{ for some } \delta > 0 \right\}.$$

A subset S of $\mathbb{R}^n$ is called *open* if it is the same as its interior, i.e. $S = \text{int}(S)$. Note that S is open when its complement is in $\mathbb{R}^n$, namely $\mathbb{R}^n \setminus S$, is closed. Also, interior of S is the largest open set contained in S. Interior-point algorithms for LPs maintain iterates in the interiors of the polyhedra defined by the inequality constraints for the underlying problem. For example, for an LP problem in SEF the inequalities describe the nonnegative orthant and interior-point algorithms keep all of the primal iterates in the interior of the nonnegative orthant for LP problems in SEF.

We consider the LP problems (P) in SEF and assume without loss of generality that $\text{rank}(A) = m$. The dual is

$$\min \left\{ b^\top y : A^\top y - s = c, s \geq \mathbb{0} \right\}.$$

Note that we have explicitly included the slack variables s. For now, we will also assume that there exist *interior* solutions for both problems, i.e. there exist $\bar{x}$ and $(\bar{y}, \bar{s})$ such that $A\bar{x} = b, \bar{x}_j > 0$ for all j and $A^\top \bar{y} - \bar{s} = c, \bar{s}_j > 0$ for all j. So, an *interior* solution is a feasible solution which satisfies all the inequality constraints strictly. Note that $\bar{x}$ and $\bar{s}$ are Slater points for (P) and (D) respectively (in suitable reformulations of (P) and (D) in the form of NLP).

By the weak duality theorem, we have

$$b^\top \bar{y} - c^\top \bar{x} \geq 0.$$

In fact, we can write the duality gap in terms of $\bar{x}$ and $\bar{s}$

$$b^\top \bar{y} - c^\top \bar{x} = \bar{x}^\top A^\top \bar{y} - c^\top \bar{x} = \bar{x}^\top (A^\top \bar{y} - c) = \bar{x}^\top \bar{s} > 0.$$

If $\bar{x}$ and $\bar{s}$ had been optimal solutions, we would have had $\bar{x}^\top \bar{s} = 0$. Recall the necessary and sufficient conditions for optimality (see, Theorem 4.6)

$$Ax = b, \quad x \geq \mathbb{0}, \quad \text{(primal feasibility)},$$

$$A^\top y - s = c, \quad s \geq \mathbb{0}, \quad \text{(dual feasibility)},$$

$$x_j s_j = 0 \quad \text{for all } j \quad \text{(complementary slackness)}.$$

Now we would like to find a new solution x, y, s so that x is feasible for (P), (y, s) is feasible for (D) and the duality gap $b^\top y - c^\top x = x^\top s$ is smaller than the current duality gap: $\bar{x}^\top \bar{s}$. Recall that the simplex algorithm keeps primal feasibility and complementary slackness and strive for dual feasibility. There are also so-called dual-simplex algorithms which keep dual feasibility and complementary slackness and strive for primal feasibility (see for instance [13]). Actually, we have seen an algorithm which behaved in this way (keeping dual feasibility and complementary slackness and striving for primal feasibility), namely the Hungarian Method for finding minimum cost perfect matchings in bipartite graphs (see Section 3.2). Here, we are keeping primal feasibility and dual feasibility and striving for complementary slackness. So, we would like to find direction dx in the primal and directions dy, ds in the dual such that:

1. $\bar{x} + dx$ is feasible in (P),
2. $\bar{y} + dy$ and $\bar{s} + ds$ are feasible in (D),
3. $(\bar{x} + dx)^{\top}(\bar{s} + ds)$ is very small, preferably zero.

The first item above implies $A dx = \mathbb{0}$, the second item implies $A^{\top} dy - ds = \mathbb{0}$, and the third one implies $(\bar{x}_j + dx_j)(\bar{s}_j + ds_j) = 0$. The first two systems of equations are linear in dx, dy, and ds; however, the third one is

$$\bar{s}_j dx_j + \bar{x}_j ds_j + dx_j ds_j = -\bar{x}_j \bar{s}_j$$

which is quadratic (we have the term $dx_j ds_j$). Of course, if we could solve these three groups of equations at once and maintain the nonnegativity of $(\bar{x} + dx)$, $(\bar{s} + ds)$, we would have optimal solutions of (P) and (D). Since we do not know how to solve these equations (without solving the linear programming problem), we will ignore the quadratic term and try to solve the big linear system with the last group of linear equations $\bar{s}_j dx_j + \bar{x}_j ds_j = -\bar{x}_j \bar{s}_j$ instead.

So, we solve the following system of linear equations. Note that the last group of equations is trying to force a solution dx, ds such that $(\bar{x} + dx)^{\top}(\bar{s} + ds)$ is close to zero.

$$A dx = \mathbb{0}, \tag{7.5}$$

$$A^{\top} dy - ds = \mathbb{0}, \tag{7.6}$$

$$\bar{s}_j dx_j + \bar{x}_j ds_j = -\bar{x}_j \bar{s}_j, \quad \text{for all } j \in \{1, 2, \ldots, n\}. \tag{7.7}$$

This is a system of linear equations and we know how to solve it efficiently (e.g., by Gaussian elimination). Under our assumptions (rank$(A) = m$ and $\bar{x} > 0, \bar{s} > 0$), this linear system always has a unique solution.

One very important condition we neglected so far is the nonnegativity constraint on x and s. Most likely, $\bar{x} + dx$ is not feasible in (P) or $\bar{s} + ds$ is not feasible in (D), possibly both. However, the above development convinced us that the dx, dy, ds are good directions and we should thus move in these directions, but not "all the way" to the boundary. So we consider the solution

$$\bar{x} + \alpha dx, \quad \bar{y} + \alpha dy, \quad \bar{s} + \alpha ds,$$

for some $\alpha > 0$. We should pick α such that $\bar{x} + \alpha dx > 0$ and $\bar{s} + \alpha ds > 0$. We can choose, for instance

$$\alpha = 0.999 \min \left\{ \frac{\bar{x}_j}{-dx_j}, \frac{\bar{s}_j}{-ds_j} : dx_j < 0, ds_j < 0 \right\}. \tag{7.8}$$

Now we show that the duality gap decreases from iteration to iteration. Here we are able to show that the decrease is proportional to $(1 - \alpha)$. That is, when the step size is large, we get a large decrease in the duality gap. Let us calculate the new duality gap

$$
\text{new gap} = \sum_{j=1}^{n} (\bar{x}_j + \alpha dx_j)(\bar{s}_j + \alpha ds_j),
$$

$$
= \sum_{j=1}^{n} \bar{x}_j \bar{s}_j + \alpha \sum_{j=1}^{n} (\bar{s}_j dx_j + \bar{x}_j ds_j) + \alpha^2 \sum_{j=1}^{n} dx_j ds_j.
$$

From (7.6), we have $ds = A^\top dy$. So

$$
\sum_{j=1}^{n} dx_j ds_j = (dx)^\top A^\top dy = (dy)^\top (A dx) = 0 \text{ (since } A dx = 0).
$$

Also note that using (7.7) we have $(\bar{s}_j dx_j + \bar{x}_j ds_j) = -\bar{x}_j \bar{s}_j$. Hence

$$
\text{(new duality gap)} = \sum_{j=1}^{n} \bar{x}_j \bar{s}_j + \alpha \sum_{j=1}^{n} (\bar{s}_j dx_j + \bar{x}_j ds_j)
$$

$$
= \sum_{j=1}^{n} \bar{x}_j \bar{s}_j - \alpha \sum_{j=1}^{n} \bar{x}_j \bar{s}_j
$$

$$
= (1 - \alpha) \sum_{j=1}^{n} \bar{x}_j \bar{s}_j
$$

$$
= (1 - \alpha) \text{ (current duality gap)}.
$$

Indeed, if $\bar{x} + dx \geq 0$ (i.e. $\alpha = 1$ yields a feasible solution in (P)) and $\bar{s} + ds \geq 0$ (i.e. $\alpha = 1$ also yields a feasible solution in (D)), then they ($\bar{x} + dx$ and $\bar{s} + ds$) will be optimal. On the other hand, unless $\alpha = 1$ at some iteration, we never reach an optimal solution. So, we might choose to stop iterating when the duality gap is below some small positive number ϵ (say 10^{-8}). Now we can state the *primal–dual affine-scaling algorithm* [50] (see Algorithm 7.9). We are given starting points x^0, (y^0, s^0) feasible in (P) and (D) respectively with $x^0 > 0$, $s^0 > 0$. Let $\epsilon \in (0, 1)$ denote the desired tolerance (we want to find x, (y, s) feasible in (P) and (D) respectively such that $x^\top s < \epsilon$).

It turns out that the solution dx, dy, and ds of the linear system (which is solved at each iteration) can be written explicitly in terms of A and the current solution $\bar{x}$ and $\bar{s}$ (note that $A\bar{x} = b$ and $A^\top \bar{y} + \bar{s} = c$). Define X to be the diagonal matrix with entries $\bar{x}_1$, $\bar{x}_2, \ldots, \bar{x}_n$ down the diagonal (and S is the diagonal matrix with entries $\bar{s}_1, \bar{s}_2, \ldots, \bar{s}_n$

Algorithm 7.9 Primal–Dual affine scaling algorithm

1: **loop**
2:   Let $k := 0$.
3:   **if** $(x^k)^\top s^k < \epsilon$ **then**
4:     **stop** x^k and (y^k, s^k) are the desired solutions.
5:   **end if**
6:   Let $(\bar{x}, \bar{y}, \bar{s}) := (x^k, y^k, s^k)$
7:   **Solve** (7.5)-(7.7) for (dx, dy, ds)
8:   **if** $(\bar{x} + dx, \bar{y} + dy, \bar{s} + ds)$ is feasible **then**
9:     Let $(x^{k+1}, y^{k+1}, s^{k+1}) := (\bar{x} + dx, \bar{y} + dy, \bar{s} + ds)$ and **stop** $(x^{k+1}, y^{k+1}, s^{k+1})$ is an optimal solution of (P) and (D)
10:   **end if**
11:   Choose step size $\alpha_k \in (0, 1)$ such that $\bar{x} + \alpha_k dx > 0$, and $\bar{s} + \alpha_k ds > 0$
12:   $(x^{k+1}, y^{k+1}, s^{k+1}) := (\bar{x} + \alpha_k dx, \bar{y} + \alpha_k dy, \bar{s} + \alpha_k ds)$
13:   $k := k + 1$
14: **end loop**

---

down the diagonal). Then the solution vectors dx, dy, and ds are given by the following formulas:

$$dy = - \left(AXS^{-1}A^\top \right)^{-1} b,$$

$$ds = A^\top dy,$$

$$dx = -\bar{x} - XS^{-1}ds.$$

To derive these formulas, we can rewrite (7.7) as $ds = -\bar{s} - X^{-1}Sdx$ substituting into (7.6), and then multiplying both sides by the nonsingular matrix XS^{-1} we obtain

$$XS^{-1}A^\top dy + \bar{x} + dx = 0.$$

Multiplying both sides from left by A, using the fact that $A\bar{x} = b$ and (7.5), we derive $AXS^{-1}A^\top dy = -b$. Since A has full row rank and XS^{-1} is an n-by-n diagonal matrix with positive entries on the diagonal, the matrix $\left(AXS^{-1}A^\top \right)$ is nonsingular and we reach the claimed solution for (dx, dy, ds).

Just as in the previous sections, in implementations of the algorithms, instead of explicitly forming the inverse of $(AXS^{-1}A^\top)$, we solve the linear system of equations $\left(AXS^{-1}A^\top \right) dy = -b$ for the unknown (dy). This is typically done by first computing a suitable decomposition of the matrix $\left(AXS^{-1}A^\top \right)$.

Remark 7.15   (i) *The idea that led us to the system of equations (7.5), (7.6), (7.7) is very closely related to Newton's method for solving a system of nonlinear equations.*

(ii) *There are many different ways of choosing the step size α; some have practical and/or theoretical advantages. In practice, we may even want to choose different step sizes in the primal and the dual problems, e.g., $\alpha_P = 0.9 \min\left\{\frac{x_j}{-dx_j} : dx_j < 0\right\}$ and $\alpha_D = 0.9 \min\left\{\frac{s_j}{-ds_j} : ds_j < 0\right\}$ and then update $(\bar{x}, \bar{y}, \bar{s})$ using $(\bar{x} + \alpha_P dx, \bar{y} + \alpha_D dy, \bar{s} + \alpha_D ds)$.*

(iii) *The algorithm can be modified so that it allows infeasible starting points. The only requirement is that $\bar{x}$ and $\bar{s} > 0$. In this combined two phase version of the algorithm, the right-hand side vectors of (7.5) and (7.6) are replaced by $(b - A\bar{x})$ and $(c - A^\top \bar{y} + \bar{s})$ and are updated in each iteration.*

7.8.1 A polynomial-time interior-point algorithm*

In our primal–dual interior-point algorithm, we can choose the step size more conservatively to guarantee that our algorithm converges in polynomial-time. Let us define the following *potential function*:

$$\psi(x, s; \rho) := (\rho + 1) \ln\left(\frac{x^\top s}{n}\right) - \ln\left(\min_j\{x_j s_j\}\right),$$

where $x, s \in \mathbb{R}^n$, $x > 0$, $s > 0$, and ρ is a positive scalar to be chosen later. In the current context, the potential function tries to balance competing strategies of reducing $x^\top s$ fast and not getting too close to the boundary of the feasible region prematurely. The first term in the potential function leads to values tending to $-\infty$ as $x^\top s$ approaches to zero and the second term goes to $+\infty$ as x or s converge to a point on the boundary of the nonnegative orthant. So, the potential function will only allow the iterates to get close to the boundary when the duality gap $x^\top s$ decreases at a suitably comparable pace.

In the algorithm, we will choose the step size by the following rule:

choose the step size $\alpha_k \in (0, 1)$ such that $\psi(x + \alpha_k dx, s + \alpha_k ds; \rho) = 1 + \rho \ln\left(\frac{1}{n\epsilon}\right)$.

We can make sure that the starting point satisfies certain technical conditions depending on the data (A, b, c). (See the exercises at the end of the section.) Then we can prove an upper bound on the number of iterations of the algorithm. We assume that we are given $\epsilon \in (0, 1)$. We define

$$\rho := 2 \ln\left(\frac{1}{\epsilon}\right).$$

THEOREM 7.16 *Suppose we are given (x^0, y^0, s^0) as in the statement of the primal–dual affine scaling algorithm. Further assume that*

$$\psi(x^0, s^0; \rho) \leq 1 + \rho \ln\left(\frac{1}{n\epsilon}\right).$$

Then the primal–dual affine scaling algorithm with the above step size strategy based on ψ, terminates in at most $O\left(n \ln^2\left(\frac{1}{\epsilon}\right)\right)$ iterations with x^k, (y^k, s^k) feasible in (P) and (D) respectively such that their duality gap is less than ϵ.

This theorem guarantees finding feasible solutions with very good objective values. The technical assumption on $\psi(x^0, s^0; \rho)$ can be satisfied by a reformulation trick. Moreover, it turns out, for LP, if we can get the duality gap below a certain threshold (for data (A, b, c) with rational entries, let t denote the number of bits required to store them in binary; then choosing $\epsilon := \exp(-2t)$ suffices), then using these very accurate approximate solutions, we can quickly compute exact optimal solutions. Using these techniques and ideas, we can prove that LP problems with rational data can be solved in polynomial-time. We work through some of the details in exercises at the end of this chapter and in the Appendix.

Exercises

1 Suppose $A \in \mathbb{R}^{m \times n}$ with $\text{rank}(A) = m$, $\bar{x} > \mathbb{0}$ and $\bar{s} > \mathbb{0}$ are given. Prove that the system (7.5)–(7.7) has a unique solution.

2 Consider the LP given in the form of (P) with

$$
A = \begin{pmatrix} 1 & 2 & 1 & 1 & 0 \\ 3 & 1 & -1 & 0 & 1 \\ 0 & 1 & 1 & 1 & 0 \end{pmatrix} \quad b = \begin{pmatrix} 5 \\ 8 \\ 3 \end{pmatrix} \quad c = \begin{pmatrix} 4 \\ 2 \\ 1 \\ 1 \\ -3 \end{pmatrix}.
$$

Starting with the primal–dual pair $\bar{x} = (1, 1, 1, 1, 5)^\top$, and $\bar{s} = (4, 9, 3, 4, 4)^\top$, apply the primal–dual interior-point method until the duality gap goes below 10^{-4}. Then identify a pair of (exact) optimal solutions for (P) and its dual. Finally, verify the correctness of your answer by utilizing duality theory.

3 Suppose that during an application of the primal–dual interior-point method (to a pair of primal–dual LPs, with the primal (P) in standard equality form), when we solve the system

$$
A d_x = \mathbb{0},
$$
$$
A^\top d_y - d_s = \mathbb{0},
$$
$$
\bar{s}_j d_{x_j} + \bar{x}_j d_{s_j} = -\bar{x}_j \bar{s}_j, \quad \text{for all } j
$$

(for given current iterate $\bar{x} > 0, \bar{s} > 0$ such that $A\bar{x} = b$, $A^\top \bar{y} - \bar{s} = c$, for some $\bar{y}$), we find that $d_x = 0$. Prove that in this case there exists $\hat{y}$ such that

$$
A^\top \hat{y} = c
$$

and that every feasible solution of (P), including $\bar{x}$, is optimal.

4 (Advanced) Consider Theorem 7.16. Prove that the same iteration bound can be achieved if we take a constant step size $\alpha = \Theta\left(\frac{1}{nt^2}\right)$ in every iteration. (Note, t is defined in the paragraph following Theorem 7.16.)

5 (Advanced) In the system of linear equations determining the search direction, replace the right-hand side of (7.7) by

$$-\bar{x}_j \bar{s}_j + \gamma \frac{x^\top s}{n}, \quad \text{for all } j \in \{1, 2, \ldots, n\},$$

where $\gamma \in [0, 1]$ a given constant.

(a) Prove that in this case

(new duality gap) = $[1 - \alpha(1 - \gamma)]$ (current duality gap).

(b) What would be a potential benefit for choosing a value for γ that is positive? (For example, if you choose $\gamma := 0.5$, can you prove a lower bound for the step size α, which is considerably larger than $O(1/nt^2)$.)

6 (Advanced) Let (P) denote an LP problem in standard inequality form and let (D) denote its dual. Suppose we are given a feasible solution $\bar{x}$ for (P) and a feasible solution $(\bar{y}, \bar{s})$ for (D) such that $\bar{x}^\top \bar{s} < \epsilon$ for some positive ϵ. Construct an efficient algorithm which computes an extreme point of the feasible region of (P), $\hat{x}$, such that

$$c^\top \hat{x} > z^* - \epsilon,$$

where z^* is the optimal objective value of (P).

HINT: Consider a related advanced exercise from Section 2.4.2 (namely, Exercise 6). If $\bar{x}$ is a basic feasible solution, then we are done (justify); otherwise, the columns of A corresponding to the positive entries of $\bar{x}$ are linearly dependent. Can you find a way of efficiently modifying $\bar{x}$ while staying feasible, reducing the number of positive entries of $\bar{x}$, and not making the current objective function value $c^\top \bar{x}$ worse?

7 (Advanced) Suppose that in our NLP, in addition to the inequality constraints $g(x) \leq 0$, we have explicit equality constraints. These can be written as $h_i(x) = 0, i \in \{1, \ldots, p\}$. If these constraints are affine, given by $h(x) = Ax - b$, then we have a way of "moving around" in the set of solutions to these equality constraints (e.g., if A has full row rank, we find a basis B and we rewrite the constraints as $x_B + A_B^{-1} A_{NXN} = A_B^{-1} b$ and consider directions of local movement from the basic solution determined by B).

Suppose that h_i is continuously differentiable at $\bar{x}$ for every i, $h(\bar{x}) = 0$ and $\nabla h_1(\bar{x}), \ldots, \nabla h_p(\bar{x})$ are linearly independent. Using the *inverse function theorem* below, derive a way of locally moving around the point $\bar{x}$ which generalizes the special case of affine h above.

Inverse function theorem: Let $F : U \subseteq \mathbb{R}^n \to \mathbb{R}^n$ be continuously differentiable on the set U which is open. Further assume that at $\bar{x} \in U$ gradients of $F_1, \ldots, F_n$ are linearly independent. Let $DF(\bar{x})$ denote the matrix of those gradients. Then there exists an open set V in U containing $\bar{x}$ and an open set W containing $F(\bar{x})$ such that $F(V) = W$ and F has a *local*, continuously differentiable inverse $F^{-1} : W \to V$. Moreover, for every $w \in W$ with $x := F^{-1}(w)$, we have

$$DF^{-1}(w) = [DF(x)]^{-1} .$$

7.9   Further reading and notes

For a further introduction into nonlinear optimization, see Peressini et al. [54], Boyd and Vandenberghe [11], and Nesterov [53]. There are also algorithms designed for linear optimization using insights gained from developing the theory of nonlinear convex optimization (see Khachiyan [37], Karmarkar [35], and Ye [72]). For a more detailed discussion of hardness of nonlinear optimization problems, see Murty and Kabadi [51], and Vavasis [67].

There are historical as well as theoretical connections to game theory. Suppose two players P and D are playing a two-person zero-sum game. Player P's strategies are denoted by a vector $x \in \mathbb{R}^n$ and player D's strategies are denoted by a nonnegative vector $y \in \mathbb{R}^m$. Players choose a strategy (without knowing each other's choices) and reveal them simultaneously. Based on the vectors x and y that they reveal, Player P pays Player D $\left[f(x) + \sum_{i=1}^{m} y_i g_i(x) \right]$ dollars (if this quantity is negative, Player D pays the absolute value of it to Player P). Player P's problem is

$$\min_{x \in \mathbb{R}^n} \max_{y \in \mathbb{R}^m_+} \{ L(x, y) \} = \min_{x \in \mathbb{R}^n} \max_{y \in \mathbb{R}^m_+} \left[f(x) + \sum_{i=1}^{m} y_i g_i(x) \right],$$

player D's problem is

$$\max_{y \in \mathbb{R}^m_+} \min_{x \in \mathbb{R}^n} \{ L(x, y) \} = \max_{y \in \mathbb{R}^m_+} \min_{x \in \mathbb{R}^n} \left[f(x) + \sum_{i=1}^{m} y_i g_i(x) \right].$$

There are very many ways of utilizing the Lagrangian in the theory and practice of optimization. For example, we can use the Lagrangian in designing primal–dual algorithms (even for combinatorial optimization problems). Consider $\bar{x} \in \mathbb{R}^n$, a current iterate of an algorithm. Suppose $\bar{x}$ violates a subset of the constraints $g_i(x) \leq 0$, for $i \in J$. We can choose $\bar{y}_i > 0, \forall i \in J$ (essentially penalizing the violations) and minimize $L(x, \bar{y})$. This gives us a new iterate $\bar{x}$ and we may repeat the process by choosing a new $\bar{y}$ depending on the current violation and repeat. In many cases, choosing a good $\bar{y}$ is also done via solving an easier optimization problem. Note that we outlined a primal–dual scheme, with alternating moves, based on relaxations, in the primal and dual spaces. Usage of the Lagrangian, as the name suggests goes back at least to the times of Lagrange, it is connected to the works of Leonhard Euler as well as Joseph Louis Lagrange during the mid-1700s. In addition to its connections to the foundations of nonlinear optimization, the work also inspired a lot of research in the areas of differential equations as well as mechanics.

For further information on convex relaxation approaches, see Tunçel [64] and the references therein. Utilization of Karush–Kuhn–Tucker Theorem in the design of algorithms for nonconvex NLP problems continues to be fruitful. For example, in nonconvex optimization problems arising from "big data" applications such as those in compressed sensing seem amenable to such approaches based on the KKT Theorem, see for instance [12]. KKT Theorem is named after Karush [36], as well as Kuhn and Tucker [44]. For a historical account, see Cottle [18].

For a review of interior-point algorithms, see Ye [72] and the references therein. There are many primal–dual interior-point algorithms in the literature. The Primal–dual affine-scaling algorithm was proposed by Monteiro, Adler, and Resende [50]. For some of the variants, including the one we discussed, see Tunçel [65] and the references therein.

Appendix A Computational complexity

An algorithm is a formal procedure that describes how to solve a problem. For instance, the simplex algorithm in Chapter 2 takes as input a linear program in standard equality form and either returns an optimal solution, or detects that the linear program is infeasible or unbounded. Another example is the shortest path algorithm in Chapter 3.1. It takes as input a graph with distinct vertices s, t and nonnegative integer edge lengths, and returns an st-path of shortest length (if one exists).

The two basic properties we require for an algorithm are: *correctness* and *termination*. By correctness, we mean that the algorithm is always accurate when it claims that we have a particular outcome. One way to ensure this is to require that the algorithm provides a certificate, i.e. a proof, to justify its answers. By termination, we mean that the algorithm will stop after a finite number of steps.

In Section A.1, we will define the running time of an algorithm; we will formalize the notions of *slow* and *fast* algorithms. Section A.2 reviews the algorithms presented in this book and discusses which ones are fast and which ones are slow. In Sections A.3 and A.4 we discuss the inherent complexity of various classes of optimization problems and discuss the possible existence of classes of problems for which it is unlikely that any fast algorithm exists. We explain how an understanding of computational complexity can guide us in the design of algorithms. (For much more information on the topic, we refer the interested reader to the recent textbooks of Kleinberg and Tardos [41], and Sipser [60].)

A.1   What is a fast (resp. slow) algorithm?

As an example consider the $0, 1$ *feasibility* problem. Here, we are given a rational $m \times n$ matrix A, a rational vector b with m entries, and we are asked if there exists a solution $\bar{x}$ to $Ax \leq b$ where all entries of $\bar{x}$ are either 0 or 1. This problem is a finite problem, as every variable can only take two possible values. Since we have n variables, there are 2^n possible assignments of values to the variables. Hence, to solve the $0, 1$ feasibility problem we could try all possible 2^n assignments of $0, 1$ values $\bar{x}$ and check for each case whether $\bar{x}$ satisfies the inequalities $Ax \leq b$. This procedure is an algorithm for the $0, 1$ feasibility problem, as correctness and termination are trivially satisfied in this case.

This is however, not a satisfactory algorithm. One drawback is that it is *very slow*. Suppose for instance that we have $n = 100$ variables. We need to enumerate 2^{100}

possible assignments of values to the variables. Assuming that we have implemented our algorithm on a computer that can enumerate a million such assignments every second, it would take over 4×10^{16} years. According to current estimates the universe is 13.75 billion years old. Thus, the running time would be nearly three million times the age of the universe! Clearly, this is not a practical procedure by any reasonable standard. This illustrates the fact that brute force enumeration is not going to be a sensible strategy in general.

There is a further shortcoming of this algorithm. Suppose that the algorithm states that there is no 0, 1 solution to the system $Ax \leq b$. How could we convince anyone that this is indeed the case? The algorithm provides no help and anyone wanting to verify this fact would have to solve the problem from scratch.

A.1.1 The big "O" notation

Before we can proceed further, we require a few definitions. Consider functions f, g : $\mathbb{N} \to \mathbb{N}$.[1] We write:

(a) $f = O(g)$ if there exist constants c_1, c_2 such that for all $n \geq c_1, f(n) \leq c_2 g(n)$,
(b) $f = \Omega(g)$ if there exist constants c_1, c_2 such that for all $n \geq c_1, f(n) \geq c_2 g(n)$,
(c) $f = \Theta(g)$ if $f = O(g)$ and $f = \Omega(g)$.

Thus, $f = O(g)$ means that for large n, some fixed multiple of $g(n)$ is an upper bound of $f(n)$. Similarly, $f = \Omega(g)$ means that for large n, some fixed multiple of $g(n)$ is a *lower* bound for $f(n)$. Hence, if $f = \Theta(g)$, then f behaves like g asymptotically.

Example 30 Suppose for instance that $f(n) = 2n^3 + 3n^2 + 2$. Then $f = O(n^3)$, as for all $n \geq 4$ we have $3n^2 + 2 \leq n^3$. Hence, in particular

$$2n^3 + (3n^2 + 2) \leq 3n^3.$$

(We apply the definition with $c_1 = 4$ and $c_2 = 3$.) Similarly, we can show that $f = \Omega(n^3)$. Hence, $f = \Theta(n^3)$. In general, we can show that if $f(n)$ is a polynomial of degree k, i.e.

$$f(n) = \sum_{i=1}^{k} a_i n^i,$$

for constants $a_1, \ldots, a_k$ where $a_k \neq 0$, then $f = \Theta(n^k)$.

A.1.2 Input size and running time

It will be important to distinguish between a problem and an *instance* of that problem. For example, if our problem is 0, 1 feasibility, an instance of that problem would be described by a matrix A, and a vector b and solving that instance means checking if

---

[1] $\mathbb{N}$ is the set of all positive integers, i.e. $1, 2, 3, 4, \ldots$.

there exists a 0, 1 solution to $Ax \leq b$ for that specific matrix A and vector b. When we say that an algorithm solves the 0, 1 feasibility problem, we mean that we have an algorithm that works for every instance of the problem.

The *input size* of a given instance is the number of bits we need to store the data in a computer. The way the data are encoded is not important[2] as long as we avoid storing numbers in base 1. We define the running time of an algorithm as the number of *primitive operations* that are needed to complete the computation where a primitive operation is one that can be completed in a short fixed amount time on a computer (i.e. a constant number cycles on the processor). This may include for instance:

- arithmetic operations like $+, -, *, /$ involving numbers of a bounded size,
- assignments,
- loops, or
- conditional statements.

Example 31 Consider a simple algorithm that multiplies two square integer matrices.

Algorithm A.10

1: **Input:** Matrices $A, B \in \mathbb{N}^{n \times n}$.
2: **Output:** Matrix $C = AB$.
3: Initialize $C := 0$
4: **for** $1 \leq i \leq n$ **do**
5:    **for** $1 \leq j \leq n$ **do**
6:       $C_{ij} := 0$
7:       **for** $1 \leq l \leq n$ **do**
8:          $C_{ij} := C_{ij} + A_{il}B_{lj}$
9:       **end for**
10:    **end for**
11: **end for**
12: **return** C

Let us assume that we are in the case where all numbers in the matrices A and B are within a fixed range. Then every one of these numbers can be stored using a fixed number of bits, say k. Thus, the total number of bits required to store the data A, B (i.e. the input size) is $2 \times n^2 \times k = \Theta(n^2)$.

The main work of the algorithm is done in two main loops in steps 4 and 5. In particular, steps 6–9 are executed n^2 times, once for each pair $1 \leq i, j \leq n$. Each such

---

[2] To store numbers that can take values between 1 and ℓ in base 1 we require ℓ bits, however, by using base p, we only require, $\log_p(\ell)$ bits, an exponential difference in storage requirements.

execution needs $n + 1$ assignments, n additions, and n multiplications. Thus, the entire algorithm (modulo the initialization) needs

$$n^2(3n + 1) = 3n^3 + n^2 = \Theta(n^3)$$

primitive operations in total.

Of course, we expect algorithms to have a longer running time as the size of the instance increases. Hence, we will express the running time as a function f of the input size s. Consider Example 31 again. For an instance A, B of the problem (where A, B are $n \times n$ matrices), the size s of the problem is $\Theta(n^2)$ and the running time is $\Theta(n^3)$. Thus, we have the following running time:

$$f(s) = \Theta\left(s^{\frac{3}{2}}\right).$$

In the case of Example 31, the running time is always going to be the same for every instance of the same size s. Thus, there is no ambiguity when we talk about running time. This need not be the case in general however. Consider for instance the simplex algorithm. We may be given as input a problem that is already in canonical form for an optimal basis, or we may be given a problem of the same input size that will require many steps of the simplex algorithm. How should we define the running time in this context? We take a *worst case* view, and define the running time function f by letting $f(s)$ be the *longest* actual running time of the algorithm for input instances of size s.

A.1.3 Polynomial and exponential algorithms

We say that an algorithm is (or runs in) *polynomial-time* if its running time function f is a polynomial in the input size s; i.e. if $f(s) = O(s^k)$ for some constant k. An algorithm is (or runs in) *exponential-time* if for some constant $k > 1$ we have $f(s) - \Omega(k^s)$; i.e. there are input instances of size s that require our algorithm to execute a number of steps that is exponential in s.

Henceforth, we will say that polynomial-time algorithms are *fast*, and exponential-time algorithms are *slow*. According to our definitions the simple minded algorithm for the $0, 1$ feasibility problem is slow and the algorithm for multiplying two matrices is fast.

To motivate the notion of fast and slow, suppose that your computer is capable of executing 1 million primitive instructions per second. Assume that you have an algorithm that has running time $f(s)$ for an input size s. The following table shows the actual running time of this algorithm on your computer for an input of size $s = 100$, depending on $f(s)$. Clearly, the slow, i.e. exponential-time algorithms (the rightmost two), will be of little use in this case.

$f(s)$	s	$s \log_2(s)$	s^2	s^3	1.5^s	2^s
Time	< 1 sec	< 1 sec	< 1 sec	1 sec	12, 892 years	4×10^{16} years

Let us consider the impact of improved hardware on the ability of fast and slow algorithms to solve large instances. Suppose that in 1970 Prof. Brown was running a (fast) algorithm with a running time $f(s) = s^3$ and that he was able to solve a problem of size 50 in one minute on a computer. Forty years later, in 2010, Prof. Brown has a new computer that is a million times faster than the one he had in 1970. What is, then, the largest instance he can solve in one minute running the same algorithm? If m denotes the number of primitive operations that the computer could run in one minute in 1970, we must have $50^3 \leq m$. We claim that Prof. Brown can solve problems of size $100 \times 50 = 5000$ with his 2010 computer. This is because for such a problem the running time is $(100 \times 50)^3 = 100^3 \times 50^3 \leq 10^6 \times m$. Hence, the size of the largest instance that can be solved has been *multiplied* by a factor 100.

On the other hand, suppose that in 1970 Prof. Brown was running a (slow) algorithm with a running time $f(s) = 2^s$ and that he was able to solve a problem of size at most 50 in one minute. As the largest instance he could solve in 1970 had size 50, we must have had $2^{51} > m$. We claim that Prof. Brown can only solve problems of size at most $50+20$ with this new computer. This is because for a problem of size $50 + 21$ the running time is $2^{50+21} = 2^{51} \times 2^{20} > m \times 2^{20} > m \times 10^6$ as $2^{20} > 10^6$. Hence, the size of the largest instance that can be solved has only increased by a small *additive* amount!

This is in fact a general phenomenon: if we multiply the number of allowed elementary operations by a constant, the largest size of a problem we can solve by a polynomial-time algorithm changes by a multiplicative constant, whereas the largest size of a problem we can solve by an exponential-time algorithm changes by an additive constant.

Exercises

1 A standard legal size paper sheet is 0.1 millimeter thick. How many times do you need to fold a legal size paper sheet into two such that the resulting folded sheet has thickness larger than the distance between the earth and the moon (384,400 km)?
HINT: It is a surprisingly small number.

2 The Planet Express manufacturing plant produces gizmos of k different types. Production of each gizmo of type i requires a_i hours of labor, b_i dollars in additional costs, and has a resale value of c_i dollars. In addition, no more than five gizmo's of any type i should be produced. You have at your disposal α hours of labor and a budget of β dollars.
(a) Formulate as an integer program the problem of deciding how many units of each gizmo to produce so as to maximize the total resale value of the items produced while not exceeding the labor and budget at your disposal. HINT: You will have an integer variable x_i to indicate how many gizmos of type i are produced.
(b) Professor Farnsworth proposes the following scheme for solving the integer program. Since each variable x_i can take only one of the following values $\{0, 1, 2, 3, 4, 5\}$, try all possible combinations of assignment of values to the variables x_i, check if the corresponding solution is feasible and pick the one which yields the maximum resale value. Discuss the merits and shortcomings of such a strategy.

(c) Captain Leela uses the company's mainframe computer to solve the integer program following the strategy outlined by the Professor. Suppose that she can now solve problem instances with 20 different gizmo types. If the computer is upgraded to a new version that is 10 000 times faster, for how many gizmo types can Leela now solve the problem?

3 Xavier is a student who finds the simplex algorithm complicated and proposes the following alternative for finding an optimal solution to,

$$\max\{c^\top x : Ax = b, x \geq \mathbf{0}\}.$$

Using linear algebra find a maximal set of linearly independent columns A_B where $B \subseteq \{1, \ldots, n\}$. Then A_B is a square matrix. Set $\bar{x}_j = 0$ for all $j \notin B$ and let $\bar{x}_B$ be a solution to the system $A_B x_B = b$. Then $\bar{x}$ is a basic solution. Among all basic solution $\bar{x} \geq \mathbf{0}$, choose the one that maximizes $c^\top x$.

Discuss the merit of Xavier's proposals. In particular, indicate if the scheme always gives a correct answer, and if it is practical.

A.2 Examples of fast and slow algorithms

In this section, we discuss the running time of the algorithms presented in this book and classify them into slow and fast algorithms.

A.2.1 Linear programming

Linear programs was the first and most basic class of optimization problems that we introduced in Chapter 1. In Chapter 2, we studied one of the most popular and successful methods for solving such problems: the simplex algorithm. Even after more than 70 years since its invention by George Dantzig in the late 1940s [21], the simplex algorithm remains *the* dominant method for solving LPs in most state-of-the-art linear programming solvers [7].

Given this, it is hard to believe that the simplex algorithm is in fact a slow algorithm, or to be more precise: we do not know of a pivoting rule for which the simplex method runs in polynomial-time. In fact, Klee and Minty [40] were the first to exhibit example instances on which the simplex algorithm (with Dantzig's pivoting rule) requires an exponential number of steps to terminate. Later such examples were found for virtually all *deterministic* pivoting rules (e.g., see [3]). If pivot rules are allowed to be randomized, we can do slightly better: the simplex algorithm can be shown to run in $O(m^{\sqrt{n}})$ time for LPs with m constraints and n variables (see [33, 47]). The work in the area of pivoting rules for the simplex method continues to date. In 2011, Friedmann [27] proved that the *least entered rule* proposed by Zadeh in 1980 cannot yield a polynomial-time simplex algorithm in the worst case (this particular problem had stayed open for 31 years).

So how can we explain the fact that simplex works so well in practice? Borgwardt [10], gave one of the first partial answers, and showed that simplex performs well on *average* instances; he showed the algorithm with the *shadow vertex* pivot rule runs in expected polynomial-time for a certain class of random instances. Also see Adler, Megiddo and Todd [1], Haimovich [31], Smale [62, 61] and the references therein for related results. However, *random* instances are not a good model for *typical* instances as Spielman and Teng [63] pointed out. The authors proposed a different argument for explaining the practical performance of the simplex method. They argued that input data for practical LP problems is often *inherently noisy*, and proposed the following *smoothed complexity* model for the analysis of algorithms. Apply a tiny random perturbation to the data of any given input instance, and bound the worst case *expected* running time of the algorithm. In this model, the authors were able to show that the running time of simplex can be bounded by a polynomial in the input size! So, if we ever encounter an instance for which simplex method requires an unexpectedly large number of iterations, then there always exists a tiny perturbation of that data so that the LP resulting from the perturbed data is much better behaved.

The quest for a pivoting rule for simplex that guarantees polynomial running time closely relates to the *Hirsch* conjecture for polyhedra. Given a polyhedron P, we define a graph $H(P)$ that has one vertex for each extreme point of P. Two vertices $x^{(1)}$ and $x^{(2)}$ are connected by an edge if the corresponding two extreme points are *adjacent*; i.e. if there exist bases corresponding to these two extreme points that differ in exactly two elements (thus could appear consecutively in the simplex algorithm). We say that two extreme points $x^{(1)}$ and $x^{(2)}$ are at distance k if there is an $x^{(1)}, x^{(2)}$-path in $H(P)$ with at most k edges. Figure A.1 illustrates these concepts: $x^{(1)}$ and $x^{(2)}$ are adjacent, $x^{(1)}$ and $x^{(3)}$ are not, and $x^{(3)}$ and $x^{(4)}$ are at distance 2.

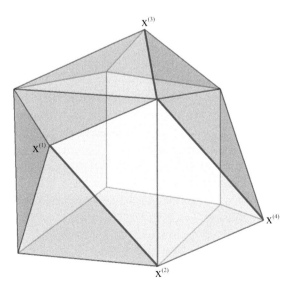

Figure A.1

The Hirsch conjecture is now as follows:

In an n-dimensional polyhedron P that is defined by m inequalities, any two vertices $x^{(1)}$ and $x^{(2)}$ are at distance at most $m - n$.

It is now not difficult to see that an upper bound $f(n, m)$ on the running time of simplex when applied to an LP with n variables and m inequalities would imply the same bound on the maximum distance between any two vertices of the polyhedron corresponding to the feasible region. The existence of short paths between any two extreme points of $H(P)$ does not imply, however, that simplex can *find* such a path. It turns out that the Hirsch conjecture as stated is *false*; more than 50 years after being initially posed, Santos [57] very recently found a complex counter example. The counter example does not preclude that the diameter of a polyhedron is polynomial (or even linear) in m and n, leaving a tantalizing open problem for the ambitious reader!

All of the above said, linear programming has polynomial-time algorithms. In 1979, the Soviet mathematician L.G. Khachiyan [38] proved that the *ellipsoid method* (originally proposed by Shor [59] as well as Nemirovski and Yudin [52] for a class of well-behaved nonlinear optimization problems) can be implemented so that it becomes a polynomial-time algorithm to solve linear programs, thereby showing that there is a fast algorithm for linear programming.

While this discovery received a great deal of attention (among others, in an article in the *New York Times*), the excitement about its practical impact quickly dissipated. The ellipsoid method can be proven to run in polynomial time for LP problems in particular (for a recent reference, see [42]) and convex optimization problems in general, under suitable conditions (see [52] and [64]). While this immediately implies that the ellipsoid method outperforms almost all popular variants of the simplex method in the worst case, the latter is much faster empirically. We note, however, that the ellipsoid method has several crucial consequences for the theory of optimization.

Since Khachiyan's discovery, many variants and other polynomial algorithms for linear programming (like *interior-point methods* introduced in Section 7.8) have been developed, and some of them are competitive with the simplex method in certain typical instances.

A.2.2   Other algorithms in this book

The shortest path algorithm developed in Section 3.1 is known as *Dijkstra's* algorithm [22]. When implemented in a straightforward way, its running time is $O(|V|^2)$, but with some care (when using so called *Fibonacci heaps*) it is possible to achieve a running time of $O(|E| + |V| \log |V|)$ [26], and this is the fastest-known running time for the shortest path problem with nonnegative weights. Better results are known for special cases.

Section 3.2 discusses the minimum cost perfect matching problem in bipartite graphs. The Hungarian algorithm presented is due to Kuhn [43], and a straightforward implementation runs in time $O(|V|^2|E|)$ (e.g., see [16]). With some care, the algorithm can be implemented to run in time $O(|V||E| + |V|^2 \log |V|)$ [24].

Equivalent to finding a minimum weight perfect matching is the problem of finding a maximum weight matching in a given graph (e.g., see [42]). If we are interested in finding the matching of largest *cardinality* in a given bipartite graph $G = (V, E)$, we may use the algorithm of Hopcroft and Karp [32] which runs in time $O(\sqrt{|V|}|E|)$. Surprisingly, Micali and Vazirani [49] showed that the same running time can also be obtained in general graphs. Maximum weight matchings in general graphs can be found in time $O(|E||V| + |V|^2 \log |V|)$ [28].

Finally, the cutting plane algorithm for integer programming presented in Chapter 6 is not known to run in polynomial-time, and is in fact widely expected to be a slow (i.e. exponential-time) algorithm. The reason for this belief lies in the fact that such an algorithm could be used to solve truly difficult, so-called *NP-complete* decision problems. We will discuss this in the next section.

A.3   The classes *NP, co-NP* and *P*

A.3.1   Decision problems

A problem where for every instance the answer is either **YES** or **NO** is called a *decision problem*. For example, given a rational matrix A and vector b, we may ask if the system $Ax = b$ has a solution. In this decision problem, an instance is described by the matrix A and the vector b. Moreover, the answer is either: **YES** the system $Ax = b$ has a solution or **NO** the system $Ax = b$ does not have a solution. A list of decision problems is given in Table A.1.

Given a connected graph $G = (V, E)$ with vertices s, t and nonnegative edge weights $w \in \mathbb{N}^E$ the *shortest path problem* asks us to find the length of the shortest st-path. (Here we are not concerned with finding the actual shortest st-path.) Clearly, if we have a polynomial-time algorithm for solving the shortest path problem, we can solve the

Table A.1 Decision problems

Input	Rational matrix A and vector b.	ⓐ
Question	Does $Ax = b$ have a solution?	
Input	Rational matrix A and vector b.	ⓑ
Question	Does $Ax \leq b$ have a solution?	
Input	Rational matrix A and vector b.	ⓒ
Question	Does $Ax \leq b$ have an integer solution?	
Input	Bipartite graph G.	ⓓ
Question	Does G have a perfect matching?	
Input	Graph $G = (V, E)$, $s, t \in V$, $w \in \mathbb{N}^E$ and $k \in \mathbb{Z}$.	ⓔ
Question	Does G have an st-path of length $\leq k$?	
Input	Graph $G = (V, E)$, $s, t \in V$, $w \in \mathbb{N}^E$ and $k \in \mathbb{Z}$.	ⓕ
Question	Does G have an st-path of length $\geq k$?	

associated decision problem ⓔ in polynomial-time. Namely, compute the length ℓ of the shortest *st*-path, if $\ell \leq k$ then the answer is **YES** otherwise the answer is **NO**.

Moreover, we claim that given a polynomial-time algorithm for solving the decision problem ⓔ can we use it to find a polynomial-time algorithm for the shortest path algorithm. In particular, this implies that the shortest path problem and the decision problem ⓔ are equivalent from the standpoint of computational complexity. Since every edge has nonnegative integer weight, the length of any *st*-path is an integer between 0 and $u = \sum_{e \in E} w_e$. A simple strategy would be to try consider every integer value $k \in \{0, \ldots, u\}$ and use the algorithm for ⓔ to check whether the length of the shortest *st*-path is less than or equal to k. We will clearly be able to establish the length of the shortest path in this manner, however, this will not result in a polynomial-time algorithm for the shortest path problem. The difficulty here is that we need to solve problem ⓔ u times while $O(|E| \log_2(u))$ bits suffice to store all the edge weights. Thus, the number of times we use solve problem ⓔ can be exponential in the size of the instance. While this simple strategy fails, the following procedure will work,

---

Algorithm A.11

1: Initialize $\ell = 0$, $u = \sum_{e \in E} w_e$
2: **while** $u > \ell$ **do**
3:     $k = \lfloor \frac{u+\ell}{2} \rfloor$.
4:     **if** length of shortest *st*-dipath is $\leq k$ **then**
5:         $u = k$
6:     **else**
7:         $\ell = k + 1$
8:     **end if**
9: **end while**
10: Return ℓ

We let the reader verify that ℓ is indeed the length of the shortest *st*-path. The procedure will terminate after at most $\log_2(u)$ steps, in particular the number of times problem ⓔ needs to be solved is a polynomial function of the size of w.

More generally, for every optimization problem that asks to minimize (resp. maximize) some function $f(x)$ there is an associated decision problem that asks for a fixed k if there exists $\bar{x}$ such that $f(\bar{x}) \leq k$ (resp. $f(\bar{x}) \geq k$). Thus, we will restrict our discussion to decision problems.

A.3.2 The class *NP*

Informally a decision problem is in NP^3 if there exists a short certificate for every **YES** instance. Consider the following metaphor. Alice wishes to organize a party that is to be restricted to UW (University of Waterloo) students. She books a venue and hires

---

[3] *NP* stands for *Non-deterministic Polynomial-time.*

a bouncer to check every person that wishes to enter the venue. The bouncer has to solve the following decision problem: Given person X at the door, is X a UW student? If the answer is **YES** then the person is accepted to the party, if the answer is **NO** then the person is denied entry. Every UW student is given an identification card and of course such IDs are only given to UW students. Thus, if person X is a UW student he or she can easily convince the bouncer of that fact by exhibiting his or her ID. In other words, for every **YES** instance X of the decision problem, there is a certificate (in this case the UW ID) that allows us to efficiently verify that it is indeed a **YES**instance.

Let us formalize these ideas. Consider a class of decision problems $\mathscr{D}$. A *YES-Checking algorithm* is an algorithm that takes as input an instance D of $\mathscr{D}$ as well as a potential certificate $C(D)$ and returns **OK** if $C(D)$ shows that D is indeed a **YES** instance and returns **NOT-OK** otherwise. (Note, when the **YES**-checking algorithm returns **NOT-OK** it does not mean that instance D is a **NO** instance, it just means that the certificate $C(D)$ is not sufficient to establish that D is a **YES** instance.) We say that $\mathscr{D}$ is in *NP* if for every **YES** instance D of $\mathscr{D}$ there exists a certificate $C(D)$ for which the **YES**-Checking algorithm returns **OK**. Moreover, we require that the running time of the **YES**-Checking algorithm be polynomial in the size of instance D. In our earlier metaphor, where the decision problem is to check if person X is a UW-student, the **YES**-checking algorithm is the bouncer, and the bouncer can efficiently verify if instance X is a **YES** instance by using the UW-ID as a certificate.

Let us consider problem ⓓ in Table A.1. An instance is a bipartite graph G. If G is a **YES** instance, then G has a perfect matching. We can use as a **YES** certificate a perfect matching M. The **YES**-checking algorithm is an algorithm that given a subset of edges B says **OK** if B is a perfect matching and **NOT-OK** if B is not a perfect matching. Clearly, it is easy to design such an algorithm with running time polynomial in the size of the graph G. Thus, by definition, ⓓ is in *NP*.

Let us consider problem ⓐ in Table A.1. An instance is a rational matrix A and a rational vector b. If A, b is a **YES** instance, then there exists $\bar{x}$ such that $A\bar{x} = b$. We can use as a **YES** certificate a vector $\bar{x}$ such that $A\bar{x} = b$. The **YES**-checking algorithm is an algorithm that given a vector $\hat{x}$ checks whether $A\hat{x} = b$. We need to make sure that the algorithm runs in polynomial time in the size of A and b however. In order to do this, we need to ensure that we can choose $\bar{x}$ whose size is polynomial in the size of A and b. The next remark indicates that this can always be achieved,

Remark A.1 *Let A be a rational matrix, let b be a rational vector. If $Ax = b$ has a solution, then it has a solution $\bar{x}$ with size polynomial in the size of A and b.*

Proof We may assume that A is an $n \times n$ nonsingular matrix, for otherwise we can reduce the system. Thus, $\bar{x}$ is the unique solution to $Ax = b$. Since A, b are rational, we may assume after scaling that both A and b are integer. Let A^j be the matrix obtained from A by replacing the jth column by the vector b. *Cramer's Rule* from Linear Algebra now tells us that

$$\bar{x}_j = \left(A^{-1}b\right)_j = \frac{\det(A^j)}{\det(A)}.$$

In particular, $\bar{x}_j$ is a rational number. We claim that in addition both the numerator and the denominator can be represented using a number of bits polynomial in the size of A, b. We consider the denominator only, as the argument for the numerator is similar. Note, that $\det(A) \leq n!\alpha^n \leq n^n\alpha^n$, where $\alpha = \max\{|A_{ij}| : i,j \in \{1,\ldots,n\}\}$. Thus, the denominator for $\bar{x}_j$ requires at most $O(\log_2(n^n\alpha^n)) = O(n\log_2(n) + n\log_2(\alpha))$ bits of storage. $\qquad\square$

Problems ⓑ, ⓒ, ⓔ, ⓕ are also in *NP*. We indicate for each problem the required certificate:

ⓑ $\bar{x}$ such that $Ax \leq b$,
ⓒ an integer $\bar{x}$ such that $Ax \leq b$,
ⓔ a set of edges B that forms an *st*-path in G of length $\leq k$,
ⓕ a set of edges B that forms an *st*-path in G of length $\geq k$.

In exercises in section A.3.4, you are asked to verify that we can pick $\bar{x}$ to be of size polynomial in the size of A, b for problems ⓑ and ⓒ.

A.3.3 The class *co-NP*

Informally speaking, a decision problem is in *co-NP* if there exists short certificate for every **NO** instance. More formally, a *NO-Checking algorithm* is an algorithm that takes as input, an instance D of $\mathscr{D}$ as well as a potential certificate $C(D)$ and returns **OK** if $C(D)$ shows that D is indeed a **NO** instance and returns **NOT-OK** otherwise. (Note, when the **NO**-checking algorithm returns **NOT-OK** it does not mean that instance D is a **YES** instance, it just means that the certificate $C(D)$ is not sufficient to establish that D is a **NO** instance.) We say that $\mathscr{D}$ is in *co-NP* if for every **NO** instance D of $\mathscr{D}$ there exists a certificate $C(D)$ such that the **NO**-Checking algorithm returns **OK**. Moreover, we require that the running time of the **NO**-Checking algorithm be polynomial in the size of instance D.

Note that the existence of short **YES** certificate does not imply the existence of a short **NO** certificate. To illustrate this fact, consider the following metaphor. Bob, wishes to organize a party that is to be restricted to everyone that is not a UW student. He hires a bouncer to check every person that wishes to join the party. As for Alice's party, the bouncer has to solve the decision problem: given person X at the door, is X a UW student? In this case, however, if the answer is **NO** , then the person is accepted into the party; if the answer is **YES**, then the person is denied entry. Suppose person X is *not* a UW student; can he or she easily convince the bouncer of that fact? There is no obvious way of doing this. The fact that person X does not carry a UW ID does not prove that X is not a UW student – he or she might have hidden his or her ID in order to gain access to Bob's party for instance. Thus, there is no readily available **NO** certificate in this case.

Let us consider problem ⓐ in Table A.1. An instance is a rational matrix A and a rational vector b. If A, b is a **NO** instance, then there does not exist $\bar{x}$ such that $A\bar{x} = b$. The *Fredholm Theorem of the alternatives* from linear algebra implies that we can then find $\bar{y}$ such that $\bar{y}^\top A = 0$ and $\bar{y}^\top b \neq 0$. Moreover, such a vector $\bar{y}$ clearly implies that $Ax = b$ has no solution. Thus, $\bar{y}$ is our certificate. Moreover, it can be readily checked that we can choose $\bar{y}$ that is of size polynomial in the size of A, b. It follows that there is a **NO**-Checking algorithm with running time polynomial in the size of A, b. Thus, ⓐ is in *co-NP*.

Let us consider problem ⓑ in Table A.1. An instance is a rational matrix A and a rational vector b. If A, b is a **NO** instance, then there does not exist $\bar{x}$ such that $A\bar{x} \leq b$. A variant of Farkas' lemma (see Lemma 4.8 and Exercise 1 in Section 4.4) implies that we can then find $\bar{y} \geq \mathbb{0}$ such that $\bar{y}^\top A = 0$ and $\bar{y}^\top b < 0$. Moreover, it can be readily checked that we can choose $\bar{y}$ that is of size polynomial in the size of A, b. It follows that there is a **NO**-Checking algorithm with running time polynomial in the size of A, b. Thus, ⓑ is in *co-NP*.

Let us consider problem ⓓ in Table A.1. Let G be a bipartite graph with bipartition U, W. If G is a **NO** instance, then G has no perfect matching. If $|U| \neq |W|$, then U, W is our certificate. Otherwise, *Hall's theorem* (Theorem 3.12) implies that there exists a deficient set S, i.e. $S \subseteq U$ such that $|S| > |N_G(S)|$. Then S is our certificate. Hence, ⓓ is in *co-NP*. In Exercise 2 in Section A.3.4, you are asked to prove that the decision problem ⓒ is in *co-NP*. It is not known whether ⓒ and ⓕ are in *co-NP*, in fact it is widely believed that they are not.

A.3.4   The class P

A decision problem $\mathcal{D}$ is in P if there is a polynomial-time algorithm that for every instance D of $\mathcal{D}$ returns **YES** if D is a **YES** instance and returns **NO** if D is a **NO** instance. Observe that if $\mathcal{D}$ is in P, then $\mathcal{D}$ is also in *NP*, as the algorithm in this case can determine efficiently whether the instance is a **YES** or **NO** instance without the help of any certificate, or equivalently the empty string is a certificate for the **YES** instances. Similarly, P is contained in *co-NP*. We can represent this by the Venn diagram below. Note, it is not known whether $P \neq NP$. In fact, this question appears among the

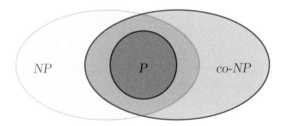

seven entries of the *Millennium Prize Problems* list of the *Clay Mathematics Institute*. A correct resolution of the status of the question entails its author to a \$1 000 000 reward! Similarly to $P \neq NP$, it is not known whether $P \neq$ *co-NP*, or whether $P = NP \cap$ *co-NP*.

We proved that problems ⓐ, ⓑ, ⓓ, and ⓒ in Table A.1 are all in $NP \cap$ *co-NP*. It follows in fact from our discussion in Section A.2 that these problems are all in P.

Exercises

1 Let A, b be a rational matrix and vector. Show that if $Ax \leq b$ has a solution, then it has a solution that is of size polynomial in the size of A, b. Deduce that problem ⓑ from Table A.1 is in *NP*.

HINT: Consider the extreme points of $\{x : Ax \leq b\}$ and use Remark A.1.

2 Show that problem ⓒ from Table A.1 in is *co-NP*.

HINT: See Proposition 3.6.

3 Suppose that we have a set of factories labeled 1 through m each producing the same car model. Suppose that we have a set of showrooms labeled 1 through n. Exactly p_i cars need to leave factory i for every $i = 1, \ldots, m$ and exactly d_j cars need to arrive at showroom j for every $j = 1, \ldots, n$. Thus, if we denote by $x_{i,j}$ the number of cars transported from factory i to showroom j, then we have to satisfy

$$\sum_{j=1}^{n} x_{i,j} = p_i, \qquad i = 1, \ldots, n. \tag{1}$$

$$\sum_{i=1}^{m} x_{i,j} = d_j, \qquad j = 1, \ldots, m. \tag{2}$$

$$x_{i,j} \geq 0, \qquad i = 1, \ldots, n, \ j = 1, \ldots, m. \tag{3}$$

The *transportation feasibility problem* is the decision problem that given m, n, $p_1, \ldots, p_m$ and $d_1, \ldots, d_n$ asks if there exists an integer $\bar{x}$ satisfying all of (1), (2), and (3).

(a) Show that this problem is in *NP*?

(b) Is this problem in *co-NP*?

(c) Is this problem in *P*?

4 (Advanced) Let A be a rational matrix and let b be a rational vector. The goal of this exercise is to show that the decision problem ⓒ in Table A.1 is in *NP*. Let P denote the polyhedron $\{x : Ax \leq b\}$. Given sets $A, B \subseteq \mathbb{R}^n$, we denote by $A + B$ the set $\{a + b : a \in A, b \in B\}$. Given vectors $x^{(1)}, \ldots, x^{(k)}$, we denote by $\text{conv}\{x^{(1)}, \ldots, x^{(k)}\}$ to be the set of all convex combinations of $x^{(1)}, \ldots, x^{(k)}$ (see Exercise 7 in Section 2.8.4).

We will make use (without proof) of the following theorem of Weyl–Minkowski:

$$P = \text{conv}\{x^{(1)}, \ldots, x^{(k)}\} + \text{cone}\{r^{(1)}, \ldots, r^{(\ell)}\} \tag{$\star$}$$

where $x^{(1)}, \ldots, x^{(k)}$ are extreme points of P and where $\text{cone}\{r^{(1)}, \ldots, r^{(\ell)}\} = \{x : Ax \leq 0\}$.

(a) Show that $x^{(1)}, \ldots, x^{(k)}$ are of size polynomial in A, b.

(b) Show that we may assume $r^{(1)}, \ldots, r^{(\ell)}$ are integer and of size polynomial in A, b.

(c) Let

$$B = \left\{ x : \sum_{j=1}^{\ell} \mu_j r^j : 0 \leq \mu_j \leq 1, j = 1, \ldots, \ell \right\}.$$

Show P has nonempty intersection with $\mathbb{Z}^n$ if and only if conv$\{x^{(1)}, \ldots, x^{(k)}\} + B$ does.

(d) Deduce that the decision problem ⓒ in Table A.1 is in *NP*.

A.4   Hard problems

A natural question is whether there exists a polynomial-time algorithm for every optimization/enumeration problem. This is not the case. Consider the following problem: given a rational matrix A and vector b, find the set of all extreme points of the polyhedron $P = \{x : Ax \leq b\}$. It is not hard to see that, in general, the number of extreme points of P can be exponential in the size of the instance A, b. Thus, to simply write down the answer to the enumeration problem, will require exponential time. The situation for problems in *NP* is markedly different, there are no problems for which we can prove that a polynomial algorithm does not exist (for otherwise we would know that $P \neq NP$). We will show instead that there exists a large collection of problems in *NP* that have the property that if any one of these problems is in *P*, then all them are in *P*.

A.4.1   Reducibility

We revisit the idea of reducibility introduced in Section 2.6.1. Consider two decision problems, say $\mathscr{A}$ and $\mathscr{B}$. If given an polynomial-time algorithm to solve problem $\mathscr{B}$, we can solve every instance of $\mathscr{A}$ in polynomial-time, then $\mathscr{A}$ is *reducible* to $\mathscr{B}$. In other words, $\mathscr{B}$ is at least as hard as $\mathscr{A}$ (from our complexity point of view, where we only care whether a problem can be solved in polynomial-time). Being reducible is a transitive relation,

Remark A.2 *Let $\mathscr{A}$, $\mathscr{B}$, and $\mathscr{C}$ be decision problems. If $\mathscr{A}$ is reducible to $\mathscr{B}$ and $\mathscr{B}$ is reducible to $\mathscr{C}$, then $\mathscr{A}$ is reducible to $\mathscr{C}$.*

In the remainder of this section, we give examples of reductions.

Let $x_1, x_2, \ldots, x_n$ denote Boolean variables. So, x_j is either TRUE or FALSE. A *literal* is either a variable x_j or its complement $\bar{x}_j$. A *clause* is a disjunction of a finite collection of literals (e.g., clause C_j can be $(x_5 \vee \bar{x}_3 \vee \bar{x}_2 \vee x_{10})$). A *formula* is a conjunction of a finite collection of clauses. For example

$$(x_5 \vee \bar{x}_3 \vee \bar{x}_2 \vee x_{10}) \wedge (\bar{x}_1) \wedge (x_2 \vee \bar{x}_4 \vee x_8 \vee x_9 \vee x_7).$$

A formula is satisfied for an assignment of TRUE/FALSE values to its variables if every clause is satisfied. For instance, in the above example we can satisfy the formula by assigning x_1 FALSE, x_2 TRUE, x_3 FALSE, and all other variables to any TRUE/FALSE. The *SAT* decision problem takes as input a formula, and the question is whether there is an assignment of values to the variables that satisfies the formula. The *0,1 Feasibility* decision problem takes as input a rational matrix A, a rational vector b, and asks if there is a vector $\bar{x}$ where all entries are in $\{0, 1\}$ for which $Ax \geq b$.

PROPOSITION A.3   *SAT is reducible to 0,1 Satisfiability.*

Proof Suppose that we had a polynomial-time algorithm to solve the 0, 1 feasibility problem. We could then proceed as follows to solve satisfiability. Given a formula with clauses $C_1, C_2, \ldots, C_m$ and variables $x_1, x_2, \ldots, x_n$, define the following 0, 1 feasibility problem

$$\sum_{j:x_j \in C_i} x_j + \sum_{j:\bar{x}_j \in C_i} (1 - x_j) \geq 1 \qquad (i \in \{1, \ldots, m\}) \qquad \text{(A.1)}$$

$$x_j \in \{0, 1\} \qquad (j \in \{1, \ldots, n\}). \qquad \text{(A.2)}$$

Then it can be readily checked that there is a solution to the formula if and only if there is a solution to (A.1), (A.2).   □

The *Quadratic Feasibility* problem takes as input a set of functions $f_i(x)$ for $i = 1, \ldots, k$ of degree at most 2 where each of f_i is described by rational coefficients. The question is whether there exists $\bar{x}$ such that $f_i(\bar{x}) \leq 0$ for $i = 1, \ldots, k$.

PROPOSITION A.4   *0,1 Satisfiability is reducible to Quadratic Feasibility.*

Proof We can encode the constraints that $x_j \in \{0, 1\}$ as the constraint $x_j(1 - x_j) = 0$, or equivalently as $x_j - x_j^2 \leq 0$ and $-x_j + x_j^2 \leq 0$.   □

The partition problem takes as input, integers $a_1, a_2, \ldots, a_n$, and the question is whether we can find sets J and K such that $J \cup K = \{1, \ldots, n\}$, $J \cap K = \emptyset$ and

$$\sum_{j \in J} a_j = \sum_{j \in K} a_j.$$

The Knapsack problem takes as input integers $\alpha_1, \alpha_2, \ldots, \alpha_n, c_1, c_2, \ldots, c_n, b$ and M. The question is whether there exists $x \in \{0, 1\}^n$ such that

$$\sum_{j=1}^{n} \alpha_j x_j \leq b \quad \text{and} \quad \sum_{j=1}^{n} c_j x_j \geq M.$$

PROPOSITION A.5   *SAT is reducible to partition.*

PROPOSITION A.6   *Partition is reducible to Knapsack.*

Proof Given integer $a_1, \ldots, a_n$ set $M = b = \frac{1}{2} \sum_{j=1}^{n} a_j$ and for all $j = 1, \ldots, n$, let $\alpha_j = c_j = a_j$. Now it is easy to check that the answer is **YES** for the Partition problem if and only if the answer is **YES** for the Knapsack problem.   □

Exercises

1   Let A be a rational matrix and b, c be rational vectors. Let (P) denote the following linear program, $\max\{c^\top x : Ax \leq b\}$ and consider the following decision problems:
(D1) Is (P) feasible?
(D2) Is (P) unbounded?
(D3) Does (P) have an optimal solution?
Show that for any $i, j \in \{1, 2, 3\}$, problem (Di) is reducible to problem (Dj).

A.4.2   NP-complete problems

A decision problem $\mathscr{D}$ is said to be *NP-hard* if *every* problem in *NP* is reducible to $\mathscr{D}$. A decision problem $\mathscr{D}$ is said to be *co-NP-hard* if *every* problem in *co-NP* is reducible to $\mathscr{D}$. It is not obvious at all that there are *NP-complete* decision problem. The following seminal result proves that this is indeed the case:

THEOREM A.7 (Cook's theorem).   *SAT is NP-complete.*

We will omit the proof of this result.
    We summarize the relations we established between a number of problems in *NP* in the following figure. An arrow indicates that the problem at the tail of the arrow is reducible to the problem at the head of the arrow. Arrows A, B, C, and D correspond to respectively Proposition A.3, A.5, A.4, and A.6. The thick arrows follow from Theorem A.7. Note, that it follows from the fact that SAT is *NP-complete* and Re-

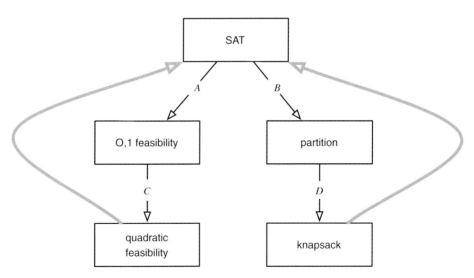

mark A.2 that all of SAT, 0,1 Feasibility, Quadratic feasibility, Partition, and Knapsack are *NP-complete*. In particular, Knapsack indicates that solving 0, 1 Integer Programs with a single constraint is difficult.

A.4.3   Complexity as guide

Suppose that your boss asks you design a polynomial-time algorithm for an optimization problem. You can use the ideas in this appendix as a guide. First define $\mathcal{D}$ to be the associated decision problem. Then investigate if there is an easy way of certifying the **YES** and **NO** answers, i.e. try to prove that $\mathcal{D}$ is in $NP \cap co\text{-}NP$. If you succeed in proving this, you should try to prove that $\mathcal{D} \in P$. If you fail to prove that $\mathcal{D} \in NP$ or that $\mathcal{D} \in co\text{-}NP$, you should try to prove that $\mathcal{D}$ is $NP\text{-}hard$ or that $\mathcal{D}$ is $co\text{-}NP\text{-}hard$. If you succeed in designing a polynomial-time algorithm for $\mathcal{D}$, then you will surely earn the gratitude of your boss; on the other hand, if you succeed in showing that $\mathcal{D}$ is $NP\text{-}hard$, it will indicate to your boss that nobody else is likely to find a polynomial algorithm and that at the very least firing would do her no good. Note, however that in practice we can solve many instances of $NP\text{-}complete$ problems quite well. Indeed Integer Programming feasibility is hard, yet countless corporations use integer programming solvers. The fact that a problem is $NP\text{-}hard$ only indicates that we cannot always expect an exact answer within a short amount of time. This is an indication that the use of heuristic or restricting the search to an approximate solution might be in order.

A.4.4   Easy versus hard problems

A striking aspect of complexity theory is that the boundary between easy (i.e. P) and hard (i.e. $NP\text{-}complete$) problems is very subtle with many pairs of problems with seemingly similar formulations falling in one category or the other. Table A.2 provides a number of examples. Problems on the left-hand side are in P and problems on the right-hand side are $NP\text{-}complete$. Consider problems ⓐ, ⓑ, ⓒ. The table shows that while solving inequalities with continuous variables or solving linear equations with integer variables is easy, combining inequalities and integer variables is hard.

   Problems ⓓ, ⓔ indicate that while finding the shortest st-path in a graph is easy, finding the longest st-path in a graph appears to be hard.

   A *walk* W is a sequence of edges

$$v_1 v_2, v_2 v_3, \ldots, v_{k-2} v_{k-1}, v_{k-1} v_k.$$

It is an st-walk if $v_1 = s$, $v_k = t$. When all of $v_1, \ldots, v_k$ are distinct, then the st-walk W is an st-path. Consider a graph G with nonnegative integer edge weights and vertices s, t, u. We can find in polynomial-time a shortest su-path P_1 and a shortest ut-path P_2. If we concatenate P_1 and P_2, we obtain a shortest st-walk that visits the intermediary vertex u. This proves together with the shortest path algorithm that ⓕ is in P as indicated in the table. However, ⓖ indicates that it is hard to find a shortest st-path that visits an intermediary vertex u.

   A graph G is *planar* if it can be drawn on the plane such that no two edges cross. A graph is k-colorable if we can find an assignment of at most k colors to the vertices with the property that adjacent vertices are assigned distinct colors. The celebrated *4-color Theorem* states that any planar graph is 4-colorable. Thus, problem ⓗ is indeed in P

Table A.2 Easy and hard decision problems

	P	NP-complete
	ⓐ	
Input	Rational matrix A and vector b.	
Question	Does $Ax \leq b$ have a solution?	
	ⓑ	ⓒ
Input	Rational matrix A and vector b.	Rational matrix A and vector b.
Question	Does $Ax = b$ have an integer solution?	Does $Ax \leq b$ have an integer solution?
	ⓓ	ⓔ
Input	Graph $G = (V, E)$, $s, t \in V$, $w \in \mathbb{N}^E$ and $k \in \mathbb{Z}$.	Graph $G = (V, E)$, $s, t \in V$, $w \in \mathbb{N}^E$ and $k \in \mathbb{Z}$.
Question	Does G have an st-path of length $\leq k$?	Does G have an st-path of length $\geq k$?
	ⓕ	ⓖ
Input	Graph $G = (V, E)$, $s, t \in V$, $w \in \mathbb{N}^E$ and $k \in \mathbb{Z}$.	Graph $G = (V, E)$, $s, t, u \in V$, $w \in \mathbb{N}^E$ and $k \in \mathbb{Z}$.
Question	Does G have an st-walk using vertex u of length $\leq k$?	Does G have an st-path using vertex u of length $\leq k$?
	ⓗ	ⓘ
Input	Planar graph G.	Planar graph G.
Question	Can G be 4-colored?	Can G be 3-colored?
	ⓙ	ⓚ
Input	A graph G and an integer k.	A hypergraph G and an integer k.
Question	Is there a matching M with $|M| \geq k$?	Is there a matching M with $|M| \geq k$?
	ⓛ	ⓜ
Input	A graph G.	A graph G.
Question	Is there a closed walk visiting each edge exactly once?	Is there a closed walk visiting each vertex exactly once?

(use the trivial algorithm that always says **YES**). On the other hand ⓘ indicates that checking if a planar graph is 3-colorable is hard.

A *hypergraph* is a pair (V, E) where V are the vertices and E are the *hyperedges* which are subsets of the vertices. Thus, a graph is a hypergraph, where every hyperedge has cardinality 2. A *matching* in a hypergraph a set of pairwise disjoint hyperedges. Checking if a graph has a matching of size at least k is easy, thus ⓙ is *P*. But ⓖ is *NP-complete*, thus checking if a hypergraph has a matching of size at least k is hard. Note, this remains hard even if all the hyperedges have cardinality three.

An st-walk is *closed* if $s = t$, i.e. it starts and end at the same vertex. ⓛ indicates that finding a closed walk that visits every edge exactly once is easy. However, ⓜ says that finding a closed walk that visits every vertex exactly once is hard. Note, such a closed walk is known as a Hamiltonian cycle.

References

[1] I. Adler, N. Megiddo, and M. J. Todd. New results on the average behavior of simplex algorithms. *Bull. Amer. Math. Soc. (N.S.)*, 11(2): 378–382, 1984.

[2] N. Alon, D. Moshkovitz, and S. Safra. Algorithmic construction of sets for *k*-restrictions. *ACM ToDs*, 2(2): 153–177, 2006.

[3] N. Amenta and G. Ziegler. Deformed products and maximal shadows of polytopes. In B. Chazelle, J. Goodman, and R. Pollack, editors, *Advances in Discrete and Computational Geometry*, pages 57–90. American Mathematical Society, 1999.

[4] A. Ben-Tal, L. El Ghaoui, and A. Nemirovski. *Robust Optimization*. Princeton Series in Applied Mathematics. Princeton University Press, Princeton, NJ, 2009.

[5] D. Bertsimas and R. Weismantel. *Optimization over Integers*. Athena Scientific, Nashua, NH, 2005.

[6] M. J. Best. *Portfolio Optimization*. Chapman & Hall/CRC Finance Series. CRC Press, Boca Raton, FL, 2010.

[7] R. E. Bixby. Personal communication. President and Co-founder, Gurobi Optimization, 2013.

[8] R. E. Bixby. The simplex algorithm: it's alive and well. Presented at the Tutte Colloquium, Waterloo, Ontario, Canada, 2013.

[9] R. G. Bland. New finite pivoting rules for the simplex method. *Math. Oper. Res.*, 2(2): 103–107, 1977.

[10] K. H. Borgwardt. The average number of pivot steps required by the simplex method is polynomial. *Zeitschrift für Operations Research*, 26: 157–177, 1982.

[11] S. Boyd and L. Vandenberghe. *Convex Optimization*. Cambridge University Press, Cambridge, 2004.

[12] X. Chen, F. Xu, and Y. Ye. Lower bound theory of nonzero entries in $\ell_1 - \ell_p$ minimization. *SIAM J. Sci. Comp.*, 32: 2832–2852, 2010.

[13] V. Chvátal. *Linear Programming*. A Series of Books in the Mathematical Sciences. W. H. Freeman & Company, New York, 1983.

[14] M. Conforti, G. Cornuéjols, and G. Zambelli. Polyhedral approaches to mixed integer linear programming. In M. Jnger, T. Liebling, D. Naddef, G. L. Nemhauser, W. R. Pulleyblank, G. Reinelt, G. Rinaldi, and L. Wolsey, editors, *50 Years of Integer Programming 1958–2008: From the Early Years to the State-of-the-Art*, pages 334–384. Springer, 2010.

[15] W. J. Cook. *In Pursuit of the Traveling Salesman: Mathematics at the Limits of Computation*. Princeton University Press, Princeton, NJ, 2011.

[16] W. J. Cook, W. H. Cunningham, W. R. Pulleyblank, and A. Schrijver. *Combinatorial Optimization*. Wiley-Interscience Series in Discrete Mathematics and Optimization. John Wiley & Sons, New York, 1998.

[17] G. Cornuéjols and R. Tütüncü. *Optimization Methods in Finance*. Mathematics, Finance and Risk. Cambridge University Press, Cambridge, 2007.

[18] R. W. Cottle. William Karush and the KKT theorem. *Doc. Math.* (Extra volume: Optimization stories): 255–269, 2012.

[19] W. H. Cunningham and J. G. Klincewicz. On cycling in the network simplex method. *Math. Program.*, 26: 182–189, 1983.

[20] G. B. Dantzig. *Linear Programming and Extensions*. Princeton University Press, Princeton, NJ, 1963.

[21] Dantzig, G. B. Maximization of a linear function of variables subject to linear inequalities. In T. C. Koopmans, editor, *Activity Analysis of Production and Allocation*, pages 339–347. John Wiley & Sons, New York, 1951.

[22] E. W. Dijkstra. A note on two problems in connexion with graphs. *Numerische Mathematik*, 1: 269–271, 1959.

[23] I. Dinur and S. Safra. The importance of being biased. In *Proceedings, ACM Symposium on Theory of Computing*, pages 33–42. ACM, New York, NY, 2002.

[24] J. Edmonds and R. M. Karp. Theoretical improvements in algorithmic efficiency for network flow problems. *Journal of the Association for Computing Machinery*, 19: 248–264, 1972.

[25] U. Feige. A threshold of ln n for approximating set cover. *J. ACM*, 45(4): 634–652, 1998.

[26] M. L. Fredman and R. E. Tarjan. Fibonacci heaps and their uses in improved network optimization algorithms. *J. ACM*, 34(3): 596–615, July 1987.

[27] O. Friedmann. A subexponential lower bound for Zadeh's pivoting rule for solving linear programs and games. In *Integer Programming and Combinatorial Optimization*, volume 6655 of Lecture Notes in Computer Science, pages 192–206. Springer, Heidelberg, 2011.

[28] H. N. Gabow. Data structures for weighted matching and nearest common ancestors with linking. In *Proceedings of the 1st Annual ACM-SIAM Symposium on Discrete Algorithms (SODA '90)*, pages 434–443. Society for Industrial and Applied Mathematics, Philadelphia, PA, 1990.

[29] G. H. Golub and C. F. Van Loan. *Matrix Computations*. Johns Hopkins Studies in the Mathematical Sciences. Johns Hopkins University Press, Baltimore, MD, third edition, 1996.

[30] M. Grötschel, L. Lovász, and A. Schrijver. *Geometric Algorithms and Combinatorial Optimization*, volume 2 of Algorithms and Combinatorics, Springer-Verlag, Berlin, second edition, 1993.

[31] M. Haimovich. The simplex method is very good!–on the expected number of pivot steps and related properties of random linear programs. Technical Report, Columbia University, NY, 1983.

[32] J. E. Hopcroft and R. M. Karp. An n**(5/2) algorithm for maximum matchings in bipartite graphs. *SIAM J. Comput.*, 2(4): 225–231, 1973.

[33] G. Kalai. A subexponential randomized simplex algorithm. In *Proceedings of the Twenty-Fourth Annual ACM Symposium on the Theory of Computing*, pages 475–482. ACM, New York, NY, 1992.

[34] D. Karger. Minimum cuts in near linear time. *J. ACM*, 47(1): 46–76, 2000.

[35] N. Karmarkar. A new polynomial-time algorithm for linear programming. *Combinatorica*, 4(4): 373–395, 1984.

[36] W. Karush. Minima of functions of several variables with inequalities as side conditions. ProQuest LLC, Ann Arbor, MI, 1939. Thesis (SM)–The University of Chicago.

[37] L. G. Khachiyan. A polynomial algorithm in linear programming. *Dokl. Akad. Nauk SSSR*, 244(5): 1093–1096, 1979.

[38] L. G. Khachiyan. A polynomial algorithm in linear programming. *Soviet Mathematics Doklady*, 20: 191–194, 1979.

[39] S. Khot and O. Regev. Vertex cover might be hard to approximate to within 2-epsilon. *J. Comput. Syst. Sci.*, 74(3): 335–349, 2008.

[40] V. Klee and G. Minty. How good is the simplex algorithm? In O. Shisha, editor, *Inequalities*, pages 159–175. Academic Press, New York, 1972.

[41] J. Kleinberg and É. Tardos. *Algorithm Design*. Pearson Studium, 2006.

[42] B. Korte and J. Vygen. *Combinatorial Optimization*. Springer, New York, 5th edition, 2012.

[43] H. W. Kuhn. The Hungarian method for the assignment problem. *Naval Res. Logist. Quart.*, 2: 83–97, 1955.

[44] H. W. Kuhn and A. W. Tucker. Nonlinear programming. In *Proceedings of the Second Berkeley Symposium on Mathematical Statistics and Probability, 1950*, pages 481–492. University of California Press, Berkeley and Los Angeles, 1951.

[45] E. L. Lawler, J. K. Lenstra, A. H. G. Rinnooy, and D. B. Shmoys. *The Traveling Salesman Problem: A Guided Tour of Combinatorial Optimization*. Wiley Series in Discrete Mathematics & Optimization, John Wiley & Sons, New York, 1985.

[46] J. Lee. Hoffman's circle untangled. *SIAM Rev.*, 39(1): 98–105, 1997.

[47] J. Matoušek, M. Sharir, and E. Welzl. A subexponential bound for linear programming. *Algorithmica*, 16(4/5): 498–516, 1996.

[48] R. R. Meyer. On the existance of optimal solutions of integer and mixed-integer programming problems. *Mathematical Programming*, 7: 223–225, 1974.

[49] S. Micali and V. Vazirani. An algorithm for maximum matching in general graphs. In *Proceedings of the 21st IEEE Annual Symposium on the Foundations of Computer Science*, pages 17–27. IEEE, New York, 1980.

[50] R. D. C. Monteiro, I. Adler, and M. G. C. Resende. A polynomial time primal–dual affine scaling algorithm for linear and convex quadratic programming and its power series extension. *Math. Oper. Res.*, 15: 191–214, 1990.

[51] K. G. Murty and S. N. Kabadi. Some NP-complete problems in quadratic and nonlinear programming. *Math. Program.*, 39: 117–129, 1987.

[52] A. S. Nemirovskii and D. B. Yudin. Informational complexity and efficient methods for the solution of convex extremal problems. *Ékonomika i Matematicheskie Metody*, 12: 357–369, 1976.

[53] Y. Nesterov. *Introductory Lectures on Convex Optimization: A Basic Course*. Kluwer Academic Publishers, Boston, MA, 2004.

[54] A. L. Peressini, F. E. Sullivan, and J. J. Uhl. *The Mathematics of Nonlinear Programming*. Undergraduate Texts in Mathematics, Springer-Verlag, New York, 1988.

[55] M. L. Pinedo. *Scheduling: Theory, Algorithms, and Systems*. Springer, New York, third edition, 2008. With 1 CD-ROM (Windows, Macintosh and UNIX).

[56] R. Raz and S. Safra. A sub-constant error-probability low-degree test, and a sub-constant error-probability PCP characterization of NP. In *Proceedings, ACM Symposium on Theory of Computing*, pages 475–484. ACM, New York, NY, 1997.

[57] F. Santos. A counter example to the Hirsch conjecture. *Annals of Math.*, 176(2): 383–412, 2012.

[58] A. Schrijver. *Theory of Linear and Integer Programming*. Wiley-Interscience Series in Discrete Mathematics. John Wiley & Sons, Chichester, 1986.

[59] N. Shor. Cut-off method with space extension in convex programming problems. *Cybernetics*, 13: 94–96, 1977.

[60] M. Sipser. *Introduction to the Theory of Computation*. Cengage Learning, 2nd edition, 2006.

[61] S. Smale. On the average number of steps of the simplex method of linear programming. *Math. Programming*, 27(3): 241–262, 1983.

[62] S. Smale. The problem of the average speed of the simplex method. In *Mathematical Programming: The State of the Art (Bonn, 1982)*, pages 530–539. Springer, Berlin, 1983.

[63] D. A. Spielman and S.-H. Teng. Smoothed analysis of algorithms: why the simplex algorithm usually takes polynomial time. *J. ACM*, 51(3): 385–463, 2004.

[64] L. Tunçel. *Polyhedral and Semidefinite Programming Methods in Combinatorial Optimization*. American Mathematical Society, 2010.

[65] L. Tunçel. Constant potential primal–dual algorithms: a framework. *Math. Programming*, 66(2, Ser. A):145–159, 1994.

[66] R. J. Vanderbei. *Linear Programming*. International Series in Operations Research & Management Science, 37, Kluwer Academic Publishers, Boston, MA, second edition, 2001.

[67] S. A. Vavasis. *Nonlinear Optimization. Complexity Issues*. Oxford University Press, 1991.

[68] V. V. Vazirani. *Approximation Algorithms*. Springer, 2001.

[69] D. P. Williamson and D. Shmoys. *The Design of Approximation Algorithms*. Cambridge University Press, 2011.

[70] W. L. Winston. *Operations Research, Applications and Algorithms*. Thomson Learning, 2004.

[71] L. A. Wolsey. *Integer Programming*. Wiley-Interscience Series in Discrete Mathematics and Optimization. John Wiley & Sons, New York, 1998.

[72] Y. Ye. *Interior Point Algorithms*. Wiley-Interscience Series in Discrete Mathematics and Optimization. John Wiley & Sons, New York, 1997.

Index

Fermat's last theorem, 209
formula, 258
formulation, 1, 4
free variable, 97
fundamental theorem of integer programming, 185
fundamental theorem of linear programming, 44, 81

gradient, 220
graph, 26
    arc, 111
    bipartite, 29
    connected, 134
    directed, 148
    tree, 134
Graph: matching, 29
Graph: perfect matching, 29
Graph: st-cut, 32

halfspace, 87
Hall's theorem, 126
Hamiltonian cycle, 204
Hirsch conjecture, 251
hyperplane, 87

incident, 27
input size, 246
input-output systems, 175
instance, 245
integer program
    IP, 14
integer feasibility problem, 25
integer program, 14
    mixed, 14
    pure, 14
interior of a set, 235
interior-point methods, 234
inverse function theorem, 241

Karush–Kuhn–Tucker theorem, 221, 224
    for nonconvex problems, 228
KKT point, 232
knapsack problem, 17
KWOil example, 7

Lagrangian, 224
Lagrangian dual, 226
largest coefficient rule, 73
largest improvement rule, 73
law of diminishing marginal returns, 175
leaf, 135
level set, 213
lexicographic rule, 97
line segment between two points, 88
line through two points, 88
linear constraint, 6
linear function, 5
linear program, 6

feasible, 44
    infeasible, 44
    unbounded, 44
linear programming relaxation, 107, 123
literal, 258
local conic convex approximation, 229
LP relaxation, 107, 123, 189

M-alternating tree, 140
M-covered, 138
M-exposed, 138
Mangasarian–Fromowitz constraint qualification, 232
Markowitz model, 43
matching
    maximum, 134
mathematical constraint, 1
maximum-weight matching, 35
Minkowski's theorem, 232

neighbors: set of, 126
nonlinear program, 37
nonlinear program:NLP, 37
nondegenerate, 174
NP, 253, 254
    Co-NP, 255
    hard, 260

objective function, 1
open set, 235
optimal value, 44
optimal solution, 4
    unique, 174

P, 256
path, 27
pivot rules, 97
pivoting, 82
polyhedron, polyhedra, 88
polynomial-time algorithm, 247
portfolio optimization, 41, 43
positive semidefinite matrix, 95
potential function, 239
primal–dual pair, 143
pruning, 198

ratio test, 65
reduced cost, 118
reducible, 258
robust optimization, 43
running time, 247

saddle point, 228
SAT, 258
separation algorithm, 203
set-cover problem, 166
shadow price, 174

Printed in the United States
By Bookmasters